New trends for innovation in the Mediterranean animal production

EAAP – European Federation of Animal Science

INRA

INIA

cirad

CIHEAM

The European Association for Animal Production wishes to express its appreciation to the *Ministero per le Politiche Agricole e Forestali* and the *Associazione Italiana Allevatori* for their valuable support of its activities

New trends for innovation in the Mediterranean animal production

EAAP publication No. 129

edited by:

R. Bouche†

A. Derkimba

F. Casabianca

Wageningen Academic
Publishers

ISBN: 978-90-8686-170-5
e-ISBN: 978-90-8686-726-4
DOI: 10.3920/978-90-8686-726-4

ISSN 0071-2477

First published, 2011

Wageningen Academic Publishers
P.O. Box 220
6700 AE Wageningen
The Netherlands
www.WageningenAcademic.com
copyright@WageningenAcademic.com

We would like to take a moment to mention that this book could not have been made without the work of our esteemed colleague Rémi Bouche who, we are sad to say, passed away in November 2009.
As researchers we grieve the passing of our colleague who was respected for his contributions to our scientific area.

This book was a project in which he, as an editor, invested a significant amount of time and effort.
We therefore also dedicate this publication
to the memory of
Rémi Bouche.

Table of contents

B. Feeding systems, nutrition, health

Round table

Conclusion

Preface

R. Bouche, A. Derkimba and F. Casabianca
INRA, Research Laboratory on Livestock Development, Corte, France-

The Mediterranean Group

The Mediterranean Contact Group is an EAAP Work Group devoted to specific scientific and development issues related to animal production in the Mediterranean area. Its aim is to encourage collaboration between scientific communities and technical sectors from Mediterranean countries, and among other activities, the organization of a Mediterranean Symposium every two years. These symposiums are always excellent tools to link Mediterranean researchers on specific subjects about livestock development in the Mediterranean area.

Topic of the symposium

The Mediterranean area shows a great diversity of livestock systems, depending mainly on local resources and traditions, but also on the space of networking where informational resources are available for producers. During the last decades, a lot of innovations have been conceived and/or introduced in Mediterranean area, conditions for livestock systems to remain competitive but also, at the same time, the way to eliminate those unable to change.

Innovation is approached as a process creating a new way of producing, organizing or selling, and inducing some appropriation of users. It must be distinguished from invention that proceeds from an inventor, even if some invention may become innovation when adapted and adopted by users. Innovate is the ability to perform on a new way what is still functioning. It must be considered as a process by which local producers or groups, researchers and advisors, administrations and policy makers are reacting when facing to these new trends and by which they are learning during these processes.

As a first level of understanding, we can identify

- Technical innovation (new technical objects within the production unit) vs. Organizational innovation (new relationships among operators).
- Process innovation (a new way to obtain the same product) vs. Product innovation (a new product).
- Incremental innovation (some change into the system) vs. Radical innovation (change the system itself).

Innovations involve, classically, production techniques, work organization, equipment and infrastructures, collective features for selection, reproduction, feeding or sanitary devices. They have concerned also (and more and more) local organizations such as product labelling, new dynamics around local breed, collective rules for supply basin or new products approaches for new markets.

More recently, some innovations have focused on environmental impacts and/or ethical issues of livestock production, due to an increasing awareness of these kinds of problems.

All these innovations are now questioned by sustainable development in economic, social and environmental perspective.

The purpose of this symposium

- Making an updated review on the main innovations that changed significantly the activities of livestock production in the Mediterranean area during the last period (since 1980). How to characterize the innovation trends through the main observed evolutions?
- Examining to what extend these innovations, when adapting or adopting, improve the efficiency, insure the socio-cultural basis or reduce the environmental impact of livestock systems. How to evaluate the positive and negative impacts of innovations?
- Reflecting about the global balance between (1) development of local systems able to adopt such innovations and (2) elimination of producers unable to adopt them. Who is supported and who is eliminated by the innovation trend?
- Inducing some new vision of innovating systems based on participatory processes between researchers and producers. How to manage the building up of innovations in a participative way?

The structure of the proceedings

In the present symposium proceedings, presentations are organised within 4 sessions and a round table preceding the conclusion.

The first session provides some conceptual input of what is innovation today and gives an overview of what is occurring in the Mediterranean areas. Then the second one focuses on innovation in production systems based on new techniques. It deals with genetics, feeding systems, products and animal identification.

The third session deals with innovation in organisation systems, valorising know how, products, taking place in collective action. The fourth one concerns environmental innovation, giving insights on natural resources management, natural treatments, as emerging changes.

Then, the report of the round table allows an access to the main discussion during this symposium, around a question addressed to all participants: Are traditions in Mediterranean livestock activities an obstacle or a source of innovations?

The last words of the proceedings are from the President of the Mediterranean Group of the EAAP in a large conclusion emphasizing the main issues for the future of livestock activities in the Mediterranean area.

Part 1.

Introduction on innovation

Agricultural systems and the innovation process

J.-M. Meynard[1] and F. Casabianca[2]
[1]INRA, Division Sciences pour l'Action et le Développement, Grignon, France; Jean-marc.meynard@grignon.inra.fr
[2]INRA, Division Sciences pour l'Action et le Développement, Corte, France; francois.casabianca@corte.inra.fr

Abstract

Why dedicate today such a symposium on animal production to innovation and the evolution of this notion in the Mediterranean area? Because innovation has become crucial in an uncertain world: innovation to anticipate, to adapt, to remain competitive or even to survive. In the near future, what would be the keys of the evolution of livestock systems? How to support the innovation process, how to guide it, how to articulate innovation with research and extension? We have shown that, in the present context of increasing globalization, the innovation process is core to multiple tensions that the actors of innovation cannot ignore. Facing such tensions, we suggest to take into account the need to prepare a range of solutions; but also to anticipate the economical, social and environmental effects of innovations; and more broadly, looking forward to the consequences of their emergence and dissemination on the development of agriculture and territories. Then, we have elaborated some proposals to stimulate capabilities of the livestock sector to innovate: how to create new ways of producing? If a model-based design is adapted for exploring breaking systems, a step-by-step design gives more room for learning process and reducing risk. So, it is suitable for a progressive involvement of breeders. However, methods for rule-based innovation (whose aims are fixed) are no more usable when aims are questioned (for instance by dramatically new environmental or social stakes). New competences are then required and new validation criteria should be invented. Such a process of exploration should be called 'innovative design' whose sources are themselves diversifying.

Keywords: innovative design, livestock system, farming system, collective learning

Introduction

In this symposium dedicated to innovation in livestock systems, many examples of innovative systems will be presented to us. The programme reveals a very great diversity of innovations, both for the processes which led to them, for their audacity and for the people involved. There is no doubt that the discussions will be very rich and varied.

But why dedicate such a symposium to innovation today? A first reason is that innovation has become vital in a world that is changing very rapidly: we must innovate to anticipate or innovate to adapt, innovate to remain competitive or innovate to survive. How then can we encourage and guide the innovation process, how can we structure innovation with research and development?

A second reason, just as important as the first, is that the globalisation of exchanges and technological developments can amplify the impacts of innovations to a frightening degree: a technological innovation can radically change production conditions, as well as the living conditions of some people involved in agriculture and territories. We saw it in South America for GM varieties of herbicide-tolerant soybean. If innovation benefits from powerful means of distribution, it can spread at very great speed. So if it has negative effects, it can soon create disasters. How do we evaluate an innovation, before its dissemination or after, to appreciate not only the expected benefits but also the resulting risks?

These are some of the questions that will be put in this symposium and for which it seemed useful to suggest some lines of thought in this introduction.

In the first part of this presentation, we will show that, in the present farming context, the innovation process is at the heart of many tensions, which those involved in innovation cannot ignore. Then we will attempt in the second part to make a few proposals aimed at developing and stimulating the capacities of the farming world to innovate: what can be done to invent new ways of producing?

Innovating to prepare a diversity of futures

The driving forces of the future evolution of farming practices

What are, and what will be in the next few years, the driving forces of the evolution of farming systems? We have identified four of them, whose importance varies according to agricultures and countries.

The degradation of the ecosystems and the recognised responsibility of agriculture

According to the *Millennium Ecosystem Assessment* (http://www.millenniumassessment.org/en/index.aspx), carried out under the aegis of the United Nations, and published in 2005, concludes very clearly:

'In the last 50 years, mankind has generated changes to ecosystems more quickly and more extensively than at any other period in the history of humanity, mainly to satisfy a rapidly-growing demand for food, fresh water, fibre and wood for building and energy. This has led to a substantial loss of the biological diversity on Earth, a high proportion of it irreversibly…and a reduction in the advantages which future generations can draw from ecosystems.'

The diagnosis identifies agriculture as responsible for:
1. The degradation of the quality of surface and ground waters, of soils (contaminations by pesticides, heavy metals, erosion...), and of the air (greenhouse gases).
2. The loss of biodiversity (disappearance of breeds or varieties, homogenisation of habitats, mortality due to pesticides and deforestation…).
3. The exhaustion of non-renewable resources (fossil fuel, fertilisers).
4. The increase in risks of catastrophes (floods, mud-slips…).

The 'Livestock's long shadow' report refines this diagnosis as far as livestock is concerned, identifying in particular the emission of methane by ruminants. The effects of farming practices on the environment are not felt only at local level: the production systems of the countries of the North and South interact via trade in goods and the establishment of world prices. As stressed by the *Ecosystem Assessment*, solutions will have to combine profound changes at political and institutional levels with a change in practices. Without waiting for political decisions to be made, it seems indispensable to anticipate changes in farming practices aimed at (1) better control of pollutions and soil degradation, (2) greater autonomy as regards non-renewable resources and (3) creating a synergy between conservation and use of biodiversity.

The necessary competitiveness of production systems, in a context of the globalisation of trade

World trade results in considerable fluctuations in prices for many agricultural products. This leads to farmers rethinking their production choices and diversifying their activities on the farm. In a context of food crisis, the imperative for world production to grow is imposed on everyone. In some countries of the South, the food-producing agricultures, historically destabilised by imports of cheap food products, now have to be rebuilt. In the North, whole territories, whose business activity was based on monoproduction that is now in crisis (wine, sheep farming…), are looking for new production systems. To ensure the competitiveness and continued existence of Mediterranean agricultures, solutions will have to be found to confront problems of climate and parasites increased by global changes. Innovations in work organization must be found (Dedieu *et al.*, 1999). Flexible, reactive farming systems must be promoted that can adapt to the instability of market prices or changes in quality standards. They must also be prepared to adapt production systems and business locations to a huge increase in the cost of energy (Boiffin *et al.*, 2004).

The evolution of consumer expectations and industrial demands

Whilst at international level, the *Codex Alimentarius* attempts to standardise minimum qualities beyond which no obstacle to free trade may be placed, competition is based more and more on the differentiation of products (Allaire, 2002). The markets, which used to be governed by sectional agreements related to standard quality, are now increasingly segmented. Processing companies and even the mass distribution multinationals are increasing the number of contracts imposing specific quality criteria on the product or restrictive specifications on crop or livestock practices. So qualification issues are evolving from the product itself towards the definition and control of how it is obtained (origin, production methods). Transforming farming systems not only involves the identification of production methods suitable for obtaining products of the desired quality (technological quality, taste, health quality…) or territorial typicity (Casabianca and Matassino, 2006), but also the drawing up of specifications in a contractual framework (how, for example, can specifications be set up without limiting technical choices and preventing innovation?). It also involves measures to coordinate farmers and industries at territory level (Moity-Maïzi *et al.*, 2001). The acute sensitivity of consumers to products health quality leads to questionings about ways of controlling plant and animal diseases: the tension between the need to minimise the use of pesticides and veterinary products and limit residues in food (and the environment, see above), and the need to guarantee, with a high level of security, the absence of toxins or pathogenic micro-organisms in food products, calls for the mobilisation of alternative methods of controlling animal and plant diseases.

A profound transformation in relationships between agricultures and territories

The transformation of farming systems cannot be conceived independently of the non-farming players in the territory: compromises, arrangements or synergies must be sought with them.

In a context where the economic and social role of agriculture is being marginalised in an increasing number of territories, particularly on the Northern shore of the Mediterranean, farmers are perceived as having a mission to manage space and natural resources, and no longer just to produce. Depending on the characteristics of the territories, specific questions are asked: what sort of agriculture is possible in an outlying suburban area, what are the opportunities and constraints of this type of territory? How can competition for water resources between agriculture and other users be managed? What crop or livestock systems are possible in areas of environmental interest, where the 'production of the environment' can become a priority over the production of food? What combinations of activities are possible in touristic areas, between farming production, product processing, direct sales, welcome to the farm, and work outside the farm? Farming activities must be organised so that they can coexist with and complement other uses of the area, other rural activities and the lifestyle quality of residents and tourists. The transformation of farming systems cannot be conceived independently of the non-farming players in the territory, and compromises, arrangements or synergies must be sought with them (Boiffin *et al.*, 2004). At the present time tools are lacking which could make it possible to anticipate and prepare the transformation of cropping or livestock systems around territorial issues: diagnostic tools to locate and give a rank order to the practices to be improved within a given objective;

tools to enable all the players in the territories to share a representation of the agri-ecological and socio-economic consequences of different scenarios of action and dialogue…

Innovation, at the heart of many tensions

Very diversified attempts at innovation

The transformations of production systems which will result from the combined efforts of these driving forces (Poux, 2006) will not occur without tension:

- The classic tension between economic and environmental requirements, illustrated in France by the tension between the high productivity, based on high yielding cultivars, and intensive use of fertilizers and pesticides, and the growing demand of society for a reduction in inputs, and pesticides in particular. The additional work load which handicaps the development of some environmentally-friendly practices can also be cited.
- The tension between individual farmer's decisions and territorial dynamics: for example, the choice of all the farmers in a region being concentrated on the most profitable activities, or on the 'best' breed of a given species, creates genetically homogeneous areas that are favourable to the spread of epidemics.
- The tension between sectors of a same territory, associated for example with competition between different productions for territorial resources (high potential soils, water…). For example, we can quote tensions between cattle farms in the Alps where contrasting productions exist in a heavily-urbanised environment: on the one hand 'cheese-producer' farms dedicated to high quality cheeses using local breeds, difficult lands and high summer pastures, and based on limited productivity per cow; on the other hand, very intensive « dairy » farms involved in mass production sectors, which are a heavy burden on the management of resources and the location of activities.

However, apart from general trends such as the inescapable and increasing influence of environmental questions, it is difficult to predict in ten or twenty years' time, evolutions in the socio-economic context, international agreements, public policies, local arrangements, movements of opinion and power struggles between pressure groups. So the relative influence of the various driving forces in transforming farming systems remains uncertain. Among the players in agriculture and Research and Development there are great divergences about the relative importance to be given to the different driving forces and therefore in the prospective definition of the innovations agriculture would need (Lémery, 2000). The diversity of possible futures is combined with the diversity of local situations (according to the soil and climate condition, the biological and cognitive resources, and the economic, social and cultural dynamics of the farming territories), endlessly multiplying the need for innovation (Meynard *et al.,* 2006).

In addition, it would be naïve to think that all the players whose decisions have an impact on farming systems (from farmers to breeders, from consultancies to public authorities, from agro-supply firms to the food business) have the same short or long term interests and objectives. The same innovation can therefore, according to the viewpoint of the many players, be considered as progress or regression; what is relevant for one can be considered by the other as inopportune, even disastrous. So there is no question of looking here for an illusory consensus on the types of desirable innovations or ideal farming systems. We propose you consider that it is necessary:

1. To prepare a diversity of solutions, to leave the choice to the players, and to be able to confront different futures.
2. To make it easier to anticipate the economic, social and environmental effects of innovations, and more widely, the consequences of their emergence and dissemination on the development of agriculture and territories. The challenge is to be able to appreciate rapidly not only what we hope to gain, but also what we risk losing.

The systemic effects of innovation

Resolving tensions and reconciling the contradictory requirements we have just mentioned will suppose that compromises are reached. In particular, new best practices will have to be sought for the use of inputs, and processes until now neglected will have to be mobilised within the agro-ecosystem (for example, several complementary species in the same field, the activity of macro-fauna in the soil, or interactions between the cultivated field and its edges), or the solution to contradictions insoluble at a given scale will have to be sought at other scales. Technological innovations (reproductive techniques, feed adjustments, veterinary products, aid to decision-taking tools…) will certainly have a role to play. However, simple (even simplistic) ways of thinking of the type 'a problem to solve, a technological innovation to solve it' have shown their limitations.

The case of pesticides is a perfect example of this. In Western Europe, in the 1970s and 1980s, people involved in agriculture, as a whole, lived with the illusion that pesticides would solve all problems of disease, pests or weeds. As other techniques no longer have authority to prevent these problems, they have seen their ways of thinking develop: to favour the expression of the production potential of crops and/or reduce working times, they have changed rotations, sowing dates, densities, the management of nitrogen or water, and methods of working the soil, and these actions are sometimes favourable to crop enemies. Thus, the systems became increasingly specialised on the most profitable species, which has increased the pressure of parasites and created a dependence on pesticides. For winter cereals, farmers brought forward the sowing dates, increased densities and doses of nitrogen fertiliser, and chose the most productive varieties (under-valuing resistance to diseases). All these evolutions are favourable to productivity, but equally to fungal diseases, pests and weeds: intensive cereal crops are very productive but, by construction, very demanding in pesticide treatments (Meynard and Girardin, 1991). So from technological innovations that were supposed to solve precise problems, pesticides became the pivots of production systems. We can now see the agronomic, environmental and health limitations of the immoderate use of pesticides. But it will only be possible to reduce their use significantly if we act on the whole of the crop system, to reduce at the same time the risks associated with bio-aggressors. Generally speaking, the strong coherence which governs farmers' technical choices, as much as the complexity of agro-ecosystem regulations, *invite us to think and act in a systemic way (Gibon* et al.*, 1999): to take into account the unintentional effects of technical choices, to consider the interactions between techniques, to reason by integrating several space and time scales, to consider the herd and the territory it uses as an ecosystem which must be well managed for production to be taken from it.*

The technical choices can also involve not inconsiderable social consequences, although, they are not usually anticipated by those who implement them, in particular for work organization (Madelrieux *et al.*, 2006). We will take the example of the mechanisation of milking, an innovation in principle useful for everyone and offering possibilities of simplification and reduction of work as well as movements (which allows farmers to take holidays). This innovation first of all poses the question of developing skills: setting the machine, disinfecting the equipment… but its adoption also supposes that it is accompanied by generalised electrification in the countryside. Failing which, setting it up can contribute to the rapid disappearance of farms unable to equip themselves with a milking machine, who are gradually marginalised and cannot find a buyer.

Similar issues of this type can be identified for the implementation of artificial insemination. Generally speaking, we must not forget to be careful to check the conditions required for an innovation, and its consequences in the diversity of situations encountered.

Developing the farming world's capacities for innovation

Research work in social sciences teaches us that innovation is not a linear process, with successive stages of research, design, development, production, and finally marketing. It results from permanent comings and goings between these 'stages': it is a collective and interactive process. The sources of innovation are potentially multiple; in fact, the innovations will sometimes be able to originate in research bodies, sometimes in development organisations, sometimes with farmers, but in all cases, the conjugation of the efforts of all will be necessary to consolidate the initial concept, to adapt it to the diversity of environments and farms, and determine its area of validity…

Organising processes of innovative design

Researchers in management sciences (Le Masson *et al.*, 2006) show that developing the innovation capacities of a structure or a collective requires its design activities to be organised and managed. Design is an active, intentional process, which simultaneously generates concepts and knowledge, in order to lead to new products and new technologies. But there is no one-to-one link between design and innovation, and every design is not necessarily innovative. In this respect, researchers in management operate a distinction between two design systems: regulated design and innovative design. In regulated design, which is the most frequent by far, the design objectives are clearly defined in advance and aim at gradual modifications to products or existing technologies. The skills necessary to innovate and the validation processes (prototypes, trials, tests, and indicators) remain unchanged from one innovation to another. This stability allows for the large-scale deployment of standardised working methods, and favours a clear division of work between research and development. In the field of livestock, regulated design could be illustrated by the essential of animal selection; or by the progressive improvement, over 30 years, in calculating livestock feed rations from feed value tables and an estimate of animal requirements.

However, regulated design is no longer suitable as soon as the design objectives are called into question (for example in agriculture by the emergence of new environmental or societal objectives), and because of this, new skills must be mobilised and validation methods revised. *So the term innovative design is used to indicate a process of exploration where it is not possible to specify in advance, the objectives, the required skills and the validation methods.*

The driving forces of agricultural development mentioned above clearly call for a considerable effort of innovative design. For example, the internalisation of environmental impacts in the design of crop and livestock systems does indeed constitute the source of a major change in evaluation methods: Evaluation scales now concern the catchment area or the landscape. In the same way, adaptation to climatic change will involve innovative design work: in fact, it calls into question the references of livestock farmers and advisors, questioning the relevance of know-how acquired in the past. To rebuild their capacity to project themselves into the future, farmers and advisors must combine knowledge acquired in their working territories and knowledge gained from climatic and agro-ecological models.

Work on the organisation of innovative design in firms shows that although the design must be closely linked both to research and to development, it must not be pledged to either of them. Two symmetrical obstacles can restrain innovative design:

- The idea, still shared by many researchers, that innovation comes naturally from the progress of knowledge. Experience of innovative design in the industrial field shows that the design activity must direct at least partially the acquisition of knowledge.
- The choice of many people in charge of agricultural development to work solely on systems which could be immediately profitable, under current technical and economic conditions. This choice curbs the exploration of potentially interesting systems.

To develop the design approach in crop and farming systems

Although, as in industry, regulated design in agriculture is widely practised and well controlled, innovative design will not be able to develop without a renewal of methods: how to integrate new objectives, define new knowledge to be incorporated into the innovation process, and reconfigure evaluation methods in a systemic approach (Donatelli *et al.*, 2007)? A review of the scientific literature leads to dividing the approaches into two large groups:

- The first group attempts *to improve the existing systems step by step*, to adapt them to the new objectives. The design work begins with a diagnosis: in what way do the present livestock or cropping systems satisfy old and new objectives? The diagnosis indicates the criteria which do not obtain a satisfactory result, the agronomic and ecosystemic functions which are in question and technical actions which should be changed to exploit these functions. On the basis of this diagnosis, evolutions of the farming systems are imagined, and implemented. Then a new diagnosis is made, new evolutions of the systems follow, etc. thus engaging in a true spiral of continuous improvement (Meynard *et al.*, 2002). These actions benefit from progress made in recent years in systemic analysis methods in situ, which make it easy to carry out precise and reliable diagnoses (Doré *et al.*, 1997; Doré *et al.*, 2006).
- The second group of approaches aims at opening the field of possibilities. The design of farming systems is based on functional models of crops, herds or farms (see for example Keating and McCown, 2001; Ingrand *et al.*, 2002; Loyce *et al.*, 2002; Duru and Hubert, 2003; De Wit *et al.*, 1988). These models are finalised representations of complex systems which include not only the plants or animals of interest, but also the living organisms associated with them, the climatic and edaphic environment, and all the human actions which apply to these various entities. Such models give a dynamic simulation of the performances of crops, herds and farms subjected to various technical and economic choices, by continuously simulating many variables (virtual experimentation). They predict the long term effects which are difficult to access by experimentation or observation, and integrate the effects of climatic variations (extreme years, climatic change ...). By using these models as a base, we can thus (1) identify, within the multitude of possible combinations, the technical acts that respect predefined requirements on criteria of production, costs, margins, work, and environmental impacts; but also (2) analyse the systemic consequences of the adoption of exogenic technological innovations, such as new varieties or livestock farming methods.

These 2 approaches are, in fact, complementary: model-assisted design is better suited to exploring systems that are breaking down, and analysing on computer their long-term risks or consequences, before they are actually implemented. Researchers are fond of this approach, which makes good use of scientific knowledge, synthesized by the models. Step-by-step design is more favourable to learning and moderate risk-taking; because of this, it lends itself better to progressive mobilisation of the farmers, and therefore to development approaches. The learning processes are here a condition and not just a consequence of the innovation process.

Favouring collective learning, to innovate at territory level

How can innovations relating to the coordination of farming systems at territory level be helped? This question often arises, for example about structures for product qualification (Geographical Indications for example), or for the collective management of remarkable environmental resources (irrigated perimeter, natural reserve).

But, now, it is also posed in ordinary rural areas where generalised coordination structures are imposed for traceability, crop protection or the prevention of risks (erosion...). It is a complex question, because the interests of the various players can be contradictory, their representations of the situation irreconcilable, or their information asymmetrical. Some cannot draw any benefit from a coordination of productive systems, or not perceive the benefits they could receive. The already existing networks of players can hinder (but also sometimes help) innovative coordination… Such questions are a special area of participatory research, which is particularly favourable to the analysis of learning processes, individual and collective, which accompany construction of standards, and the evolution of farming systems.

Among the participatory methods, 'Integrated assessment' (Bland, 1999; Pahl-Wostl, 2005) and Companion modelling (Collectif ComMod, 2004) are based on the use of computer models as mediator between scientists and stakeholders. Simulating the expected evolutions of the system according to constraints which are applied to it is a great source of learning. Comparisons of scenarios, debates and dialogues are organised around the results of the model, which lead to improving the scenarios, and favouring convergences between the stakeholders. The modelling of support and guidance is in the same group of methods, used for the collective management of territories with environmental issues. The model, of the Multi-Agent type, is constructed with the users, developing both their knowledge and that of the scientists. A role play, based on the model, makes it possible for the players to share their representations, without necessarily making them converge towards a single representation, thus favouring discussions on their actions or projected actions.

Conclusion

Innovative design must mobilize a collective and distributed intelligence, more effective, faster and less hazardous than individual approaches, that are dependent on dominant discipline or local representations (Lançon *et al.,* 2008). This supposes the organisation of complementarities and an increase in communication between the various players in the innovation: researchers and R&D engineers (collectives, largely represented in this symposium); farmers and breeders; the authorities…

The development of innovative design thus begins by inviting researchers and R&D engineers to use their expertise, their collaborations, and their practices. This evolution is illustrated by animal production research in livestock systems (Dedieu *et al.,* 2008). Until the mid-nineties, research at international level was above all situated at the interface with physiology (animal and plant) and genetics, leading to innovations relating to the methods of calculating feed rations, directing reproduction or evaluating breeding values. Today, research has explicitly moved towards interfaces with ecology (ecological management of grasslands and rangelands, soils and landscapes), and social sciences (methods of participative design for innovative systems, socio-technical systems). To successfully conclude this change, livestock farming system studies have a major asset: it is at the same time (1) a scientific discipline which produces generic knowledge, taught and shared with an international community, and (2) an engineering composed of a multitude of know-how controlled and perfected by livestock farming system experts engaged in the action.

Far from isolating research in its improvements, we must collectively recognize the major role played now and in the future by farmers in the design and development of innovative farming systems. Our job is to identify the key innovations for the future, analysing them, evaluating them, and suggesting adjustments to widen their field of validity. But it is also to develop co-design processes where the complementarity of various types of know-how is placed at the service of a joint project.

Finally, the authorities can play an important role in stimulating the design of innovative systems and their development: they must ask themselves how to encourage agro-supply companies to be involved in the collective activities of design (Aggeri, 2000), how to encourage a growing number of advisers, livestock farmers and researchers to become involved in these activities. To act on economic and social dynamics of great inertia, such as those which govern farming systems, will suppose consistency in the messages addressed to the people concerned, consistency of public policies and transparency in the evolution of regulations: isn't one of the obstacles to the development of sustainable farming the fact that public policies are not themselves sustainable?

References

Aggeri F., 2000. Les politiques d'environnement comme politiques d'innovation. Annales des Mines, Gérer et comprendre. Juin 2000, 31-43.

Allaire G., 2002. L'économie de la qualité, en ses secteurs, ses territoires et ses mythes. Géographie, Économie, Société, vol. 4 n°2, 155-180.

Bland, W.L., 1999. Toward integrated assessment in agriculture. Agricultural Systems, 60 157-167.

Boiffin J., Hubert B., Durand N., 2004. Agriculture et Développement Durable: enjeux et questions de recherche, INRA, 92 p.

Casabianca F. and Matassino D., 2006. Local resources and typical animal products, Livestock farming systems: product quality based on local resources leading to improved sustainability, Editors: R. Rubino, L. Sepe, A. Dimitriadou and A. Gibon, EAAP publication, N°. 118, 9-26.

Collectif ComMod (Companion Modelling, CIRAD-INRA-IRD), 2004. La modélisation comme outil d'accompagnement. Natures, Sciences et Sociétés, 13 (2) 165-168

Dedieu B., Laurent C., Mundler P. 1999, Organisation du travail dans les systèmes d'activité complexes: intérêts et limites de la méthode Bilan Travail. Economie Rurale 253: 28-35

Dedieu B., Chia E., Leclerc B., Moulin C.-H., Tichit, M. 2008. L'élevage en mouvement: Flexibilité et adaptation des systèmes d'élevage; Editions QUAE, Paris, 296 pages

Donatelli M., Hatfield J., Rizzoli A., 2007. Farming System Design. Int. Symposium on Methodologies on Integrated Analysis on Farm Production Systems, Catania, Italie, La Goliardica Pavese, 2 tomes, 230 et 270 p.

De Wit C.T., Van Keulen H., Seligman N.G., Spharim I., 1988. Application of interactive multiple goal programming techniques for analysis and planning of regional agricultural development. Agric. Syst. 26, 211-230.

Doré T., Meynard J.M., Sebillotte M., 1997. A diagnostic Method for Assessing Regional Variations in Crop Yield. Agric. Syst.; 54,169-188.

Doré T., Le Bail M., Martin P., Ney B., Roger-Estrade J., (Eds.) 2006. L'agronomie aujourd'hui, Editions Quae, Paris, 367 p.

Duru M. and Hubert B., 2003. Management of Grazing Systems from Decision and Biophysical Models to Principles for Action. Agronomie, 23: 689-703.

Gibon A., Sibbald A.R., Thomas C., 1999. Improved sustainability in livestock systems, a challenge for animal production science Livestock Production Science 61 107-110

Ingrand S, Dedieu B, Agabriel J, Pérochon L, 2002. Modélisation du fonctionnement d'un troupeau bovin allaitant selon la combinaison des règles de conduite. Premiers résultats de la construction du simulateur SIMBALL. Renc Rech Rum, 9, 61-64

Keating A., McCown R. L. 2001. Advances in farming systems analysis and intervention. Agricultural Systems, Volume 70, Issues 2-3, 555-579

Lançon J., Reau R., Cariolle M., Munier-Jolain N., Omont, B., Petit M.-S., Viaux P., Wery J., 2008.Elaboration à dire d'experts de systèmes de culture innovants; in Systèmes de culture innovants et durables: quelles méthodes pour les mettre au point et les évaluer ? R. Reau et T. Doré ed. Edicagri, Dijon, 91-108

Le Masson P., Weil B., Hatchuel A. 2006. Les processus d'innovation: Conception innovante et croissance des entreprises. Hermes Lavoisier, Paris, 471 p.

Lémery B., 2000. Les agriculteurs dans la fabrique d'une nouvelle agriculture, Sociologie du travail, 45 (1) 9-25

Loyce C., Rellier JP., Meynard JM. 2002. Management planning for winter wheat with multiple objectives (1): The BETHA system. Agricultural Systems, 72 (1), 9-31

Madelrieux S. Dedieu B. Dobremez L. 2006, ATELAGE: un modèle pour qualifier l'organisation du travail dans les exploitations d'élevage. Productions Animales, 19 (1) 47-58

Meynard, J.M., Aggeri, F., Coulon J.B., Habib R., Tillon J.P., 2006. Recherches sur la conception de systèmes agricoles innovants. Report for the INRA national head, 55 p.+ annex.

Meynard J.M., Cerf M., Guichard L., Jeuffroy M.H., Makowski D., 2002. Which decision support tools for the environmental management of nitrogen? Agronomie, 22, 817-829.

Meynard J.M., Girardin P., 1991. Produire autrement, Courrier de la Cellule Environnement de l'INRA, 15, 1-19.
Moity-Maïzi P., Sainte Marie C. (de), Geslin P., Muchnik J., Sautier D. (eds) 2001. Systèmes agroalimentaires localisés: Terroirs, savoir-faire, innovations. Etud. Rech. Syst. Agraires Dév. 32, 216 p.
Pahl-Wostl C, 2005, Actor based analysis and modelling approaches. The integrated assessment Journal, 5, 97-118
Poux X. (coord.) 2006. Agriculture, environnement et territoires; quatre scénarios à l'horizon 2025. Un exercice de prospective du Groupe de la Bussière. La Documentation Française, Paris, 222 p.

Broad-view research in the Mediterranean system

A. Rosati
European Association for Animal Production

Abstract

Animal research in the Mediterranean Area, while sharing the global trends, has specific characteristics determined by the dynamics of demand for animal products and consumers' requirements as well as by the natural conditions and production systems in the area. Demand for animal products in the Mediterranean region will be determined by the population growth, urbanization, increased purchasing power and special requirements of consumers. It is expected that the future demand will be met by imports from more efficient producers, local industrial landless systems and local traditional pasture-based and crop-livestock production systems. Research for the future animal production in the Mediterranean systems should focus on increased efficiency of pasture and crop-livestock production systems. This includes the efficient utilization of local genetic resources and local feed resources, innovation and improvement of traditional processing technologies and guarantee of quality and safety of local products. Priority research agendas at national and regional levels, such as EU technological platforms, are conducive to the attainment of these objectives in joint research programmes encompassing both banks of the Mediterranean.

Keywords: production systems, genetic resources, research agenda, risk analysis

Drivers of change in demand for animal products in the Mediterranean

Throughout history, the Mediterranean area has been characterised by a higher density of population in comparison with the central parts of continents surrounding the Mediterranean Sea. It is expected that the number of people living in the Mediterranean area would continue to grow on the both banks of the Sea. In the South, it is expected that the trend would follow patterns observed in other developing regions[1], while in the North, migration – both internal and external – would be the main factor of population growth. Naturally, increase in population creates increase in demand.

Animal product consumption rises also along with purchasing power. The effect of increased income on diets is greatest among lower and middle-income populations. (Delgado *et al.*, 2002). In the South, the volume of consumption of animal products will grow faster than in the North, where there are strong signals of the saturation of demand. However, changes in the structure of consumption and demand are following global trends.

Urbanization and urban lifestyles have been the historic characteristic of the Mediterranean area. However, recent and actual trends have resulted in the appearance of large, still growing agglomerations of people with millions of inhabitants (Istanbul, Cairo, Athens, Naples, Barcelona and others). Urbanization is accompanied by changes in habitual food consumption patterns and shift from cereal-based diets to diets with high animal protein and fat content. Changes in lifestyle also influence demand for new type of products, such as convenience food, functional products, pre-packed and pre-cooked products (FAO, 2007).

Consumer tastes and preferences of the Mediterranean people have some common and shared elements, typical of the famous Mediterranean diet. However, food preferences vary among and within countries and regions, mainly under the influence of cultural factors, including religion and

[1] In 1980, three quarters of the world population lived in developing countries. It is estimated that by 2030 developing countries may account for 85% of the world population (FAO, 2007).

availability of specific local products. Nowadays, consumer tastes and preferences are influenced by globalization, availability and offer of imported products via distribution channels and changes in the lifestyle. Concerns about health, environment, animal welfare, local communities – specific local products – and other factors that are influencing food preferences are also present in the Area, mostly among urban populations in the North.

Farm animal production systems

All three large groups of farm animal production systems, landless industrialized and small-scale production systems, grassland-based systems and mixed farming systems exist in the Mediterranean.[2] Industrial livestock systems account mainly for poultry meat and egg production, and, in the North, for pig production. There are also some medium to large dairy farms and beef fattening units. Industrial models never completely succeeded in imposing themselves as an absolute reference, due to cultural reasons, but also for reasons of adaptation of practices to local characteristics.

The Mediterranean areas were much less concerned by the homogenization of livestock systems than the northern regions of Europe. Mediterranean livestock farms have a strong diversity from the point of view of the species reared, the production, and the levels of intensification (forage in particular) and livestock management practices (de Rancourt *et al.*, 2006). They are characterized by a combination of grassland-based and mixed crop-livestock production. Livestock in cereal production systems is the special feature of the Area.

This diversity results from various forms of adaptations developed by farmers to adjust herd management to the territory they have at their disposal. They contribute to maintaining a mosaic of natural and cultivated areas, which contributes to the landscape and environmental quality of these areas (Riedel *et al.*, 2007), e.g. in the Mediterranean region 80% of the biodiversity of species. Livestock farming thus represents a structuring activity both from the point of view of its economic importance and its role in the structuring of space. At the same time, livestock production networks have preserved for their products strong images related to the local 'territoire', and emphasize the link between farming practices and the territory.

Livestock farming under such conditions in a way predisposes farmers to program their activities with regard to local and territorial dynamics, and in a convergent way high demands are made of them in this direction. The question is therefore relatively open concerning the forms of livestock farming it would be desirable to develop for the future, and the synergies and complementarities between different work management systems. Farming practices are strongly influenced by the demand for specific products. Reacting to market signals, in less than 40 years farmers used to switch from traditional pastoral systems to the mixed one or vice versa. (Napoleone, 2008).

Farm animal genetic resources

The Mediterranean is characterized by a marked diversity of farm animal genetic resources. Country reports prepared by the four countries of the South on the state of AnGR refer that there are 18 local breeds of cattle, 31 breeds of sheep, 12 breeds of goats, 7 breeds of camels and 5 breeds of buffaloes. In addition, there are also farms keeping exotic and mainstream breeds and their crosses with local breeds.

In accordance with country reports, in eight countries of the North there are 116 breeds of cattle, 183 breeds of sheep, 51 breeds of goats, 7 breeds of buffaloes and 52 breeds and lines of pigs. (FAO, 2007)

[2] FAO classification, FAO 2007.

Although many breeds reported by one country are present also in neighboring countries (trans-boundary breeds), these data confirm the biodiversity as a great and un-tapped development potential of the animal production in the area.

Options for research agenda

Forecasting the future of animal production and science is based on the knowledge and accurate evaluation of past and current developments and trends. This affirmation, apart from being an axiom in every science, implies a high responsibility for scientists and for all those who are involved, directly or not, in animal science and production: breeders, industrialists, policy-makers, even consumers. Our decisions and actions today, either individual or collective, contain a projection for the future, and that is something we should always keep in mind. Short-sighted thinking has proved to be harmful, if not devastating, for every field of science – even more in our science that has such close links with the environment and the human factor.

To visualize science and animal production in 2020 means identifying the major issues that will have to be dealt with in the following 10 years, evaluating the long-term outcome of our current actions and decisions and forecasting the internal and external context of animal science/production: the character and dynamics of animal science itself and the nature of the technological, environmental, industrial and social framework 10 years from now.

Research has been largely responsible for the remarkable progress of livestock production in the second half of the past century. The main beneficiary has been the consumer enjoying the constant decrease of costs of food and of the share of expenditures for food in the total costs of living. The quality and safety of animal products have been significantly improved. At the end of the century, there was a shift from output enhancing research towards broader objectives, including environmental protection, improved food quality and safety and animal welfare. (Cunningham *et al.,* 1993).

The future strategic orientation of the animal research in the Area needs to take into account both aspects: enhancing outputs, mainly from local animal genetic resources and feed base, and improving quality and safety of products. Taking into account the fragility of the Mediterranean ecosystems, research needs to find solutions for the future development of technologies of farm animal production adjusted to the conditions of the area. The applicability of production practices and technologies applied in other regions and ecological zones needs to be critically examined.

Development and utilization of farm animal genetic resources

Countries of the Northern Mediterranean coasts have established structured breeding programmes for almost all breeds of economic importance as well as small and structured programmes for conservation covered endangered breeds. In countries of the South, there is a lack of information and data on phenotypic, functional and molecular characteristics of some local breeds as well as data on performance of crosses in the given environment and production systems. To fill this gap, it would be necessary to give priority to the full characterization of local breeds as the basis for structured and market oriented breeding programmes and strategic decisions regarding the future development of animal production in these countries. Research should find new and innovative methods of farm animal characterization. Available techniques should be adjusted to the local conditions and production systems. Regional cooperation in development of new techniques and application of up-to-date methods and techniques are crucial for the success of such an endeavour.

Conservation of AnGR

In the short to medium period, many local breeds, particularly in the South, will prove to be economically inefficient in supplying the growing demand for animal products and will run the risk of becoming obsolete and extinct. To prevent this, a new approach to in vivo and in vitro conservation is required. This may include a search for new techniques as well as improved regional cooperation in establishing joint projects and conservation programmes, starting with trans-boundary breeds. The efficiency of the existing programmes in the North based on conservation through sustainable utilization for productive and non-productive purposes needs to be evaluated.

Management of collective lands and pastures

Collective lands and pastures in the Mediterranean, in spite of the introduction of intensive production systems for some species, still represent the basis for small ruminant and, to a lesser extent, for cattle and buffalo productions. Their potential productivity under the optimal management conditions, nutritional value of the flora and methods to increase the production of biomass still needs to be ascertained. There is a need for new technologies in processing locally available biomass so as to increase its palatability and nutrition value.

In some countries the existing legal system of management of collective – communal – lands represents an impediment to the rational management and full utilization of pastures. Legal and socio-economic aspects need to be analysed and evaluated in order to propose viable solutions to policy makers and legislators.

Quality and safety of all types of animal products

Consumers require quality and safe animal products. The existing techniques and practices in industrial operations, in general, guarantee that the product is safe and processed in accordance with standard practices. Traditional local products, in addition to their appeal to consumers stemming from origin, taste and methods of processing, need to improve their image of quality and safe products. There is a need for development of a multi-disciplinary approach and confrontation of the traditional methods of processing with safety requirements and implementation in practice. In the North, a critical review of EU regulations and experiences in their implementation is required. In the South, there is a need for adjustments of traditional technologies to food safety requirements.

Animal health

Methods of protection of animals from epizootic diseases and zoonoses need to be further improved. This includes development of cheap and efficient methods of animal identification, registration and control of movements, new biological products and means of immunization as well as development of production technologies, which would incorporate sanitary and preventive measures in daily operations. In the globalised world, the highest level of protection could be guaranteed only by the simultaneous and harmonized policies and practices in all parts of the world. However, more efficient regional cooperation in research and in the implementation of measures could be the great step towards safeguarding and protecting animal industry in the Mediterranean.

Innovations

Without innovation in science there can be no step forward: the lack of innovation means stagnation and decline. These general statements, universally accepted, do obviously apply to animal industry. Animal science is the source of applied innovation for the livestock industry. Therefore, the question is about how to stimulate innovation through new ideas and new developments in the field of livestock science. The minimum requirement, that is good education, must be provided by universities. Are university courses in Europe and in the Mediterranean area at the same level as in America, in Asia or more generally in the rest of the world? Of course, all European university courses are not at the same level. There is a large variability within and among EAAP member countries, making it difficult to establish if our universities provide better or poorer education in relation to the USA universities, for instance. Anyhow, it can be said that the average level of education is good, and there are also some cases of excellence among Mediterranean universities having nothing to envy to the best faculties of the academic world.

The second crucial factor encouraging innovation is the possibility to establish the proper environment for stimulating many different points of view. Up to now, the weakness of some old structure of EC-funded project– like Networks of Excellence and Integrated Projects – was that they allowed funding only a few (often only one) projects for a specific scientific field. Such formats thus sharply reduce the number of possible approaches to the same problem, consequently diminishing the opportunity to stimulate innovation. Wisely enough, such projects are not included anymore in the Seventh Framework Program.

Diversity is the right attitude in order to promote innovation. A strongly standardized culture does not allow 'others' to exist or to produce new ideas. Any society – even the scientific society – characterized by homogeneity and the will to maintain achieved standards will not stimulate the exploration of new ideas. On the contrary, a heterogeneous society will encourage and accept new ideas. The EAAP scientific group is clearly characterized by the diversity of national scientific communities; diversity which is sometimes evident also within each national community. When we work as 'Europe' or as Mediterranean and not as separate national groups we do have the best conditions to build up an 'innovative environment'. In Europe, thanks to the European Commission's research funding schemes, the exchange of ideas and the mobility of scientists became part of our professional lives. Nowadays, we think more often in terms of Europe as a whole than in terms of single countries. We should consider that the first European treaty was signed more than fifty years ago and, as a consequence, some activities within the continent have inevitably become more 'European' than others. Science is one of the fields where the European Union has invested more. The same did not happen in the southern side of the Mediterranean area, there is lack of unification of efforts although the cultural links are stronger.

Another prerequisite for innovation is allowing for the possibility to make mistakes. If scientists are, in one way or another, not allowed to fail, they will constantly choose the way of small but safe innovation. The rule of 'taking one step at the time' is easier and safer than 'taking the elevator'; however, it slows down the progress of science and does not allow exploring all possible prospects. We certainly need to learn from our mistakes, but first we should be allowed to make mistakes. In this respect, other cultures – especially Asian but also North American and North European – are generally much better than the Mediterranean ones. In the culture of this area, the possibility to take risks is almost unknown, especially as regards scientific research in the animal sector. Only few scientists who enjoy significant international acknowledgment can present innovative ideas, with the possibility to make mistakes, and have their ideas accepted in scientific meetings (where selection is sometimes done for the name of the author than for the idea of the research). However, these people are almost exclusively senior scientists. The consequence is that young scientists are hardly ever encouraged to take the risk of presenting unusual and innovative ideas. Unfortunately, experience is more important than imagination for our culture. There should preferably set aside the prestige attributed to age and create instead a better environment to welcome innovative ideas by young scientists.

The possibility of sharing information is another key factor for innovation. In a time when every possible patentable research is protected, exchange of information is often poor. Luckily enough, modern means of communication like Internet offer the possibility of learning about research activities and results from every part of the world in real time, thus offsetting largely the limiting effect of patenting.

Another major element in creating innovation is interdisciplinarity. However, for the Mediterranean and, more general European culture, the structure of the universities includes a large number of small departments, specialized in specific research fields and with few points of contact with other units. There are cases of scientists participating in extremely specialized scientific meetings, which are always attended by the same small groups. Such scientists, who enjoy being part of a restricted circle of 'best researchers', will rarely produce innovation. Great ideas can only be born through the debate with scientists of other disciplines and by sharing knowledge about similar problems solved with apparently revolutionary methods. This certainly does not happen in 'specialised meetings'. Here lies one of the key success factors of the 'large spectrum' animal science meetings organized by EAAP and by the American Society for Animal Science. In the era of extremely specialized disciplines, the increasing number of people attending the EAAP and ASAS meetings shows that many scientists have understood the importance of interdisciplinary debate.

The recent economic restrictions on funds in support of research are also creating a vast damage to innovation. The first cuts in investments in science were directed to those research subjects with the largest risks of failure. Such funding policy helps development rather than pure innovation. Another obvious major risk is the career policy in many of the Universities of the area. Researchers move fast up the career ladder mainly on the basis of the number of scientific articles they produce. However, what about the time to achieve continuous learning? Many scientists complain about the impossibility to be constantly updated with papers and books due to lack of time. Fortunately, this is not always the case, but the problem is common and real. Lacking adequate knowledge for planning future research is a major risk: if research is not tightly connected to knowledge, the consequence is poor research planning.

The great fragmentation of animal industries is also an additional factor holding back innovation in animal science. The inadequate possibility of animal industries to invest money in 'blue-sky thinking' is another impediment to improvement. In the fields where animal industries are larger in size, like specific sectors of animal nutrition or genetics, the speed of innovation is higher.

Historically, those who have had the ability to make great progress in science have always been people equipped with an interdisciplinary approach, great erudition and ability to take risks. Today, if we want animal science to experience considerable innovation, we must create the proper environment and reward those young researchers who have these qualities.

Animal science and risk analysis

One of the main risks for animal science is to give more importance to what will have to be done next than what was left behind. It lacks the culture of considering the importance of the consequences of the discoveries and their application. To find a proper solution, the research activities in the Mediterranean region, but is the same for every situation, should be able to develop a field of animal science based on the long term consequences of its application. Of course this is even truer for fragile environments like those of Mediterranean area. The application of risk analysis methods is necessary. Those methods are already often used for other fields of science (like engineering, for instance) and animal scientists just have to change the attitude and to comprehend the importance of the application of their results for livestock system, for the environment, the economy and human health. The example of what happened for the 'mad cow disease' should teach that a better applied risk analysis would have been obviously very useful.

Conclusions

Animal research for the Mediterranean systems as a part of the global science and research should rely on achievements, methods and approaches developed both in the local research units and abroad. Specific tasks of animal research in the Mediterranean area encompass themes and problems specific for the Mediterranean environment and ecosystems, trends in demand for animal products and consumers' requirements. Research must aim at innovative approaches in attaining multiple objectives in line with economic and social functions of animal production in the area: increasing local animal production, improving safety and quality of products, adjusting production systems and technologies to broader social objectives, such as poverty reduction, rural development and employment.

References

FAO 2007, The State of the World's Animal Genetic Resources for Food and Agriculture.

Delgado, C., Rosengrant, M. and Meijer, S. 2002. Livestock to 2020: the revolution continues. World Brahman Congress. Rockhampton.

Napoleone, M. and Lasseur, J., 2008. The contribution of livestock farm diversity to ensure the sustainability of small ruminant systems in the Mediterranean region. EAAP 2008, Vilnius

Van der Zijp, A.J., Boyazoglu, J., Renaud, J. and Hoste, C., 1993. Research strategy for animal production in Europe in the 21st century, EAAP Publication No. 64.

Innovations arising from applied research on a new on-line milk analyzer and a behavior meter

A. Arazi[1], N. Pinski[1], T. Schcolnik[1], E. Aizinbud[1], G. Katz[1] and E. Maltz[2]
[1]S.A.E Afimilk, Kibbutz Afikim 15148, Israel
[2]Institute of Agricultural Engineering, ARO, The Volcani Center, P.O. Box 6, Bet Dagan, 50250, Israel

Abstract

Two new sensors recently developed by S.A.E. Afikim, an on-line milk analyzer (Afilab™) and a behavior meter (Pedomer+™), provide new data at higher resolution than was previously possible. Applied research was performed to identify beneficial tools for dairy herd management from the new data that became available. A controlled large-scale field trial was carried out by the S.A.E. Afikim Applied Research Team in a commercial dairy herd of 800 milking cows in Israel from May/07 through July/07. Milk analyzer data (milk solids and SCC) were recorded three times daily for all the milking cows, and were compared to reference laboratory data. At the group and herd level, the analyzer can be used as a reliable detector of nutritional problems. Small differences in the daily calculated bulk tank fat and protein were found between the analyzer and the laboratory data (-0.05% - +0.28% and +0.01% - +0.05%, respectively). These results show that the analyzer is a useful tool to estimate payment returns to the dairies. Under milk formulas corrected for the Israeli economy, the difference between the results of the reference laboratory and the on-line analyzer was less than 1.1%. At the level of the individual cow, milk analyzer data can be used for several purposes. Real time on-line milk separation based on milk protein content can be performed on the farm. A difference of 5.2% in high quality (protein>3.2%) milk volume was found between the analyzer and laboratory data. This opens new options regarding individual feeding by using the milk components to formulate precise rations suited to the needs of individual cows. Other applications concerning detection of metabolic diseases and clinical and sub-clinical mastitis are currently under study with promising initial results involving multivariable models. These models include low fat and fat-to-protein ratio as indicators for metabolic disorders, and decreases in lactose for the detection of mastitis. Applied studies for uses of the behavioral meter were conducted by researchers from the *Israeli Agricultural Research Organization - The Volcani Center*. The results of these studies together with observations made on commercial farms revealed several useful applications derived from behavioral data (lying down time, lying down frequency and activity). At the group and herd level it was found useful for early detection of stressful events such as extreme climate and noise disturbance. It is also a promising tool for defining bedding conditions and to signal changes in farm routine. At the level of the individual cow, the data can be used to predict calving time, and will be examined for heat detection in tie stalls. The authors believe that the derived applications will improve management of commercial dairy herds. Further research will allow better understanding of complex behavior of dairy cows which will certainly reveal more applications for this new data.

Keywords: milk analyzer, milk components, behavior meter, lying behavior, animal welfare

Introduction

The milk industry has undergone extensive changes in recent decades. The number of farms is dropping while the average number of animals on each farm is increasing sharply. Automated recording of animal data is becoming essential for precision dairy farming. In the last three decades, new technologies have enabled acquisition of new data at significantly higher resolution than was previously possible. The advantages of these new data and the benefits they offer farmers have not yet been fully clarified, and require more applied research. Thus, firms developing new technologies not only wish to cooperate with academic institutions but also set up applied research groups of their own.

In this paper, two new sensors for automated data collection developed by S.A.E. Afikim are described. A milk analyzer (Afilab™) and a behavior meter (Pedometer+™). These sensors are already in use in several commercial and research farms.

Here we present the preliminary results of a study carried out by S.A.E. Afikim Applied Research Team (AART) in a commercial herd. Results of supplementary research conducted at the Israeli Institute of Agricultural Engineering, Agricultural Research Organization - The Volcani Center are also described

The development and assimilation of the new technologies called for multidisciplinary cooperation between veterinarians (epidemiologists and clinicians), animal scientists, nutritionists, economists and experts in physics, chemistry and technology (engineers and computer programmers, statistics and data processing). The model studies and clinical research conducted by multidisciplinary research groups yielded the results presented here.

Milk analyzer (Afilab™)

The Afilab™ (Figure 1) is a novel device that performs real-time analysis of individual cow milk solids (fat, protein and lactose) and gives an indication of blood and somatic cell counts (SCC). The technology is based on spectroscopy and thus does not interfere with milk flow through the line nor alter the milk in any way. An Afilab™ device is installed in each stall and analyzes each individual cow's milk at each milking session. The acquisition of milk component data takes place in the same configuration and time frame as milk weight data measured by the electronic milk meter. Currently, information on milk components is only available (if at all) from periodical milk tests.

Maltz *et al.* (personal communication) demonstrated significant fluctuations in individual cow milk component values during one milking session, between daily milking sessions, between days, and throughout lactation. These results emphasize the importance of frequent measurements of milk components and open a new horizon for applications.

A comprehensive field study was performed by the AART in a commercial dairy farm milking 800 cows three times daily from May/07 through July/07 (Schcolnik *et al.*, 2008, unpublished data).

In this study, milk components (% fat, % protein, % lactose, SCC) data as measured by the analyzer were recorded for all cows at each milking session. As a reference, milk samples were collected from each cow weekly during the noon milking, and analyzed for the same components at the Israeli Central Milk Laboratory (Combi MilkoScan™, Foss Analytical A/S, DK-3400, Hillerød, Denmark).

Figure 1. The Afilab™ (indicated by arrow) installed in the parlor on the milk-line following the milk meter.

Blood samples from each cow in the fresh cows group (cows up to ~21 days postpartum) were collected three times a week and B-hydroxybutyrate (BHBA) levels (mmol/liter) were measured as a reference for energy balance (EB)/ketosis status. Statistical analysis was performed with statistical software (JMP® 6, SAS Institue Inc., NC 27513-2414, USA). Partial results of the applications are described below.

Derived applications

Control of nutritional status in the herd. Evaluating the calculated ration fed to the cows is very easy to perform but evaluating the ration actually consumed is much more complicated. 'If milk components are utilized in conjunction with a herd health program, it allows for rapid detection of nutritional changes which may result in metabolic and/or reproduction problems' (Eicher, 2004).

With the analyzer, milk solids (fat, protein and lactose) are measured and used for group or whole herd analysis. This could be used to reveal any exception to the norm, and represents a powerful tool for early detection (at each milking) of nutritional problems (fat depression, protein or fluctuations in the fat-to-protein ratio). Until now, these changes could only be detected at the herd level after a delay of several days, when the results of the bulk milk tank test were received.

The comparison of fat and protein percentages calculated by the analyzer with the results of the milk test data from the whole herd showed differences of (-0.05%)-0.28% and 0.01%-0.05% respectively. These results allow the use of the analyzer as a detector of nutritional problems (such as energy deficiency or moldy feed) at group and herd levels.

*Follow up of dairies' milk payment*s. In many countries the price paid for milk is affected by the amount of fat and protein as well as by volume. Penalties or extra payments are also based on the level of SCC. Data on the fat, protein and SCC concentrations of individual cows are only provided today, if at all, by DHI or private milk laboratories through routine periodic milk tests. With the analyzer this could be performed for every milk shipment. In our study, the payment according to the Israeli formula for economy corrected milk (ECM) calculated using data from the analyzer and from the milk tests showed differences of -1.1% to -0.0% in three monthly tests.

Milk separation by quality. Milk composition (especially percentage protein) is of great importance for the cheese industry. Dairies spend a lot of effort and money on channelling milk according to its composition. Until now there was no option for real-time separation of milk according to its composition on the farm.

On-line milk analysis allows the separation of milk during milking, enabling it to be sold in different categories based on its contents.

In milk separation based on milk protein content (high protein>3.2%, low protein≤3.2%) during nine noon milking sessions, the difference in high protein milk volume between analyzer and laboratory results was 5.2% (range of -4.8% to 14.3%).

Individual feeding. Individual feeding enables a precise ration to be formulated for each cow according to its nutritional needs. This feeding method is particularly important when concentrate supplements are needed (pasture management, robotic milking, fresh cows or other non homogenous groups). In addition, with the global rise in the prices of feedstuff, the economic importance of precise formulation of individual feed is increasing.

Currently, individual feeding is based on the results of periodic milk tests, or on milk production regardless of its composition. Using the analyzer, rations can be adjusted periodically as desired.

We applied the NRC 2001 formula to predict dry matter intake (DMI) and nutritional needs comparing results of the analyzer with those of the milk laboratory incorporating other available data (milk yield and body weight recorded by AfiFlow™ and Afiwighet™, respectively).

When we performed the calculations for an average milking cow, 21 to 150 days in milk (DIM), in three different months, the differences between the analyzer and the laboratory results were 0.6 MgCal/day (1.3%) and 4 grams metabolic protein (0.1%) per day per cow in the energy and protein balances, respectively.

Figure 2 gives an example of decision making in allocating concentrates when the cow is fed forages *ad libitum* in the common trough, and all the concentrates are allocated by computer controlled self feeders. The decision regarding the amount of concentrates to allocate was made according to the 4% fat corrected milk (4% FCM) yield. It can be seen that the 4% FCM calculated from daily data on milk fat content (three days running average), measured by the milk analyzer was significantly lower than the one calculated using periodic data on milk fat.

The decrease recorded in milk fat by the analyzer that would normally not be detected until the next milk test would have resulted in an overestimation of 3-5 kg/day of 4% FCM based solely on milk test data. Allocation of concentrates according to the periodic milk test amounted to a higher ration of almost 2 kg DM/day. This would have caused a waste of about 60 kg DM of concentrates over a period of 30 days (between days 14 and 45 days after calving), most likely leading to a further decline in milk fat (E. Maltz, personal communication).

Diagnosis of metabolic diseases. Associations between ketosis and Sub Acute Ruminal Acidosis (SARA), with milk components have been suggested.

SARA is defined as temporary periods of depressed ruminal pH below 5.0-5.5, and is common in modern dairy herds. Clinical diagnosis is based on primary symptoms (reduced or erratic feed intake, reduced rumination, mild diarrhea, foamy feces), and on direct diagnosis of reduced ruminal fluid pH. Nordlund *et al.* (2004) suggested that when more than 10% of the milking cows (in Holstein herds) show milk fat percentages lower than 2.5%, SARA should be suspected. Tomaszewski and Cannon (1993) suggested that rumen acidosis is present when the fat-to-protein ratio (FPR) in the milk is below 1.0. The associations between SARA and milk solids could not be tested in the present study because the observed herd did not contain cows with milk fat lower than 2.5%. A possible association will be explored when suitable commercial herds become available

FPR was suggested as a potential indicator for EB (Grieve et. al. 1986), and FPR values of 1.35-1.5 were found in different studies as an indication of negative EB (Heuer et. al., 1999 and Heuer and Pflug, 1993). Friggens and co. argued that automated sampling and on-line milk analysis may

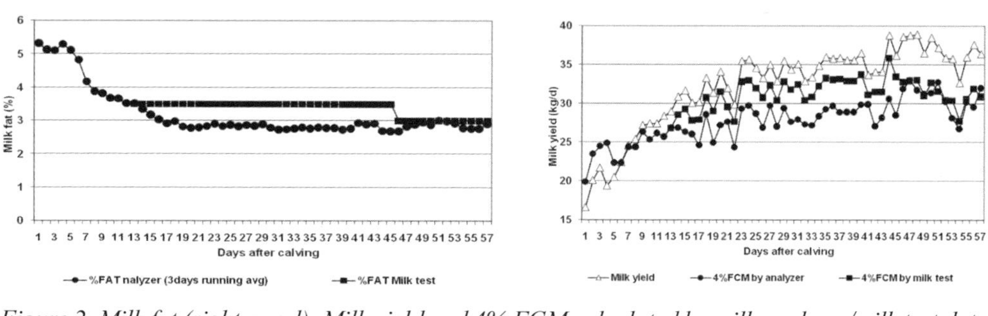

Figure 2. Milk fat (right panel), Milk yield and 4% FCM calculated by milk analyzer/milk test data (left panel) of cow 2823. E. Maltz (personal communication).

improve the accuracy of EB determination (Friggens et. al., 2007). FPR is also used successfully to indicate negative EB in routine Israeli Herd Health analysis (Nir-Markusfeld, 2003).

We compared the agreement between BHBA blood levels and FPR in samples derived from the analyzer and the laboratory respectively, for various cut offs. Some of the results are presented in Table 1.

The sensitivities we established were not satisfactory as a diagnostic tool either using data from the milk analyzer or from the laboratory. Nevertheless the linear correlation between BHBA, FPR and additional automated recorded parameters (such as milk yield, activity, conductivity, BW and DIM), suggest that future models based on other Afimilk data will improve these results.

Retrospective analysis – Data from the analyzer (fat and protein) was used to evaluate the influence of some risk factors (such as EB status) on production parameters in retrospective analysis models (Markusfeld-nir, 1996) and found to be satisfactory.

Applications under research and development

Early detection of mastitis. Mastitis affects milk quality and cow productivity. It has a significant impact on the dairy farm (economically and labor wise). Early detection of mastitis, clinical or sub-clinical, and early treatment prevents milk production loss (quantity and quality) and saves labor time.

Many studies have described the correlation between milk components and udder health (Bansal *et al.*, 2005; Leitner *et al.*, 2004;Nielsen *et al.*, 2005). Pyorala (2003) suggested that the lactose level and electric conductivity of the milk are two of the most promising parameters for monitoring sub-clinical mastitis.

Our observations in a number of commercial dairy farms support these results. Initial analysis of 21 clinical coli mastitis events on one large commercial farm revealed that a decrease in lactose had high specificity to coli-related mastitis. However the degree of sensitivity was not satisfactory.

Using multivariable models with additional data from other sources (milk meter) is a promising way to improve these initial results. Thus, a field study needs to be conducted in order to perform a statistical validation and to develop an application based on data from the analyzer.

Table 1. Analyzer fat-to-protein ratio (FPR) versus B-hydroxybutyrate (BHBA)[1] (>1.4 mmole/l) in blood (n=345).

FPR	Afilab data			Laboratory data		
	Sensitivity (%)	Specificity (%)	Accuracy (%)	Sensitivity (%)	Specificity (%)	Accuracy (%)
>1.2	59.3	56.1	57.1	90.3	37.4	47.1
>1.4	33.3	82.7	67.2	45.2	75.5	70.0
>1.6	11.1	92.4	67.0	25.8	92.8	80.6
>1.8	2.8	98.3	68.4	6.5	97.8	81.2

[1] 31.3% of cases with BHBA >1.4 mmole/l.

Behavior meter (Pedometer+™)

The Pedometer+™ is a new leg tag that continuously records activity (number of steps), lying down time and lying down frequency. It is based on a 3-dimensional sensor that detects the position of the animal's leg. The concept is to determine the routine behavior of the animal (at individual, group and herd levels) and to define deviations from the daily routine. Such deviations are potential indicators of welfare, health, fertility, production and stressful events.

The results presented here are based on two studies conducted by groups led by E. Maltz, from the *Agricultural Research Organization - The Volcani Center*, and observations performed by AART in a commercial farm equipped with the Pedometer+™

Study 1. Lying down behavior of dairy cows under different housing systems

Animal welfare and comfort is an emerging issue in modern agriculture driven by consumers and animal organizations. Poor welfare usually leads to greater susceptibility to disease (Broom and Corke, 2002). Weary and Tucker (2003) suggested that monitoring standing and lying down behavior may be a suitable approach when cow comfort is assayed. Moreover, with the aim of maintaining good health, welfare and high levels of productivity, it is essential that dairy cows have enough time to lie down and rest (Tucker et. al., 2004). Furthermore, blood flow to the mammary gland is increased by 24%-28% when cows lie down (Cook 2003 based on the studies of Metcalf *et al.*, 1992 and Rulquin and Caudal, 1992), and this may improve the delivery of nutrients to the udder and hence increase milk yield (Cook, 2003)

Study 1 was performed in a commercial dairy farm ton which 550-600 cows are milked three times a day. In the first trial, the lying down behavior of 12 multiparous cows was recorded in a roofed barn with loose stalls, under comfortable thermal conditions. In the second trial, the lying down behavior of 16 first lactation cows was monitored in two adjacent barns: a completely roofed barn with loose stalls (8 cows) and a completely roofed barn with free stalls (8 cows). In the third trial, four cows from each group in the second trial switched groups. Activity and performance were monitored by the AfiFarm management system (S.A.E. Afikim, Israel). Significant differences were evaluated using Student t-tests, unpaired – between groups, paired – between periods and treatments of the same cows (Livshin *et al.*, 2005)

Results

The results of the study and observations made in commercial farms showed that lying down behaviour has a constant pattern during the day (data not shown), the longest periods of lying down being during the night, and the shortest being at noon (Livshin *et al.*, 2005).

Different patterns of lying down behaviour were found in the two different housing systems examined. It was found that cows rested longer in barns with no stalls than in barns with free stalls (Table 2). When cows were switched from the barn with no stalls to the barn with free stalls, they spent less time lying down in all three time intervals (morning, noon and night) and the reverse was found when cows were switched from the barn with free stalls to the barn with no stalls (Livshin et. al., 2005).

We suggest that such changes could prove to be powerful indicators of changes in cow comfort and can therefore be used to evaluate housing management, and improve farm routine or facilities.

In addition to the above studies, on several commercial farms, we observed that the lying down behaviour of the cows changed due to stress events such as extreme climatic conditions (very hot days or strong cold wind), noise disturbance, and bedding quality. Figure 3 shows an increase of

Table 2. Cow lying down time (mean±SD) in between-milkings diurnal intervals of two groups of four cows that were transferred from one type of barn (period 1) to the other type (period 2) (from Livshin et al.*, 2005).*

Time interval	Period 1		Period 2	
	No stall	Free stall	No stall	Free stall
04:30-12:30	153±41[*a]	120±39[*]	178±55[a]	126±55
12:30-20:30	110±45[*]	113±48[*]	148±46[a]	78±43
20:30-04:30	254±51[*a]	180±56[*]	259±58[a]	200±75

[*] Significant difference ($P<0.01$, paired t-test) in lying down behaviour of the same cows when transferred from one barn to the other in different periods.

[a] Significant difference ($P<0.01$, unpaired t-test) in lying down behaviour between groups (4 cows) inhabiting different barns within the same period.

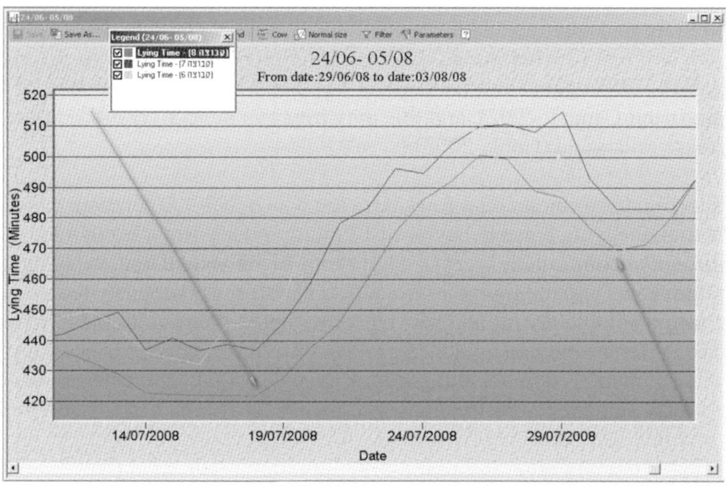

Figure 3. Average lying down time of two groups (milking cows in a free stall barn) in a large dairy farm (~600 milking cows) the arrows indicate dates at which the bedding was renewed.

~20% in average lying down time in two groups (milking cows) in a free stall barn immediately after the bedding was renewed.

Study 2. Detecting calving time of dairy cows by analyzing activity and feeding behavior

Predicting calving time may help to plan the daily farm routine, attend to expected difficult calving and decrease calf mortality.

In this study, 14 cows approaching calving were allocated 5 kg of concentrates a day each dispensed in equal portions of 1 kg in five even feeding windows, using computer controlled self feeders. Activity and feeding behavior were recorded and analyzed for the last seven days prior to calving. Activity variables that were measured included number of steps, number of lying down periods and accumulated lying down time. The data from each day were compared to those from the previous day and differences were analyzed by paired t-test for significance. Analysis was performed for day time (06:00-19:00), night time (19:00-06:00) and diurnally (Maltz and Antler, 2008)

Results

It was found that the cows' behaviour changed prior to calving. The cow's activity increasedd and the lying down time decreased significantly one day prior to calving (Figure 4), these changes can thus be used to predict calving time (Maltz and Antler, 2008). The present results follow those of an earlier study by Maltz and Antler (2007) who found that the most sensitive determinant of approaching calving was the ratio of activity to lying down (number of steps/lying down time).

These two studies conducted by researchers from the *Agricultural Research Organization The Volcani Center*, were performed on small numbers of cows, but open a new horizon for applications. To transform these ideas into useable tools for the farmer, a large quantum of data is being processed by AART to define specific outliers that could evaluate the exceptional behavior of individuals and groups (data not shown).

Applications currently under research and development

Heat detection in unfavorable conditions. Detection of cows in heat is essential for a successful breeding program. Cows in heat are usually restless, and tend to increase their activity and decrease lying down time. In some management conditions such as tie stalls, the cows do not have the option to display high activity and heat cannot be detected with regular pedometers. Monitoring changes in lying down behavior might enable the detection of heat in such unfavorable conditions. Studies evaluating this application are currently underway.

Defining the optimal population density of milk cows in barns. with the universal growing demand for milk, dairy farms are increasing the number of milking cows and the stocking density. Such overcrowding might increase SCC and decrease milk production (Nir (Markusfeld), personal communication). A study aimed at testing the possibility of using monitoring of lying down and activity to determine the optimal stocking density for each barn and its influence on milk production is currently underway.

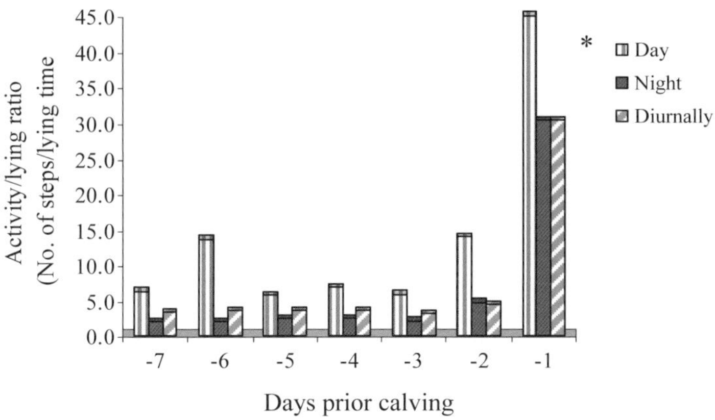

Figure 4. Activity/Lying down ratio in the last 7 days prior to calving. Ratio is displayed for the following daily periods: day time (06:00-19:00), night time (19:00-06:00) and diurnally (Maltz and Antler, 2008). * P<0.01 *day (-1) compared to day (-2).*

Discussion

Current technologies supply new data and at higher resolution than in the past. The advantages of the new data that has become available and the benefits they offer farmers are not yet fully clarified.

Companies developing new technologies are required to propose applications derived from the new technologies for the benefit of the farmers. New challenges face these companies as they need to become knowledgeable about their customer's business as well as their own. The work presented here demonstrates new management tools emerging from applied research employing new data.

The data acquired by the two new sensors presented here introduce new management tools. On-line daily milk composition analysis and cow behavior measurement are potentially useful tools to increase production, fertility, welfare, improve health, and increase the profitability of the modern dairy farm.

Some of the applications presented are ready for use: control of nutritional status of groups, monitoring the dairy's milk payments, milk separation according to quality, individual feeding, diagnosis of metabolic diseases, retrospective analysis, cow welfare and comfort assessment and predicting calving time of dairy cows. Others, such as early detection of mastitis, detection of heat in unfavorable conditions, definition of the optimal density of milking cows in barns, remain to be tested and proven.

There is a great potential for the future of dairy farm management in multivariable probabilistic models, using data collected from all available sources.

References

Bansal B.K., Hamann J., Grabowskit N.T. and K.B. Singh, 2005. Variation in the composition of selected milk fraction samples from healthy and mastitic quarters, and its significance for mastitis diagnosis. J. Dairy Sci. 72, 144-152.

Broom D.M. and M.J. Corke, 2002. Effects of disease on farm animal welfare. Acta. Vet. Bron. 71, 133-136.

Cook N.B, 2003. The influence of cow comfort on lameness and production. Proceeding of 16th annual full symposium on recent advanced in Clinical Vet. Med. U.C. Davis, 61-70.

Eicher R, 2004. Evaluation of the metabolic and nutritional situation in dairy herds: Diagnostic use of milk components. 23th World Buiatrics Congress, Québec, Canda.

Friggens N.C., Ridder C. and P. Løvendahl, 2007. On use of milk composition measures to predict the energy balance of dairy cows. J. Dairy Sci. 90, 5433-5467.

Grieve D.G., Korver S., Rijpkmea Y.S. and G. HOF, 1986. Relationship between milk composition and some nutritional parameters in early lactation. Liv. Prod. Sci. 14, 239-254.

Heuer, C., Schukken, H. and P. Dobbelaar, 1999. Postpartum body condition score and results from the first test day milk as predictors of disease, fertility, yield, and culling in commercial dairy herds. J. Dairy Sci. 82, 295-304.

Litner G., Merin U. and N. Silanikove, 2004. Changes in milk composition as affected by subclinical mastitis in goat. J. Dairy Sci. 87, 1719-1726.

Livshin, N., Antler A., Zion, B., Stojanovski, G., Bunevski, G. and E. Maltz, 2005. Lying behaviour of dairy cows under different housing systems and physiological conditions. Proc. Implementation of Precision Agriculture, 2nd ECPLF, Uppsala, Sweden, pp. 305-311

Maltz, E. and A. Antler, 2007. A practical way to detect approaching calving of the dairy cow by a behaviour sensor. Proc. Precision livestock farming '07, Edited by: S. Cox, WageningenAcademic Publishers, pp. 141-146

Maltz, E. and A. Antler, 2008. Detecting calving time of dairy cows by analyzing activity and feeding behavior in computer controlled self-feeders management. Presented at CIGAR -international conference of agricultural engineering XXXVII congress Brasilerio de Engenharia Agricola – Conbea 2008.

Markusfeld-Nir, O., 1996 Integrated herd health programs - the Israeli experience. Proceedings of the meeting 27th-29th of March 1996 of The Society for Veterinary Epidemiology and Preventive Medicine. Glasgow. Ed. M.V. Thrusfield and E.A. Goodall. pp 126-135.

Nielsen t., Larsen M., Bjerring M. and K.L. Ingvarsen, 2005. Quarter health, interval, and samplin time during milking affect the concentration of milk constituents. J. Dairy Sci. 88, 3186-3200.

Nir (Markusfeld), O., 2003. What are production diseases, and how do we manage them? Acta Vet Scand. Suppl. 98. 21.

Nordlund K.V., Cook N.B. and G.R. Oetzel, 2004. Investigation strategies for laminitis problems herds. J. Dairy Sci. 87 E27-E35.

Nutrient requirement of dairy cattle. 7th rev. ed. National Research Council, Washington, DC.: National Academy Press

Pyörälä S, 2003. Indicators of inflammation in the diagnosis of mastitis. Vet. Res. 34, 567-578.

Tomaszewski M.A. and T.J. Cannon, 1993 Using records for large herd management. Western Large Herd Management Conference, Las Vegas Nevada, USA. pp 137-140.

Tucker C.B, Weary D.M., Jeffrey Rushen J., and A. M. de Passillé, 2004. Designing Better Environments for Dairy Cattle to Rest. Advances in Dairy Technology 16, 39-53.

Weary, D.M. and C.B. Tucker, 2003. The scince of cow comfort. Proceeding of the joint meeting of the Ontario Agri. Business Association and the Ontario Association of Bovine Practitioners, Guelph, Ontario.

Part 2.

Innovations improving the efficiency of livestock systems in the Mediterranean area

A. Genetics, breeding, reproduction

Conception of new bioeconomic indexes to evaluate caprine genotypes productivity

A. Gaddour, S. Najari and M. Ouni
Laboratoire d'Elevage et de la Faune Sauvage des Régions Arides et Désertiques, Institut des Régions Arides, Médenine, Tunisie; gaddour.omar@yahoo.fr

Abstract

The comparison of different caprine genotypes' productivity remains limited by the technical and zootechnical parameters used to evaluate the individual animal performances such as kid's growth, goat lactation or herd reproduction. Data issued from 16 years of periodical animal survey of a herd compound from local goats, pure breeds and crosses was used to elaborate new bio-economic indexes allowing an accurate evaluation of the productive potentialities per genotype. The bio-economic indices have been developed to correct dairy performance and growth by the genotype reproductive parameters and kids' mortality rates. These indexes of overall productivity of genetic groups were subsequently adjusted by the metabolic weight of each group to take into account the production charges per animal which vary with the adult body size. The application of bio-economic indexes led to a genotypes classification quite different from this based on the comparison of individual performances. Some breeds, having high lactating performances or heavy kids weights at weaning, seem severely penalized by infertility or kid's mortality rates. Hence, the integration of the components of profitability, such as fertility, mortality and metabolic weight, in the evaluation criteria is essential for the reasonable herd management and genetic improvement. However, the development of bio-economic indexes must take into account all the components and production factors affecting the final animal phenotype and the farmer incomes.

Keywords: goat, kids' growth, dairy production, bioeconomic indexes

Introduction

In the Tunisian arid region, local goats are essentially raised in pastoral mode and ambulant herds valorise the vegetal resources of range lands under harsh conditions (Gaddour, 2005). Local goat populations are often considered genetically adapted to arid climate and is able to reproduce throughout the whole year, as well as for other caprine rustic populations (Najari et *al.*, 2006).

In this study we will conduct an evaluation of the profitability of different genetic groups, applying corrections to the raw performance parameters of reproductive and metabolic adult weight of each group (Akinsoyina et *al.*, 1997). Indeed the profitability of genetic groups can be estimated taking into account the efficiency of reproduction and conduct, and charges the animals infertile, to abortions and the number of kids died before weaning. As for charges of livestock, the main difference between the genetic groups may be represented by the metabolic adult weight (Caja, 1990).

Material and methods

Correction by the performance of reproduction

Index meat production

N1= number of kids produced by female at birth= (fertility rate-mortality rate)/100.

N2= number of female kids produced during the mating period at the age of 30 days= (fertility rate - mortality rate)/100.

N3= number of female kids produced during the mating period at the age of 90 days= (fertility rate - weaning mortality)/100.

Calculation of meat production corrected

V1= meat production per female at birth= N1×average weight of kids at birth.

V2= meat production per female at the age of 30 days= N2×average weight of kids to 30 days.

V3= meat production per female at the age of 90 days= N1×average weight of kids to 90 days produced per goat during the mating period on the age of 90 days= (fertility rate-weaning mortality rate)/100

Correction by weight metabolic

Index meat production

V'1= quantity of meat produced per female at birth= V1/metabolic weight.

V'2= quantity of meat produced per female at the age of 30 days= V2/metabolic weight.

V'3= quantity of meat produced per female at the age of 90 days= V3/metabolic weight.

Index of milk production

L'1= total production per female during the mating period= L1/metabolic weight

Results and discussion

Performances of meat corrected by the parameters of reproduction

The identification numbers of kids by genetic groups was calculated by subtracting the rate of mortality and fertility (Table 1).

For pure breeds, the local goat can be distinguished by the higher number of kids produced. A local goat gives kids at an average of 1.41 per goat worn by the mating period and 1.32 kids per female who have reached the age of 90 days at the mating period (Table 2).

Amongst pure-breeds, the Damascus breed confirms his reputation as a race excelling in the area of meat production. We noticed a change in the classification of the Alpine goat, it presented a lower production of meat than the local population. Only the Crossed Damascus, F1D and F2D present a higher meat production with respect to the local population after the correction process.

Table 1.Number of kids, 1 and 3 months after correction of reproductive and mortality ratios.

Genetic groups	Numbers of kids at:		
	birth (PN)	30 days (P30)	90 days (P90)
Alpine (A)	1.23	1.23	1.06
Damascus (D)	1.28	1.28	1.12
Murciana (M)	1.18	1.18	1.05
Local (Lo)	1.41	1.41	1.32
F1A	1.22	1.22	1.11
F2A	1.07	1.07	1.03
F1D	1.54	1.54	1.36
F2D	1.39	1.39	1.27
F1M	1.36	1.36	1.22
F2M	1.05	1.05	1.05

F1A, F2A: crossed alpine × local at first and second generation; F1D, F2D: crossed damascus × local at first and second generation; F1M, F2M: crossed murciana × local at first and second generation.

Table 2. Quantity of meat produced per female.

Genetic groups	Quantity of meat produced per female (kg) at:		
	birth (PrVNai)	30 days (PrVP30)	90 days (PrVP90)
Alpine (A)	4.44	9.99	13.89
Damascus (D)	4.69	10.49	16.21
Murciana (M)	2.80	7.49	11.59
Local (Lo)	4.04	9.44	15.67
F1A	3.73	9.95	14.82
F2A	3.58	9.15	14.57
F1D	5.36	11.99	21.47
F2D	4.71	11.48	18.51
F1M	3.56	8.80	14.36
F2M	2.86	7.57	12.59

F1A, F2A: crossed alpine × local at first and second generation; F1D, F2D: crossed damascus × local at first and second generation; F1M, F2M: crossed murciana × local at first and second generation.

Performances corrected by the metabolic weight

The objective of this correction is to include feeds for each genotype, to express their differences through metabolic weights of different genetic groups (Table 3).

Note that the local goat presents the highest return relative to pure breeds imported. While the performance of F1D is the highest of all genetic groups present, kids were heavier than the pure and the crossed alpine.

Table 3. Quantity of meat produced by kg of female metabolic weight.

Genetic groups	Quantity of meat produced by kg metabolic weight at:		
	Birth (PrVPNaiPM)	30 days (PrVPM30)	90 days (PrVPM90)
Alpine (A)	0.23	0.52	0.72
Damascus(D)	0.26	0.58	0.90
Murciana (M)	0.20	0.54	0.84
Local (Lo)	0.30	0.71	1.18
F1A	0.25	0.67	1.00
F2A	0.21	0.54	0.85
F1D	0.34	0.76	1.37
F2D	0.30	0.73	1.18
F1M	0.26	0.65	1.06
F2M	0.21	0.55	0.91

F1A, F2A: crossed alpine × local at first and second generation; F1D, F2D: crossed damascus × local at first and second generation; F1M, F2M: crossed murciana × local at first and second generation.

Performances dairy corrected by the parameters of reproduction

We have made a correction by the fertility rate, to reflect the charges goats sterile (Table 4). We note that following these corrections for pure breeds, no change has been made when ranking the gross average performance. The conduct of reproduction has not affected his superiority in milk production compared to pure breeds. However, there is some change in the paternal breed classification.

Performances dairy corrected by the metabolic weight

Through this correction, we plan to take account of feedstock for each genetic group (Table 5). This correction has helped to distinguish a change compared to previous rankings. Among others, F1A is the first in the ranking of all genotypes, with a yield of 14.4 kg of milk/kg metabolic goat at mating period, there is also a performance comparable to that of Alpine in purebred for F1M.

Table 4. Quantity of milk produced per female.

Genetic groups	Quantity of milk produced (kg)
Alpine (A)	244.80
Damascus (D)	192.72
Murciana (M)	168.19
Local (Lo)	97.01
F1A	212.49
F2A	209.69
F1D	180.54
F2D	174.66
F1M	170.65
F2M	152.78

F1A, F2A: crossed alpine × local at first and second generation; F1D, F2D: crossed damascus × local at first and second generation; F1M, F2M: crossed murciana × local at first and second generation.

Table 5. Quantity of milk produced in kg per 1 kg metabolic.

Genetic groups	Quantity of milk produced in kg per 1 kg metabolic
Alpine (A)	12.72
Damascus (D)	10.70
Murciana (M)	12.16
Local (Lo)	7.31
F1A	14.40
F2A	12.29
F1D	11.50
F2D	11.15
F1M	12.60
F2M	11.10

F1A, F2A: crossed alpine × local at first and second generation; F1D, F2D: crossed damascus × local at first and second generation; F1M, F2M: crossed murciana × local at first and second generation.

Conclusion

Following the analysis, one can conclude that the assessment of the productivity of genetic groups pure and crossed led to different results depending on the approach of the comparison. Indeed, the use of bio-economic index led to the classifications of different genetic groups of those from studies of individual performances. Therefore, integration of the components of profitability, such as fertility, mortality and metabolic weight, is essential for the reasonable choice of race improvement. For milk production, the highest level of productivity is the Alpine race, while for the production of meat, Damascus race and the local population has allowed the best levels of profitability. However, the development of bio-indicators must take into account all the components and factors of production to achieve assessments with the objectives of economic and livestock crossing absorption of the local goat.

References

Akinsoyinu, A. O., Mba, A.U., And Olubajo, F.O.: Studies on milk yield and composition of the West African Dwarf goat in Nigeria. J. Dairy Res., 1997, 44: 57-62.

Caja, G.: Evolution des systèmes de production ovin-lait dans le bassin Méditerranéen Options Méditerranéennes, 1990, 12: 31-38.

Gaddour, A.: Performances de croissance et de production laitière des groupes génétiques caprins issus d'un croisement d'absorption de la chèvre locale dans les oasis du sud tunisien, Mastère en génétique et bio ressources. Faculté des Sciences de Tunis, 2005.

Najari, S. Caractérisation zootechnique et génétique d'une population caprine. Cas de la population caprine locale des régions arides tunisiennes, Thèse de doctorat d'Etat, 2005, 214 p.

Najari, S., Gaddour, A., Ben Hamouda, Djemali, M., M., Khaldi, G. Growth model adjustment of local goat population under pastoral conditions in tunisian arid zone. Journal of Agronomy., 2007, ISSN 1812-5379.

Najari, S., Gaddour, A. Abdennebi, M. Ben Hamouda, M. Et Khaldi, G. Caractérisation morphologique de la population caprine locale des régions arides tunisiennes, Revue des Régions arides., 2006, 17: 23-41.

Genetic variability and phylogenetic relationships of ovine Algerian breeds using microsatellites

S. Gaouar[1], N. Tabet Aoul[2], O. Khaib dit Naib[3], L. Hamouda[3], N. Boushaba[3], N. Brahami[3], M. Aouissat[4], L. Dhimi[3,5] and N. Saïdi-Mehtar[3]
[1]Department of biology, University of Tlemcen, Tlemcen, Algeria
[2]Department of biotechnology, University of Oran Es-sénia, Oran, Algeria
[3]Laboratory of Molecular Genetics, University of Science and Technology of Oran (USTO), Oran, Algeria
[4]Technical Institute of Breeding (ITElv) of Aïn El-Hadjar, Saïda, Algeria
[5]Technical Institute of Breeding (ITELv) of Aïn M'lila, Constantine, Algeria

Abstract

The genetic variability and phylogenetic relationships were studied on a group of six ovine Algerian breeds: Hamra, Ouled-Djellal, Sidaoun, Taâdmit, D'men and Rumbi. These studies were carried out using 4 microsatellites: OarFCB128, OarCP34, INRA63 and BM1824, which were analyzed by the PCR-SSCP method followed by silver staining DNA. After the determination of the genotypes, we used statistical methods to determine the allelic frequencies and the rates of heterozygoty for each breed. In addition, the phylogenetic relationships between the six breeds were analyzed using the estimated genetic distances. The preliminary results reported in this study, showed a great genetic variability for most of these breeds. The phenograms showed that these breeds were dispersed according to their geographical region of origin.

Keywords: molecular characterization, phylogenetic relationships, microsatellite, statistical analysis, sheep breeds

Introduction

In Algeria, sheep breeding accounts for 80% of the global domestic animal production. The Algerian ovine livestock is composed of eight main breeds: Hamra, Ouled-Djellal, Taâdmit, Rembi, Sidaoun, Barbarine, Bèrbère and D'men. Currently, we observe a decrease of the number of heads and a change of the geographical repartition of some of them by contraction of the surfaces dedicated to these whole breeds.

In this study, we analysed the genetic variability of six Algerian sheep breeds and their phylogenetic relationships using four microsatellites.

Material and methods

Samples of DNA

We used DNA from 158 unrelated animals: 30 for Hamra, 30 of Ouled-Djellal, 30 of Taâdmit, 30 of Sidaoun, 27 of Rembi and 13 of D'men.

Markers

We chose 4 microsatellites which were located on different chromosomes, starting from recommended markers by FAO's program. The localization of these microsatellites on sheep chromosomes is indicated between brackets: BM1824 (Oar 1), OarFCB128 (Oar 2), OarCP34 (Oar 3) and INRA063 (Oar 14).

Methodology

The DNA was extracted using NaCl technique (Miller *et al.*, 1989). PCR was performed with 36.7 ng of genomic DNA in a 10 µl reaction volume. The PCR procedure involved initial denaturation at 94 °C for 15 min, 30 cycles of denaturation at 94 °C for 30 seconds, annealing temperature of primer: 57 °C (BM1824), 64 °C (OarFCB128), 61 °C (OarCP34) and 55 °C (INRA063) for 1 min, elongation at 72 °C for 1 min, and finally extension at 72 °C for 7 min.

The PCR products were analysed on 6% polyacrylamide gels and sized using 3 previously genotyped DNA by an automatic sequencer, as standard for sizing. The gels were silver stained.

Statistical analysis

The allelic frequencies for each microsatellite are calculated for each breed using Genetix software (Belkhir *et al.*, 2002) which allowed to estimate observed and expected heterozygosis. For the analysis of loss in heterozygosis within studied populations F_{is} indice was computed (Wright, 1965).

The genetic relationships between the studied breeds were established from Nei's genetic distance (Nei *et al.*, 1983) using POPULATION computer package. The obtained matrix was used for the construction of the dendrogram by UPGMA method (not weighted even group method analysis).

Results and discussion

A total of 41 alleles were found for the four loci. The highest genetic diversity was observed for INRA063 locus (15 alleles) and the least one was BM1824 (6 alleles). The mean rate of observed heterozygosis, averaged over the 4 loci, was 0.6450. This value was lower than the expected heterozygosis (Table 1). The analysis of these parameters showed an important variability.

The analysis of Wright, F_{is} indice, showed that all breeds have lost heterozygosis except Hamra, Taadmit and Rembi which were in Hardy-Weinberg equilibrium (Table 2).

The dendrogram in Figure 1 showed three groups. The first one included Hamra and Ouled-Djellal. This result agrees with the fact that Ouled Djellal breed has invaded the Hamra geographical area. Sidaoun, Taâdmit and Rembi, constituted the second group; despite the fact that they were from different geographical areas, D'men made an outgroup, probably due to its isolation.

Table 1. Number of alleles and rate of heterozygosis for the studied markers.

Locus	Number of alleles	Heterozygosis	
		Observed	Expected
OarFCB128	13	0.6233	0.7920
INRA063	15	0.6271	0.8120
BM1824	6	0.6667	0.6542
OarCP34	7	0.6630	0.6732
Mean		0.6450	0.7328

Table 2. F_{is} indices for the studied markers in the six Algerian sheep breeds.

Microsatellites	F_{is}					
	Hamra	Ouled-Djellal	Taadmit	Rembi	Sidaoun	D'men
OarFCB 128	0.184	0.418	0.155	0.110	0.219	0.314
BM1824	-0.103	-0.068	0.032	-0.149	0.451	-0.081
OarCP34	0.030	0.032	-0.197	-0.135	0.202	0.393
INRA63	0.100	0.523	0.130	0.133	0.373	0.249
Mean F_{is}	0.052	0.226	0.03	-0.01	0.311	0.218

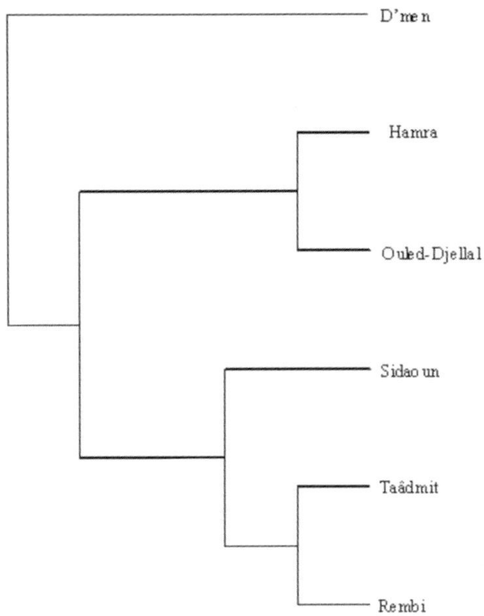

Figure 1. UPGMA dendrogram constructed from Nei's genetic distance.

Conclusion

This study was the first molecular characterization of Algerian sheep breeds. These results remain preliminary ones considering the low number of studied microsatellites. An important genetic variability was observed for the six sheep breeds with a total of 41 alleles across all the four loci. Three breeds (Ouled Djellal, Sidaoun and D'men) presented loss of heterozygosis. Further studies, using a more significant number of microsatellites, will allow to determine the level of their genetic variability and to precise their phylogenetic relationships.

References

Belkhir, K., P. Borsa, L. Chikhi, N. Raufaste and F. Bonhomme, 2002. GENETIX 4.03, logiciel sous Windows TM pour la génétique des populations. Laboratoire Génome, Populations, Interactions. CNRS UMR 5000. Université de Montpellier II, Montpellier.

Miller S.A., D.D. Dykes and H.F. Polesky, 1989. A sample salting-out for extracting DNA from human nucleated cells. Nucleic acids Res. 16: 1215.

Nei M., F. Tajima and Y. Tateno. 1983. Accurancy of Estimated Phylogenetic Trees from molecular Data. Journal of Molecular Evolution 19: 153-170.

Wright, S. 1965. The interpretation of population structure by *F*-statistics with special regard to systems of mating. Evolution.

In vitro conservation and preliminary molecular identification of some Turkish domestic animal genetic resources

M. İhsan Soysal
Department.of Animal Science, Namık Kemal University, Faculty of Agriculture, Tekirdağ, Turkey; misoysal@ttmail.com

Abstract

Changing environmental conditions (rapid increase in population, global warming, disordered structuring, and environmental pollution) change the flora and fauna of the world irreversibly and negatively. Studies carried out have showed that 27,000 animal and plant species per year disappear from nature forever. Today production of farm animals verges on an uniform pattern in the world. These lineages known as culture lineages are just animals with improved efficiency properties. However, no one has emphasized the resistance of them to environmental conditions or their disease resistances so far. Besides changes in environmental conditions, bioterrorism targets not only the life of people but also animals' life. These unfavourable factors have threatened the future of many species and lineages seriously. However, endemic domestic animals that preserve genetic variation are recognized as insurance of the future with these properties. The subject of this project has been discussed on various platforms in Turkey for a long time and declared as a priority area. Various action plans have been prepared and put into development plans but this issue has not been solved so far. This national issue that should be solved immediately would be resolved to some extent with a big project by the contribution of Turkish Ministry of Agriculture and Rural Affairs, TUBITAK and 10 universities from all regions of Turkey. This project aims to establish banks (Gene banks) which preserve animal genetic resources, to start characterization of endemic domestic animal species to use gained knowledge in registration studies, to build national researcher capacity in animal genetics and biotechnology areas and to combine and disseminate knowledge. These banks would guarantee not just five years, a hundred years of our country. Samples from 1,500 individuals belong to total 5 species would be preserved.

Keywords: domestic animals, genetic characterisation, molecular identification, conservation

Introduction

Animal genetic resources are considered as a part of the present biodiversity. Unfortunately; our native animal genetic resources have not been appreciated as it should have been; some are already gone extinct and some are at the verge of extinction. Yet, animal genetic resources must be considered as the insurance of the future, since they may have an important potential to improve social and economic life.

Conservation of animal genetic resources is important; because, there might be a need of recovering the lost/decreased genetic variation and they might be possessing special genes or important genetic information. Conservation strategies must be planned (1) not to lose the genes or gene combinations that might be of use, (2) to make use of the advantages of heterosis, (3) to use the present resources as the insurance of the future, (4) to preserve the cultural products, and (5) to use the present material in diverse research fields. In fact, animal genetic resources have been being conserved upon national conservation projects in economically developed countries (Hiemstra and FAO, 2003). Conservation of animal genetic resources includes *in situ* and *ex situ* programs (Ertugrul, 2005; İhsan, 2004; Hiemstra and FAO, 2003).

The animals are conserved alive at their native place in context of an *in situ* program. *Ex situ* conservation programs include both conserving the animals alive in a place different than their native place like a zoo or breeding farm; and conserving the material taken from the animal like gamete, embryo, sperm, tissue or cell, and DNA. Cryopreservation technique makes possible the long term storage of high number of samples in small volume containers. Moreover, rapidly developing technology enables to apply cryopreservation techniques to different species like cattle, sheep, goat, horse, swine, poultry (Piltti *et al.,* 2004; Stachecki *et al.,* 2004; Bağış *et al.,* 2004; Aller *et al.,* 2002; Leibo and Songsasen, 2002; Dobrinsky JR, 2001, 2002). Cryopreserved embryo and sperm might enable us to bring back the individuals of a lost breed. Cryopreserving healthy germ cells and embryo may help to eradicate animal health problems that might emerge in the future, as well (Piltti *et al.,* 2004; Stachechi *et al.,* 2004).

Embryo cloning techniques might also be employed in conservation of animal genetic resources for different purposes. As the technology develops, there will be an increase in the application ways of these techniques. Recent advances in science more specifically in nuclear transfer (NT) technology are promising. There are successful studies on cloning of an adult individual using NT as well as on NT between closely related species, which indicated the possibility of restoring the populations having few members left and saving the species facing extinction (Chen *et al.,* 2002; Loi *et al.,* 2001, 2002; Kitiyanant *et al.*, 2001; Arat *et al.,* 2001, 2003, 2004). That is why, several US biotechnology companies, agricultural organisations and zoos have started to construct tissue and cell banks (Ryder *et al.,* 2001).

Turkey, which has a great number of endemic species, has an important share in the genetic resources of the earth. However, the increase in the population size as well as the economic pressure speed up the changes in traditional farming techniques and causes the decrease in biodiversity.

As in other countries, the breeding and hybridization practices are being conducted and highly productive animals are being imported to increase the productivity of local breeds of Turkey. Consequently, this brought about the decrease in biodiversity of farm animal genetic resources and endangered their existance. Thus, the importance of preserving the balance between breeding studies and conservation of the animal genetic resources was revealed.

Our native breeds, being the closest relatives of the first domesticates, must be given priority in conservation (Bruford *et al.,* 2003). Molecular genetics studies have shown that the European cattle, sheep and goat breeds have originated and spread from Anatolia (Loftus *et al.,* 1999; Troy *et al.,* 2001, Bruford and Towsend, 2004; Lenstra *et al.,* 2005). The information lost during this stage might still be present in Turkish native breeds.

The importance of our native genetics resources is widely accepted and *ex situ* conservation studies must be started immediately. Moreover, genetic studies must be carried out and their economic value must be estimated, the priority of the populations and/or breeds must be identified for *in situ* conservation. It has been shown that the gene pool of the populations in breeding farms can easily differentiate from the breed's gene pool (Koban, 2004) due to genetic drift and wrong management strategies. Therefore, the results of the proposed project will guide breeding farms in both choosing the individuals to conserve and constructing proper management strategies.

Several establishments in the world have been carrying out studies on the conservation of endangered species (for example: Millennium Seed Bank, Wildlife Conservation and Monitoring Center, Conservation on Biological Diversity, Japanese Ministry of Agriculture, and Ryder, 2001; Mariante *et al.,* 2002; Matsas *et al.,* 2004; Hiemstra and FAO, 2003). In Turkey, the establishment of a central bank or an institute with affiliated banks for conservation, thus enabling the sharing of the samples and knowledge among the country has been emphasized within the priority subjects in context of TÜBİTAK (Turkish Scientific and Technological Research Council)'s Vision-2023 Science and Technology Prevision Project. In addition, 'Biodiversity Conservation Program' within the area of 'Biotechnology and Genetic Technologies' was listed among the priority action topics by the Turkish Academy of Sciences (TUBA) Molecular Life Sciences and Technology Foresight Project (2003-2023). Moreover, the importance of conservation of animal genetic resources was emphasized in the 1st Comission Report of the II Agriculture Assembledge (2004). Furthermore, Turkey took part as contractor in the international agreements on conservation of the biodiversity and genetic resources. One of these agreements is the UN Convention on Biological Diversity. In the meantime, a commission to work on conservation of animal genetic resources was formed by FAO (UN Food and Agricultural Organisation) and started its studies. The proposed project will contribute to fulfil the commitments resulted from these agreements.

In conclusion, the reasons stated above and the 'commonwill' on the conservation of native animal genetic resources resulted from the preliminary studies have lead the preparation of the proposed study. The project aims at conserving some of the native animal genetic resources and construction of banks to hand these on next generations.

Material and methods

The preparation of studying material

This project comprises to *ex situ - in vitro* preservation methods. This WP-I consists of assurance of living materials and identification of the breeds to be conserved in context of the project. Based on reports of national and international organizations and population records, the species and the breeds to be included in conservation studies were determined.

Technological development about preservation of live material

In order to develop and ameliorate the technologies of live material preservation techniques, WP-2 consists of several sub work packages. These sub packages employ techniques like tissue cryopreservation, using of alternative cell types and alternative auxiliary reproduction techniques

Preservation of genetic resources

This work package consists of cryopreservation in order to protect genetic resource of identified animals. Those resources are genomic DNA, somatic cells, tissue, embryos and germ cells. From each individual 300 dose sperm and 300 embryos should be cryopreserved. The *ex situ* and *in vitro* protected samples of those species and breeds will be kept in gene banks constructed in LHMAE/ TAGEM and GMBE.

Characterization of species

A comprehensive sample collection will be planned in the context of the proposed project. However, in genetic diversity and phylogenetic studies, since establishment of the necessary infrastructure and acquiring enough number of qualified staff will take remarkable time, most frequently used two molecular methods (microsatellite markers and mtDNA sequencing) are going to be employed.

Distribution and standardization of data

WP-5 consists of protocol standardization, distribution of data by constructing a database and a web page, getting in touch with several national and international organizations. Moreover, WP-5 includes arrangement of seminars, congresses and media declarations to represent the project.

Results

Culture breeds have high productivity traits, but, their resistance to harsh environmental conditions and diseases were not taken into account during selective breeding. Those negative factors seriously threaten the future of lots of species and breeds.

The owner of the project is the Turkish Ministry of Agriculture and Rural Affairs. The project is directed by TÜBITAK- GEBI. Furthermore, there are 10 Turkish universities and TAGEM involved. The aims of the proposed project are:

- Construction of DNA, cell, tissue, embryo and sperm banks for animals facing extinction.
- Characterization of entire animal species within the context of the project.
- The implementation of new scientific technologies and developments on animal biotechnology and animal genetics, into the researches carried out in Turkey and the ameliorations of extant technologies. Furthermore, educating qualified staffs on those fields.
- Gathering and distribution data.
- Coupling infrastructures and human resources for a specific purpose.

The project comprises 13 sheep, 6 cattle, 5 goat, Anatolian water buffalo and 5 horse breeds. The most important result of this project is animal gene banks since it will be a pioneer in Turkey in conservation of most of native animal gene resources. Two different gene banks are going to be established in conformity with FAO proposals. Studied DNA samples and stored tissues and cells of conserved animal species as part of this project will also be the sole resource both in Turkey and in the world. In genetic diversity and phylogentic studies, highly used molecular marker types (microsatellite markers and mtDNA sequencing) are going to be used. For the first time a remarkable number of breeds are going to be genetically characterized in such extensive ways. Besides, generated data will contribute to registration of studied breeds.

Different cryopreservation conditions (DNA, somatic cell, germ cell, embryo) are going to be used at the same time. In Turkey, on that field this project is the most comprehensive and well attended project with 2 governmental organizations and 10 universities.

- The funds otherwise allocated for researchers' education abroad are going to be utilized within the country in context of the project. Because the project is highly comprehensive, the duration was determined as 4.5 years and the budget of the project is: 9,126,532 TRL.
- Owner institution: Turkish Ministry of Agriculture and Rural Affairs, administrative institution: TUBITAK Marmara Research Center GEBI, project coordinator: Assoc. Prof. Dr. Sezen Arat (sezen.arat@mam.gov.tr).
- Partner institutions: Adnan Menderes University, Ankara University, Ataturk University, Istanbul University, Lalahan Livestock Central Research Institute, Middle East Technical University, Mustafa Kemal University, Namık Kemal University, Ondokuz Mayıs University, Selçuk University, Uludağ University.

References

Aller, J.F., G.E. Rebuffi, A.K. Cancino and R.H. Alberio, 2002. Successful transfer of vitrified llama (Lama glama) embryos. Anim Reprod Sci. 73 (1-2): 121-7.

Arat, S., 2001. Bovine cloning using adult donor cells treated with roscovitine. Biol Reprod Supplement. 1;173.

Arat, S., 2004. Cold storage of tissue as source for donor cells does not reduce the in vitro development of bovine embryos following nuclear transfer. Reprod Fertility Dev 16(1, 2):135.

Arat, S., 2003. Gene expression and in-vitro development of interspecies nuclear transfer embryos. Mol Reprod Dev. 66:334-342.

Arat, S., S.J. Rzucidlo, J. Gibbons, K. Miyoshi and S.L. Stice, 2001. Production of transgenic bovine embryos by transfer of transfected granulosa cells into enucleated oocytes. Mol Reprod Dev; 60: 20-26.

Bagıs, H., H. Sağırkaya, H. Odaman and D. Andras, 2004. Vitrification of pronuclear stage mouse embryos on SSV versus in cryotube comparision of the effect of ecuilibration time and different sugers in the vitrification solution. Mol Reprod Dev; 67:186-192.

Bruford, M.W. and S.J. Townsend,, 2004. Case studies in the genetics of animal domestication: sheep. In: Zeder, M., Decker-Walters, D., Bradley, D., Smith, B.D. (eds.): D ocumenting Domestication: New Genetic and Archaeological Paradigms, California University Press.

Bruford, M.W., D.G. Bradley and G. Luikart, 2003, DNA markers reveal the complexity of livestock domestication. Nature Genetics 4: 2-12.

Chen, D-Y., 2002. Interspecies implantation and mitochondria fate of panda-rabbit cloned embryos. Biol Reprod 67: 637-642.

Conservation on Biological Diversity (www.biodiv.org/chm/conv/default.htm).

Dobrinsky, J.R., 2002. Advancements in cryopreservation of domestic animal embryos. Theriogenology Jan 1; 57 (1): 285-302.

Dobrinsky, J.R., 2001. Cryopreservation of swine embryos: a chilly past with a vitrifying future. Theriogenology Nov 1; 56 (8): 1333-44.

Ertuğrul, M., G. Dellal, C. Elmacı, A.O. Akın, O. Karaca, T. Altın and İ. Cemal, 2005. TMMOB Ziraat Mühendisleri Odası, Türkiye Ziraat Mühendisliği VI. Teknik Kongresi, I. Cilt 275-290.

Geijsen, N., M. Horoschak, K. Kim, J. Gribnau, K. Eggan and G.Q. Daley, 2004. Derivation of embryonic germ cells and male gametes from embryonic stem cells. Nature 427:148-154.

Hall, S.S.G., 2004. Livestock biodiversity: genetic resources for the farming of the future. Blackwell Publishing Company.

Hiemstra, S.J., 2003. Guidelines for the constitution of national cryopreservation programmes for farm animals (FAO).

Japanese Ministry of Agriculture, Forestry and Fisheries (http://bank.dna.affrc.go.jp).

Kitiyanant Y., 2001. Somatic cell cloning in buffalo: effect of interspecies cytoplasmic recipients and activation procedures. Cloning Stem Cell 3:97-104.

Koban, E., 2004. Genetic Diversity of native and crossbreed sheep breeds in Anatolia. PhD Thesis, Orta Doğu Teknik Üniversitesi, Ankara, Turkey.

Lacham-Kaplan, O., 2004. *In vivo* and *in vitro* differentiation of male germ cells in the mouse Reproduction. Aug;128(2):147-52.

Leibo S.P. and N. Songsasen, 2002. Cryopreservation of gametes and embryos of non-domestic species. Theriogenology Jan 1; 57 (1):303-26.

Lenstra, J.A., 2005. Evolutionary and demographic history of sheep and goats suggested by nuclear, mtDNA and Y-chromosome markers. Symposium: The role of biotechnology in Turin, Italy 5-7 March.

Loftus, R.T., O. Ertuğrul, A.H. Harba, M.A.A. El-Barody, D.E. MacHugh, S.D.E. Park and D.G. Bradley, 1999. A microsatellite survey of cattle from a centre of origin the Near East. Molecular Ecology 8: 3-8.

Loi P., 2002. Nuclei of nonviable ovine somatic cells develop into lamps after nuclear transplantation. Biol Reprod. 67: 126-132.

Loi P., 2001. Genetic rescue of an endangered mammal by cross-species nuclear transfer using post-mortem somatic cells. Nature Biotech 2001; 19:962-964.

Mariante Ada, S. and A.A. Egito, 2002 Animal genetic resources in Brazil: result of five centuries of natural selection. Theriogenology. 2002 Jan 1; 57 (1): 223-35.

Matsas, D, V. Huntress, H. Levine, S. Ayres, J. Amini, R. Duby, P. Borden, G. Saperstein, E., Overstrom, 2004. Preservation of heritage livestock breeds:integrated program to cryopreserve germplasm from Tennessee myotonic goats. Reprod Fertil Dev. 2004; 17 (1,2).195.

Maudet, A., E. Beja-Pereira, H. Zeyh, A. Nagash, D. Kence, M.-P. Özüt, S. Biju-Duval, D.W. Boolormaa, P. Coltman, G. Taberlett and A. Luikart, 2004. A standard set of polymorphic microsatellites for threatened mountain ungulates (Caprini, Artiodactyla). Molecular Ecology Notes 2004; 4, 49-55.

Millennium Seed Bank (www.rbgkew.org.uk/seedbank/msb.html).

Oskam, A., A. Burrell, T. Temel, S. van Berkum, N. Longworth and I.M. Vilchez, 2004. Turkey in the European Union, Consequences for Agriculture, Food, Rural Areas and Structural Policy. Rural Areas and Structural Policy. Final Report. Wageningen University.

Piltti, K., H. Lindeberg, J. Aalto and H. Korhonen, 2004. Live cubs born after transfer of OPS vitrified-warmed embryos in the farmed European polecat (Mustela putorius). Theriogenology. 2004 Apr 1; 61 (5): 811-20.

Reist-Marti S.B., H. Simianer, J. Gibson, O. Hanotte and J.E.O. Rege, 2003. Weitzman's approach and conservartion of breed diversity: an application to African cattle breeds. Conservation Biology 17(5):1299-1311.

Ryder O.A., 2001. DNA banks for endangered animal. Science; 288 (5464):275.

Shivaji, S., S.D. Kholkute, S.K. Verma, A. Gaur, G. Umapathy, A. Singh, S. Sontakke, K. Shailaja, A. Reddy, S. Monika, V. Sivaram, B. Jyotsna, S. Bala, M.S. Ahmed, A. Bala, B.V. Chandrashekar, S. Gupta, S. Prakash and L. Singh, 2003. Conservation of wild animals by assisted reproduction and molecular marker technology. Indian J Exp Biol; 41(7):710-723.

Soysal I., 2004. Türkiye Yerli Hayvan Genetik Kaynaklarımız, Tekirdağ Ziraat Fakültesi.

Stachecki, J.J. and J. Cohen, 2004. An overview of oocyte cryopreservation. Reprod Biomed Online. Aug; 9 (2): 152-63.

Toyooka, Y, N.Tsunekawa, R. Akasu and T. Noce, 2003. Embryonic stem cells can form germ cells in vitro. PNAS; 100:11457-11462.

Troy C.S., D.E. MacHugh, J.F. Bailey, D.A. Magee, R.T. Loftus, P. Cunningham, A.T. Chamberlain, B.C. Sykes and D.G. Bradley, 2001. Genetic evidence for near-eastern origins of domestic cattle. Nature 2001; 410: 1088-1091.

Wakayama, S., S. Kishigami, N. van Thuan, H. Ohta, T. Hikichi, E. Mizutani, R. Yanagimachi and T. Wakayama, 2005. From The Cover: Propagation of an infertile hermaphrodite mouse lacking germ cells by using nuclear transfer and embryonic stem cell technology. PNAS, January 4; 102 (1): 29 - 33.

Wildlife Conservation and Monitoring Centre (www.wcmc.org.uk).

Zwaka T.P. and J.A.A. Thomson, 2005. Germ cell origin of embryonic stem cells? Development January 15, 2005; 132(2): 227-233.

Red deer (*Cervus elaphus*) farming in Greece: predicting the rate of inbreeding and estimating the effective population size

G. Vatzias[1], E. Asmini[2], D. Papavasiliou[1], G. Maglaras[1], S. Nagy[3] and Z. Zamborszky[4]
[1]T.E.I of Epirus, Kostaki, Arta, z.c. 47100, Greece
[2]T.E.I of Larisa,Trikalon ring Rd, z.c. 41200, Greece
[3]University of Pannonia, Veszprem, Hungary
[4]University of Kaposvar, Kaposvar, Hungary

Abstract

The Greek deer industry made its appearance recently, with the establishment of a few small private deer farms with minimal (basic) infrastructures and imported breeding stock. The Department of Forestry owns three State farms where indigenous red deer are raised. The deer population used in this study was composed of 70 hinds and 17 stags, raised in Northern Greece, at the public red deer farm of Chrisopigi, near the city of Serres. Due to the lack of introducing new bloodlines in the herd for more than 20 years and indications of genetic drift, based on the low birth rate of the herd in addition to abnormal phenotypic characteristics (antler deformation), the inbreeding rate (ΔF) and the effective population size (N_e) over 1 generation of the population, was estimated. The average generation interval (L) was estimated to be four years, while 48 and 16 new hinds and stags respectively, enter the herd each generation. The results indicated that the expected rate of inbreeding in each generation was approximately 0.91% while the calculated effective population size was 45.6 animals. According to the literature, the effective population size is slightly below of the critical threshold required number of animals (50), indicating that there is a build up of mutational load. The above results lead to the conclusion that immediate measures should be taken in order to reduce inbreeding rate and increase the effective population size.

Keywords: red deer, inbreeding, conservation

Introduction

Red deer farming is a new livestock venture in Greece. There are three state farms with deer populations, two of which are fenced and the other one is within a National Park. Within the last five years there were livestock producers interested to establish red deer farms for venison production. In the process of looking for the source of the red deer breeding stock, it was clear that the indigenous population raised in the state farms, was the best proposition.

However, the quality of the breeding stock of the three State farms was in question. In order to evaluate the genetic merit of the red deer breeding stock, the State farm of Chrisopigi was chosen. Inbreeding rate and the minimum effective population size were the main two parameters that were evaluated. There are different approaches to calculate the inbreeding coefficient, especially if the number of males and females within a population differ dramatically or if the evaluated population is known to express a high inbreeding rate (Stansfield, 1983). The effective population size is an important factor which indicates the effectiveness of live conservations schemes (Meuwissen, 2007). According to the literature, (Woolliams and Toro, 2007; Meuwissen, 2007), a minimum number of 50 animals is required in order to keep inbreeding rate at low levels, but there are other theories that increase this number to 500 animals in order to avoid deleterious mutations which will lead the population to extinction (Lynch *et al.*, 1995). Knowing the inbreeding rate, the effective population size, is then calculated to give a better picture of the population's genetic status. Another important factor is the generation interval which can be managed in order to reduce genetic drift (Meuwissen, 2007).

Therefore, the aim of this study was to evaluate the genetic status of the Chrisopigi State farm's breeding stock, by estimating the effective population size based on four different models. From the results it will become clear which model fits better to the genetic status of the specific red deer population. The results also will determine whether or not these animals can be sold and be the breeding stock of private farms.

Material and methods

The red deer farm, where all the observations took place, is a state farm under the authority of the Forestry Department of the prefecture of Serres. It is located in the area of Chrisopigi and covers more than 300 acres. The farm is fenced. There is no reproductive management or genetic improvement program that is applied on the breeding stock of the red deer population.

The population of the breeding stock is composed of seventeen mature males and seventy mature females. For calculation purposes the following standards were considered:

- four young males and 12 young females enter the reproductive population each year;
- the minimum effective size of a live population is 50 animals per generation;
- the generation interval of the specific population is 4 years based on the following equation:

$$L = 1/2\ (1m + lf) \qquad (1)$$

Where, L = generation interval, m = male, f = female.

The models used to calculate the effective population size are the following:

- Based on the number of newly entered males and females in the reproductive population, per generation (G) or per year (Y):
  $Ne(g) = 1/[2L\Delta F(y)],$ (2)
  $\Delta F = [8mL^2]^{-1}+[8fL^2]^{-1}$ (3)
  $Ne(y) = 1/[2\Delta F(y)],$ (4)
  $\Delta F = [8mL^2]^{-1}+[8fL^2]^{-1}$ (5)
- Based on the total number (n) of the breeding stock (males: M, females: F) per generation:
  $Ne(g) = 4\ (Nm)\ (Nf)/(Nm+Nf)$ (6)
- In case of high inbred population
  $Ne(g) = N/(1+F),$ (7)
  $F = \Delta F,$
  $\Delta F = 1/(2N)$, or (8)
  $\Delta F = (Nm+Nf)/[8(Nm)(Nf)]$ per generation (9)

Results

The results indicated the following:

- Based on the number of newly entered males and females in the reproductive population, per generation (G):
  $Ne(g) = 1/[2L\Delta F(y)],$ (10)
  $\Delta F = [8mL^2]^{-1}+[8fL^2]^{-1}$ (11)
  Where:
  M: 4/year or 16/generation;
  f: 12/year or 48/generation;
  L^2: 16;
  $\Delta F = 0.0026$ ή 0.26%/year (≈ 0.20% m, male effect);
  $Ne(g) = 1/2.08 = 0.48 \times 100 = 48;$
  $Ne(g) = 48 < 50.$

The above result indicates that the population effective size is marginal (48<50) and the danger of further genetic drift is real.

- Based on the number of newly entered males and females in the reproductive population, per year (Y):
  $Ne(Y) = 1/[2\Delta F(y)]$, (12)
  $\Delta F = [8mL^2]^{-1}+[8fL^2]^{-1}$ (13)
  Where:
  M: 4/year or 16/generation;
  f: 12/year or 48/ generation;
  L^2: 16;
  $\Delta F = 0.0026$ ή 0.26%/year (≈ 0.20% m, male effect);
  $Ne(Y) = 1/0.52 = 1.92 \times 100 = 192$;
  $Ne(Y) = 192>50$.

The above result indicates that the population effective size is large (192>50) and there is no danger to the population of genetic erosion and eventually extinction.

- Based on the total number (n) of the breeding stock (males: M, females: F) per generation:
  $Ne(g) = 4\ (Nm)\ (Nf)/(Nm+Nf)$ (14)
  Where:
  m: 17/generation;
  f: 70/generation;
  $Ne(g) = 4{,}760/87 = 54.7$;
  $Ne(g) = 54.7>50$.

The above result indicates that the population effective size is marginal (54.7>50) on the positive side, but the danger of further genetic drift is possible.

- In case of high inbred population,
  $Ne(g) = N/(1+F)$, (15)
  $F = \Delta F$, (16)
  $\Delta F = 1/(2N)$ or (17)
  $\Delta F = (Nm+Nf)/[8(Nm)(Nf)]$ per generation (18)
  Where:
  m: 17/generation;
  f: 70/generation;
  $\Delta F = 0.0091$ ή 0.91%/per generation;
  $Ne(g) = 87/1.91 = 45.6$;
  $Ne(g) = 45.6<50$.

The above result indicates that the population effective size is lower (48<50) compared to the minimum number of animals, and the danger of further genetic drift and possible extinction of the population is real.

Discussion

The inbreeding coefficient and the size of population, are both important factors which determine whether or not a conservation program and further along a genetic improvement program can be applied on a population (Simianer, 2005; Simm, 2000; Woolliams, 2007). In this study, the selection of the appropriate calculating model of the size of population, is very crucial, because as the results showed, four different models indicated different results. In order to avoid misleading interpretations, important factors and indicators, based on observations or other measurements of the population, should be taken into account and then a decision could be taken for the use of the appropriate model (Bennewitz *et al.*, 2007).

The red deer population of the Chrisopigi State farm, has shown indications and characteristics (low fertility rate, abnormal antler development etc.) of genetic erosion, due to high inbreeding rate. To the same conclusion leads the fact that there is not any reproductive management program in the farm and for the last 20 years there was no introduction of new sires in the breeding stock. Therefore, it is mandatory to take measures that will lead to lower inbreeding rate and then apply a genetic improvement program on the population of the Chrisopigi State farm.

Red deer farming in Greece has the potential to be the best option for new farmers, especially for those who live in remote mountainous regions. Besides farming many other activities can be related to that, such as, environmental education, agrotourism, etc. and be a good motive for young people to remain in their towns, establish their own enterprises and support the local economy.

References

Bennewitz, J., H. Eding, J. Ruane and H. Simianer, 2007. Selection of breeds for conservation. In: Utilization and conservation of farm animal genetic resources, K. Oldenbroek (editor), Wageningen Academic Publishers, Wageningen, The Netherlands, 131-146.

Lynch, M., J. Conery and R. Burger, 1995. Mutation accumulation and the extinction of small populations. American Naturalist 146: 489-518.

Meuwissen, T., 2007. Operation of conservation schemes. In: Utilization and conservation of farm animal genetic resources, K. Oldenbroek (editor), Wageningen Academic Publishers, Wageningen, The Netherlands, 167-193.

Simianer, H., 2005. Decision making in livestock conservation. Ecological Economics, 54: 559-572.

Simm, G., 2000. Genetic Improvement of Cattle and Sheep, Farming Press,Tonbridge, United Kingdom, 433 pp.

Stansfield, W.D. (ed.), 1983. Schaum's outline of theory and problems of Genetics. McGraw-Hill, Inc., New York, USA, 266-287.

Woolliams J., 2007. Genetic contributions and inbreeding, In: Utilization and conservation of farm animal genetic resources, K. Oldenbroek (editor), Wageningen Academic Publishers, Wageningen, The Netherlands, 147-165.

Woolliams, J. and M. Toro, 2007. What is genetic diversity? In: Utilization and conservation of farm animal genetic resources, K. Oldenbroek (editor), Wageningen Academic Publishers, Wageningen, The Netherlands, 55-74.

The use of crossbreeding programs for increasing domestic beef production

M. Ben Salem[1] *and H. Kélifa*[2]
[1]*Institut National de la Recherche Agronomique de Tunisie, Laboratoire de Production Animale et Fourragère, Rue Hédi Karray, 2049 Ariana, Tunisie*
[2]*Office de l'élevage et des Pâturages, 30, rue Alain Savary, 1002 Tunis Belvédère, Tunisie*

Abstract

Beef production in Tunisia comes from native cows, exotic Holsteins and dual purpose breeds and their respective crossbreds. However, given the limited meat potential of these animals, domestic beef production does not meet the total demand of the population. Therefore efforts have been directed to identifying appropriate crossing breeds for increased production. First, exotic dual purpose breeds such as the Brown Swiss and the Tarantaise were introduced to determine their possible use in crossbreeding with dairy animals. However, although some improvement was made in meat production, crossbreeding with these breeds still did not help in achieving the desired production level. Meanwhile, the population increased and the improved revenue of households have served to emphasize the need for more beef production. Recently, alternative crossbreeding programs with beef breeds were implemented. The objective of this work is to evaluate, trough continuous field studies, the potential of using the White Blue Belgium (BBB) and the Charolais breeds in industrial cross breeding programs for increasing beef production. Data on breeding, birth weight (BW), calving ease, average daily gain (ADG) and carcass weight (CW) and value of crossbred animals were collected for the 2000-2006 period. A total of 467 producers and 24,000 cows were included in the study. Results showed that crossbreeding with beef breeds increased BW, ADG and CW 5 kg, 217 g and 77 kg, respectively. Carcass quality and market value were also improved. This suggests that crossbreeding with beef breeds is one of the successful methods which can be used in the country to reach self-sufficiency in domestic beef and improve farmers' revenue, while better meeting the consumer preferences in terms of meat quality.

Keywords: beef, breed, crossbreeding

Introduction

Cattle ensure nearly half of the red meat production in Tunisia, with the two thirds coming from fattened calves and the reaming part from cull cows (GIVLAIT, Tunisian Ministry of Agriculture, 2007). In 2007, total domestic red meat production reached 120,000 tonnes. However, such production level does not cover the population total need of red meat despite all the deployed efforts to improve animals' productivity and to raise producer's revenue. Field studies showed that growing performance of animals and profitability of feedlots remain low because of many limiting factors. Major factors include the low nutritive values of predominant cereal forages, inadequate feeding management strategies, and most importantly the lack of beef breeds. Indeed, like in most developing countries, beef production in Tunisia is carried out at the smallholder level where most feeding systems are based on imported concentrate feeds and locally available low quality roughages. Moreover, cattle population in the country is composed of the native cow, imported dairy (Holstein-Friesian) and dual purpose (Brown Swiss and the Tarantaise) breeds, and their respective crossbreds, with the Holstein being by far the dominant one with more than 90% of the total herd. All these animals are primarily used to produce milk. They supply beef as a by-product when surplus calves and cull cows are slaughtered for meat. Local animals have a low aptitude for meat production. Pulished data (Ben Dhia and Antic, 1971; Rondia *et al.*, 1984; Atti and Ben Dhia, 1987, 1990) showed ADG of only 750 g for local young bulls between the age of 3 and 12 months when raised under favorable conditions. The ADG for the Brown Swiss, the Tarantaise and the Holstein were 1,036, 1,016 and 1,043 g/day, respectively.

Meanwhile, the continuous population increase and the improved revenue of households have served to emphasize the need for more beef production. Efforts have been therefore directed to identifying appropriate crossing breeds for increased meat production. First, exotic dual purpose breeds such as the Brown Swiss and the Tarantaise were introduced to determine their possible use in crossbreeding with dairy animals. However, although some improvement was made in meat production, crossbreeding with these breeds still did not help in achieving the desired production and profitability levels. Recently, an alternative crossbreeding program with beef breeds was implemented as a new alternative to previous programs. The objective of this work is to evaluate, trough long term field studies, the potential of using beef breeds, the White Blue Belgium (BBB) and the Charolais(CH) as compared to the use of dual purpose breeds (Brown Swiss and Tarentaise), in crossbreeding programs with local animals for increasing beef production in Tunisia.

Methodology

Most of the data used in this work is collected from field studies we conducted as a part of the regular follow up and evaluation of the national crossbreeding program implemented in late 2000 by the Office of Livestock production and pasture (OEP) which aims to improving beef production efficiency and meat quality using crossbreeding of local animals and their crossbreds with dairy breeds with beef breeds. The project involved cattle producers in different parts of the country (North East, North West, and the Center). It consisted first of identifying potential candidate beef producers to be included in the program, then in supplying them with the appropriate semen for artificial insemination. Insemination was performed following the synchronization of the animals by qualified inseminators either from the private business of from OEP. Genotypes used in the program are the Brown Swiss and the Tarantaise as dual purpose breeds and the CH and the BBB as the beef breeds. Crossbreeding genotype (CH vs. BBB) was chosen according to the body conditions of the cows so that to ease calving and avoid dystocia.

Data on breeding, birth weight (BW), calving ease, average daily gain (ADG) and carcass weight (CW) of crossbred animals were collected for the 2000-2006 period through regular visits and surveys conducted by the survey team of the project. A total of 24,000 cows were included in the study, with a significant increase in the number of involved animals over time going from 1000 in 2000 to more than 13,000 heads in 2006.

To evaluate the effect of crossbreeding with beef type breeds on major production indices, two representative samples of both groups (local × dual purpose crossbred animals and local by beef type crossbreds) were used for data collection and comparisons. All animals were raised under the same feeding and management conditions. Almost all used diets were mainly based on low quality cereal hays and/or silages and straw. They were supplemented with a commercial concentrate containing about 14 to 15% of crude protein. Effects on birth weight, average daily gain, growing period, slaughter age, carcass weight, and carcass value were examined.

Information on feeding diets and managements, feedlot conditions, calving ease, birth weight, animal behavior, percentage of adults and calves death were collected. This collection was based on planned comprehensive survey conducted among the participating producers.

Impact of crossbreeding on animal performances

Major animal performances are given in Table 1. All indicators were improved when using crossbreeding with beef breeds as compared to the use of either dairy or dual purpose animals in crossbreeding with the local cattle.

Results showed that the average carcass weight of local × dual purpose breeds was only 243 kg, whereas average carcass weight reached 320 kg for local × beef type crossbred calves with the highest value (329 kg) being observed for the local × charolais crossbreds. That is an average increase of 77 kg (or about 30%) per carcass. Such an improvement was achieved under the same conditions of management. However, when compared to dual purpose animals, local × beef type animals had similar average carcass weights despite their higher meat potential. This suggests that the nutritional requirements in terms of energy and proteins of the latter animals were not fully met. This is particularly true given the low nutritional values of common forages used in the basal diets along with their inadequate protein supplementation almost in all fattening operations regardless the genetic type of the animal.

In addition, and given the higher dressing percentage of beef animals, the 77 kg increase in the carcass weight can be considered quite important not only because of the weight per se, but also given the higher quality of the meat and the specific butchers and consumers preference for such type of meat. Moreover, we also have to consider the higher sale price for these animals' meat. Indeed, the price for each kg of these animals' carcasses is 0.5 TD higher than that from other animals, which amounts to an average additional gain of about 400 TD per carcass.

In terms, of average daily gains, we found that crossbreeding with beef breed resulted in higher daily gains under similar fattening conditions and at the same slaughter age. Beef breeds resulted in average daily gains of 950 g per day, against 733 g when crossbreeding with dual purpose breeds. That is an improvement of about 217 g per head per day, or the equivalent of 29%. However, despite such improvement, the observed gains were quite low compared to the genetic potential of the animals. Higher values of 1000 g/head/day were reported for dual purpose animals when fed better quality forages (Lahmar *et al.*, 2005).

Table 1. Average animal performances according to the genotype.

Genotype	Calving ease (%)	Birth weight (kg)	Growing period (month)	Slaughter weight (kg)	Daily gain (g)	Carcass value (TD[1])
Local × Charolais	98.5	45	20	329	950	2,500
Local × BBB	90	40	20	310	950	2,350
Average 1	94.2	42.5	20	320	950	2,425
Holstein	-	35	-	326	850	2,282
Brown Swiss	-	-	-	304	800	2,128
Average 2	-	-	-	315	825	2,205
Local × Holstein	94	-	21	238	800	1,700
Local × Brown Swiss	96.2	38	22	250	700	1,750
Local × Tarantaise	100	37	20	240	700	1,680
Average 3	96.7	37.5	21	243	733	1,710

[1] TD: Tunisian Dinar; 1TD=0.58 euros as of September 2008.

Results also showed that average weight at birth for dual purpose crossbred animals was 37.5 kg whereas the average for beef crossbred animals was 42.5 kg. That is an increase of 5 kg which confers to these animals a better growth start and a higher aptitude for fattening. Similar weights were reported for dual purpose animals. Such increase in birth weight did not result in any significant decrease in the percentage of calving ease (Table 1). This is probably the result of the adequate initial choice of cows to be used in the crossbreeding program. A choice which was primarily based on body size, lactation number and health status of cows.

Conclusion

Previous production programs dealing with the development of beef relied mainly on the use of dairy animals and to a less extent on dual purpose breed. They have focused on improving management practices in feedlots. A limited number of genetic studies have involved only crossbreeding of dairy and dual purpose breeds with the native cattle. Limited improvement was made in meat production and beef type of animals best able to produce beef efficiently need to be identified. Results in this work showed that using the CH and the BBB beef breeds in crossbreeding programs with local and/or their crossbred with dairy breeds improves birth weight, daily growth gain, carcass weight and carcass quality and market value. This suggests that crossbreeding with beef breeds is a new method which can be used in the country to reach self-sufficiency in domestic beef while improving farmers' revenue, particularly under current economical conditions where production efficiency is needed more than ever before. However, for this to be successful a significant improvement also need to be done in terms of the production of good quality silages and hays forage and forage diversification which enables beef type animals and their crossbreds to achieve their genetic potential for meat.

References

Atti, N. and M. Ben Dhia, 1987. Performances des bovins croisés en Tunisie. Annales de l'INRAT, vol. 60 Fasc. 13 pp: 24.

Atti N. and M. Ben Dhia, 1990. Performances des bovins croisés en Tunisie. Amélioration génétique des bovins sous climat sud-méditérranéen. EAAP publication N°47 pp: 196-199.

Ben Dhia M. and A. Antic, 1971. Résultats préliminaires de croisement d'absorption des bovins de race locale. Séminaire sur l'élevage et la production de viande bovine. Tunisa.

Lahmar, M., M. Djemali and H. Khémiri, 2005. Cattle production in the mountain regions of North West Tunisia: present situation and prospects for increased productivity. Proceedings of the international symposium 'animal production and natural resources utilisation in the Mediterranean mountain areas'. EAAP Publication N° 115. pp: 395-400.

Rondia G., A. Deker., M. Jabari and A. Antoine, 1984. Produire plus de grain et de lait en Afrique du Nord. Projet ferme modèle de frétissa. Publication Agricoles N°5.

GIVLAIT, Tunisian Ministry of Agriculture, 2007. The red meat sector in Tunisia.

Proposal of some alternatives to simplify a MOET programme

I. Mayorga[1,2]*, F. Masia*[2]*, L. Mara*[2]*, S. Casu*[2] *and M. Dattena*[2]
[1]*Department of Veterinary Clinical Sciences, University of Padova, 35100 Padova, Italy*
[2]*DIRPA-AGRIS, 07040 Olmedo(SS), Italy*

Abstract

The aim of this study was to propose some alternatives to simplify a sheep MOET program looking at hormonal reduction, cost-effective and less time consuming treatments. Twenty-nine animals were divided into 3 groups: natural oestrus (NT) (n=10), double cloprostenol injection (DC) by 7 days apart (n=10) and double sponge (DS) 40 mg FGA for 14 days; the sponge was replaced by a new one, maintained from day 7 to 14 (n=9) as a control group; 350 I.U. of porcine FSH per sheep was administered in eight decreasing doses twice daily starting on day 4 after onset of oestrus (day 0) in the NT group and 48 h before second cloprostenol injection or sponge removal in the DC and DS groups respectively. A single dose of 125 µg cloprostenol was injected at the same time of first FSH dose in the DS group and on day 6 after oestrus in the NT group to induce luteolysis. All ewes were naturally mated 24 h after cloprostenol injection or sponge removal. Embryos were recovered surgically by flushing at Day 7 after mating. Among all recovered embryos (NT+DC+DS) 129 were vitrified, but only 22 vitrified and 20 fresh embryos from NT group were transferred in pair into 21 synchronised natural oestrus recipients using a direct transfer technique with the help of open pulled straw (OPS). Corpora lutea (CL), embryos recovered (ER), embryos fertilized (EF) and high quality embryos (EQ_1) per ewe were analysed by ANOVA; while recovery (RR), fertility (FR), embryo quality (Q_1R) rates by Chi Square analysis; lambing rate (LR) of fresh and vitrified embryos was similarly analysed. No statistical differences were founded between NT, DC and DS groups in the number of CL/ewe (10.7±3.4, 8.8±5.9, 7.6±5.1), ER/ewe (7.2±3.9, 5.0±6.6, 4.7±4.1), EF/ewe (7.2±3.9, 5.0±6.6, 3.0±3.7), EQ_1/ewe (6.2±3.8, 4.5±6.5, 2.6±3.6), RR% (67.2, 56.8, 62.0) and Q_1R% (86, 90, 88) respectively. However, statistical differences were founded on FR% (100, 100, 62) between NT, DC and DS groups respectively. LR after transfer was similar for vitrified (59%) and fresh (60%) NT embryos. Results suggest the suitability of natural oestrus and cloprostenol for simplify some steps during a MOET program in alternative to traditional protocols avoiding intravaginal sponges, reducing costs and length of treatments.

Keywords: superovulation, FSH, cloprostenol, natural oestrus, embryo transfer

Introduction

In the last 30 years the biotechnologies of multiple ovulation and embryo transfer (MOET) have been developed into animal breeding and husbandry with the aim to speed up the genetic improvement and preserve genetic resources of endangered livestock.

MOET in sheep is now developed and pregnancy rate is over 60%. However, a reduced number of this animals undergo to MOET programme, due to the cost, long hormonal treatments, surgeries and embryo handling procedures. For these reasons, cost reduction and simplification of any stage of embryo production or transfer is likely to widespread the use.

Recently with information provided by ultrasonography new protocols of superovulation have been proposed and the results have been improved (Mayorga *et al.*, 2008; Menchaca *et al.*, 2007). Moreover the use of vitrification combined with the direct embryo transfer technique allow the reduction of the cost in ovine MOET programme, avoiding the use of any special equipment (microscope-free) (Dattena *et al.*, 2004; Isachenko *et al.*, 2003).

In our day sheep offer the potential of making an important contribution to provide food and fibre for a growing world population thus a simplified MOET programme can be useful especially for animal production in developing countries. Having this in mind, this paper proposes some alternatives to simplify a MOET programme avoiding the use of progesterone devices during superovulation and the uses of vitrification technique combined with direct embryo transfer procedures.

Material and methods

In vivo embryo production

Twenty nine Sarda ewes were selected as donors and divided into 3 groups: natural oestrus (NT) (n=10); double cloprostenol injection (DC) (n=10) and double sponge (DS) (n=9) (control group). The group NT was selected throughout oestrus detection (day 0) with the introduction of a vasectomized ram in the flock. The ewes were considered to be on oestrus when they show oestrus behaviour and were mated by the vasectomised ram. In the case of DC and DS groups, the animals were selected at random and the oestrus cycles were synchronised with a double dose of 125 μg cloprostenol injection by 7 days apart in the DC group and by insertion of a intravaginal sponge containing 40 mg fluorogestone acetate for 14 days; the sponge was replaced by a new one, maintained from day 7 to 14 in the DS group.

Superovulatory treatment consisted of 350 I.U. of porcine FSH administered in eight decreasing doses at every 12 h starting on day 4 after oestrus detection (day 0) in the NT group and 48 h before second cloprostenol injection or sponge removal in the DC and DS groups respectively. A single dose of 125 μg cloprostenol was injected at the same time of first FSH dose in the DS group on day 6 after oestrus detection in the NT group to induce luteolysis. All ewes were naturally mated 24 h after cloprostenol injection or sponge removal.

Recovery of embryos

Seven days after mating an inguinal laparotomy of 5 cm was performed and the reproductive tract was exteriorised with minimal manipulations to assess the number of corpora lutea (CL). Embryos were recovered by flushing each uterine horn with 20-40 ml flushing media (HTCM 199 + BSA 0,4%), at flushing the recovered embryos were evaluated morphologically using a stereomicroscope and their quality was scored on a scale of 1 to 3 (Niemann *et al.,* 1981); only embryos quality 1-2 were kept less than 30 min at room temperature in HTCM 199 + BSA 0.4% and considered for to be transferred or vitrified.

Vitrification of blastocysts

The vitrification procedure employed was based on the method originally designed by Yang *et al.* (1992) and Vajta *et al.* (1998) for cow embryos and modified for ovine embryos by Dattena *et al.* (2004). Briefly, all vitrification solutions were prepared using Dulbecco's PBS supplemented with 0.3 mM sodium pyruvato, 3.3 mM glucose and 20% FBS. Expanded blastocysts were exposed at room temperature to equilibration solution (V1) 10% ethylene glycol (EG) and 10% dimethylsulfoxide (DMSO) for 4-5 min, then to the vitrification solution (V2) 20% EG; 20% DMSO and 0.5 M sucrose for $\leq$ 45 sec., then were loaded into open pulled straws (OPS) and immediately plunged into LN2 (2 blastocysts per straw).

Embryo transfer

Because the reduction of experimental animals only embryos of NT group (n=42) were transferred into 21 naturally synchronised recipients ewes. Twenty were utilised for fresh indirect embryo transfer

(IET group) and 22 were utilised for vitrified direct embryo transfer with the help of OPS (DET group). Seven days after natural onset of oestrus in the recipients an inguinal laparotomy 4-5 cm was performed and the number and quality of corpora lutea (CL) was assessed, after this procedure in the IET group 2 fresh embryos per recipient were transferred into the top of the uterine horn ipsilateral to the ovary showing at least one functional CL using a TomCat™ catheter connected to a 1-ml syringe, while, at the moment of transfer in the DET group, the OPS containing the vitrified embryos were warmed by holding for 6 sec. in air and then dipped into a falcon tube contain HTCM199+ 20% serum+ 0.5 M sucrose in a water bath at 37 °C for aprox. 15 s after this procedure 2 vitrified, thawed embryos per recipient were transferred into the top of the uterine horn ipsilateral to the ovary showing at least one functional CL using the OPS as the catheter for transplantation and a TomCat™ catheter was used to connect the straw to a 1 ml syringe (Isachenko *et al.*, 2003). Pregnancy diagnose was performed 45 days after transfer by abdominal ultrasonography.

Analysis of data

Data on number of corpora lutea (CL), embryos recovered (ER), embryos fertilized (EF), and high quality embryos (EQ_1) per ewe were analysed by ANOVA (GLM SAS procedure), whereas, data on recovery (RR), fertility (FR) and embryo quality (Q_1R) rates per treatment were analysed by a Chi Square analysis; pregnancy rate and lambing rate of fresh and vitrified embryos was similarly analysed. Statistical significance was denoted as $P<0.05$.

Results

No statistical differences were found between NT, DC and DS groups in the number of CL/ewe, ER/ewe, EF/ewe, EQ_1/ewe, RR% and Q_1R%. However, statistical differences were founded on FR% (100, 100, 62) between NT, DC and DS groups respectively (Table 1). Pregnancies and lambing rates after transfer were similar for vitrified and fresh NT embryos (Table 2).

Table 1. Length and cost of the treatments, number of corpora lutea (CL), embryo recovery (ER), embryo fertilized (EF), embryo quality (EQ_1) per ewe, recovery (RR), fertility (FR) and embryo quality (EQ_1) rates in natural oestrus, double cloprostenol, double sponge groups.

Group	(n)	Days	Cost/ewe	CL/ewe	ER/ewe	EF/ewe	EQ_1/ewe	RR	FR	Q_1R
NT	10	14	€ 101.8	10.7±3.4	7.2±3.9	7.2±3.9	6.2±3.8	67.2	100^a	86
DC	10	16	€ 107.5	8.8±5.9	5.0±6.6	5.0±6.6	4.5±6.5	56.8	100^a	90
DS	9	22	€ 109.3	7.6±5.1	4.7±4.1	3.0±3.7	2.6±3.6	62	62^b	88

Different superscripts indicate treatments with significant differences ($P<0.05$).

Table 2. Pregnancy and embryo lambing rates after transfer in fresh indirect embryo transfer (IET) and vitrified direct embryo transfer (DET) groups respectively.

Group	Source of embryos	Recipients	Embryos transferred (n)	Pregnancy rate (day 45)	Lambing rate
IET	Fresh	10	20	70	60
DET	Vitrified	11	22	82	59

Conclusion

Results suggest the suitability of natural oestrus and cloprostenol to simplify some steps during a MOET program. In fact, although significant differences between the alternative treatments and the control were detected only in the FR rate the other variables tended to be improved. This suggests that it is possible to use a MOET programme to avoid the use of intravaginal progesterone sponges, reducing cost and time of treatments. In addition the vitrified direct embryo transfer technique provides a useful tool in attempt to adapt a MOET programme also in field condition.

References

Dattena, M., C. Accardo, S. Pilichi, V. Isachenko, L. Mara, B. Chessa and P. Cappai, 2004. Comparison of different vitrification protocols on viability after transfer of ovine blastocysts *in vitro* produced and *in vivo* derived. Theriogenology 62: 481-493.

Isachenko V., J.L. Alabart, M. Dattena, F. Nawroth, P. Cappai, E. Isachenko, M. J. Cocero, J. Olivera, A. Roche, C. Accardo, A. Krivokharchenko and J. Folch, 2003. New technology for vitrification and field (microscope-free) warming and transfer of small ruminant embryos. Theriogenology 59: 1209-1218.

Mayorga, I., F. Masia, L. Mara, F. Chessa, S. Casu, N. Juyena and M. Dattena, 2008. Superovulation fsh-p protocol in sarda ewes without progestagen synchronization treatment. Reprod. Dom. Anim. 43: suppl 3-pag. 193 (abstract).

Menchaca A., M. Vilariño, M. Crispo, A. Pinczak and E. Rubianes, 2007. Day 0 Protocol: Superstimulatory treatment initiated in the absence of a large follicle improve ovarian response and embryo yield in goats. Theriogenology 68: 1111-1117.

Niemann, H., E. Schilling, B. Sacher, D. Smidt, 1981. Der FDA - and DAPPI-TEST; Ein Beitrag zur Vitalitatsbeurteilung von Rinderembryonen mit Hilfe der flureszenzmikrokopie Berel. Muh. Tierarztl. Wochenschr 9: 441-445.

Vajta, G., P. Holm, M. Kuwayam, P. J. Booth, H. Jacobsen, T. Greve and H. Callesen, 1998. Open-Pulled Straw (OPS) vitrification: a new way to reduce cryoinjuries of bovine ova and embryos. Mol. Reprod. Dev. 51: 53-58.

Yang, N.S., K. H. Lu, I. Gordon and C. Polge. 1992. Vitrification of bovine blastocysts produced in vitro. Theriogenology 37:326 (abstract).

Adaptation and reproductive performances of exotic saanen breed introduced in Lebanon

S. Abi Saab[1] *and J. Daher*[1,2]
[1]*Lebanese University, Faculty of Agricultural and Veterinary Sciences, Beirut, Lebanon; Sabisaab@yahoo.com*
[2]*Universite Saint Esprit, Faculty of Agriculture, Kaslik, Lebanon*

Abstract

Introducing exotic goat breeds in the Eastern Mediterranean region aims to improve milk and meat production. The objectives of this study were to determine the capacity of adaptation and the reproductive performances of high yielding goat breeds under the prevailing conditions of Lebanon. Eighty primiparous 7 months old Saanen goats were imported from Europe, kept under an intensive management at a 900 m altitude, and fed wheat straw with a commercial concentrate. Adaptation parameters (respiration rate, pulse rate and body temperature) were measured upon arrival at day 1 (D1) and D2, on a weekly basis during the first month and then on the D45 and D75. Reproductive performances (estrus cycle length, fertility, gestation length, and prolificacy) were also studied. Results showed that the mean values of respiration and pulse rates were high just after arrival (32.4±8.5 breaths/min and 88.5±14.8 beats/min, respectively) then decreased significantly ($P<0.05$) to reach 17.5±2.8 breaths/min and 64.0±6.9 beats/min at D45, respectively and 17.8±1.9 breaths/min and 64.1±1.9 beats/min at D75, respectively. Body temperature (38.9±0.3 °C) was not affected by transportation. These finding indicate the capacity of adaptation of the newly introduced goats to the stressful conditions of management and climate of Lebanon. Estrus cycle length was 21.42±4.9 days and fertility at first mating was 76.4%, reaching 93.5% at the end of the reproductive season. Gestation length (152.49±3.6 d) and prolificacy 1.45±0.5 kids/doe, were not affected by the adaptation trial. Saanen goats introduced and raised under Lebanese conditions could increase goat farming profit if good management practices were provided.

Keywords: breed testing, Saanen goats, adaptation, reproduction, Lebanon

Introduction

The majority of goat distribution in the world is observed in developing countries where harsh climatic environmental and feeding conditions prevail. In fact, among other domestic ruminants, goats could be considered the most competitive under hard conditions given their high capacity for adaptation (Devendra, 1990; King, 1983). In Lebanon, about 500,000 goats are grazed by about 7,000 farmers (Ministry of Agriculture, 2000). The Baladi goat, the dominant breed (95%), is characterized by its rusticity and adaptation to the prevailing conditions; milk yield is about 120-140 kg/head/lactation (180 d); prolificacy reaches 1.3 kids/doe. However, the applied extensive grazing benefits on these goats are tangible for only 2 months (Abi Saab *et al.*, 2008). Therefore the introduction of select exotic breeds in order to increase productivity and insure durability of goat farming might be of interest. The Saanen breed is one of the most performing dairy breeds, characterized by a high milk yield (>750 kg of milk/head/lactation) over a long period of milk production (300 days). Thus, raising Saanen goats under an intensive management system in Lebanon could improve the overall goat milk production without being affected by the environment. Previous trials of introducing this specific breed to Lebanon with efforts from multiple entities have not yet proven efficient. The present work aims to evaluate the adaptation and the reproductive performance of the Saanen goats when introduced to Lebanon and raised under an intensive commercial system.

Material and methods

Eighty primiparous Saanen goats (7-8 months old) were imported through air freight from Europe to Lebanon in October 2006. 77 female and 3 male goats were kept in a commercial farm located at Annaya-Jbeil at a 900 m altitude. Intensive husbandry conditions were applied with a zero grazing system. Clean water, trace mineral and salt blocks were always available. Wheat straw were offered *ad libitum*, concentrate and pellet alfalfa were distributed twice daily (morning and evening); diets were formulated to achieve consistency of supply of energy and protein, according to the physiological state of animals (NRC, 1981). Goats were vaccinated against FMD and enterotoxaemia and were treated against internal and external parasites.

Adaptation parameters

Respiration rate, pulse rate and body temperature were evaluated upon arrival at day 1 (D1) and day 2 (D2), on a weekly basis during the first month at D7, D14, D21, and D28, then on D45 and D75. Respiration and pulse rate were taken through a stethoscope, while rectal temperature was taken by a medical thermometer.

Reproductive performance

The reproductive management consisted of controlled mating. As the period of importation coincide with the natural breeding for goats in Lebanon, females were divided into 3 groups to be bred by a separate buck each. Twice daily, heat was detected and does in estrus were exposed to bucks. Estrous cycle length was recorded on does returning to estrus. Fertility percentage by non-estrus returns, gestation length from mating to kidding and prolificacy by number of kids per doe were all calculated. Lineage and daily climatic temperatures were tracked. Data were analyzed using Sigmastat 2.0 (Aspire Software). Means were compared using tukey's test.

Results and discussion

Adaptation parameters

Stressful conditions during transportation were very obvious on the goats where respiration and pulse rate were at their highest (Figure 1). These values decreased significantly ($P<0.05$) with time. In fact, the respiration rate decreased and reached 25.3±5.2 breaths/min at D7 and 21.3±4.0 breaths/min at D21; then stabilized around 17 breaths/min after D28. Simultaneously, pulse rate dropped to 68.5±10.7 beats/min at D7 and 65.0±7.5 beats/min at D 21. Limited fluctuations were observed after D28 and plateau around 64 beats/min after D45. Body temperature showed little fluctuation within the normal range (38.9±0.3 °C; $P>0.05$). Climatic temperature during the experimental period was 18.1±2.5 °C.

These findings show that goats react against the stress of transport by increasing pulse and respiration rates. These rates decrease after a short period readjusting the physiological parameters to fit the new prevailing climatic conditions in Lebanon. However, body temperature is not affected by transportation. After adaptation, values of physiological traits were acceptable due to the favorable climatic temperature that was within the thermal comfort zone. Higher values were obtained in Turkey (Darcan and Güney, 2008) where climatic temperature reached more then 30 °C increasing simultaneously goat's respiration and pulse rate to more than 70 breaths/min and 100 beats/min, respectively. Previously, our findings (Abi Saab and Sleiman, 1995; Sleiman and Abi Saab, 1995) have shown that the adapted local Awassi sheep breed had lower respiration and pulse rate than other exotic sheep breeds. These findings possibly indicate a longer adaptation requirement for foreign goat and sheep breeds without any compromise to the performance of these exotic goats.

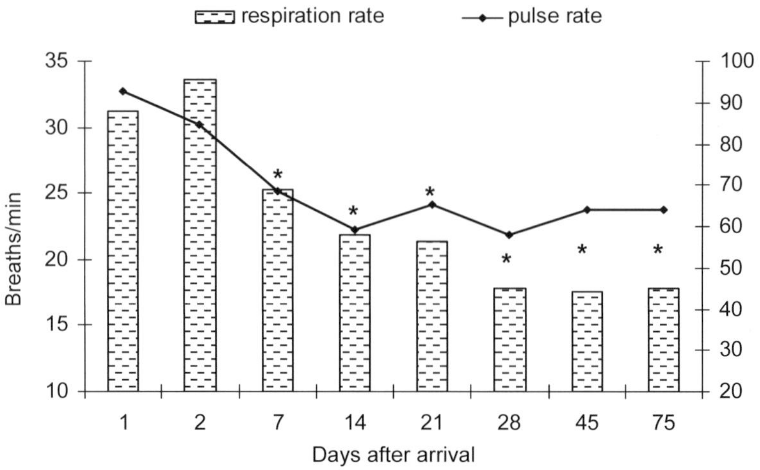

Figure 1. Respiration and pulse rate of newly introduced primiparous Saanen goats to Lebanon.
** Respiration rate and pulse rate significantly different from D1 and D2 (P<0.05).*

Reproductive performance

Mean reproductive performances are shown in Table 1. Estrous cycle length was 21.4±4.9 d for primiparous goats under Lebanese conditions matching earlier reports in the region (Khouri, 1997). 76% of the mated goats did not return to estrus after the first mating whereas only 22% showed a second estrus from which only 1.6% accepted a third mating. These observed performances are comparable to local goat and sheep operations in Lebanon and the region (Abi Saab and Sleiman, 1986; Khouri, 1997). These values indicate a highly adapted reproductive efficiency of the Saanen goat breed when exposed to the local prevailing Lebanese conditions. Gestation length for the first parity was 152.4±3.6 d, higher than gestation length during the same period during the same parity under similar intensive management in Egypt (Mourad, 2001). Prolificacy reached 1.45±0.5 kids/doe higher than that obtained others with an average of 1.22 (Crepali *et al.,* 1999; Mourad, 2001; Silva *et al.,* 1998).

Overall, these reproductive parameters indicate a potential for fast adaptation of these Saanen goats. In fact, this exotic breed showed a higher prolificacy rate than other similar breeds in the same region.

Table 1. Mean reproductive performances of primiparous Saanen goats under Lebanese conditions.

Parameter	Mean±SD
Estrous cycle length	21.4±4.9 days
Fertility	93.5%
Gestation length	152.4±3.6 days
Prolificacy	1.45±0.5 kids/doe
Sex ratio	0.58

Conclusion

Saanen goats newly introduced to Lebanon increase their pulse and respiration rate as response to the stress of transportation. Physiological traits become normal after a short period of adaptation. Reproductive performances were not affected by their import. The potential for introduction of the Saanen breed to Lebanon seems promising when given the right management and provided the required feed and sanitary environment.

References

Abi Saab, S. and F.T. Sleiman, 1986. Reproductive performance of artificially inseminated Awassi, finnlandrace, Texel sheep and their crosses. Lebanese Science Bulletin, 2(2): 12-23.

Abi Saab, S. and F.T. Sleiman, 1995. Physiological responses to stress of filial crosses as compared to local Awassi breed. Small Ruminant Research 16: 55-59.

Abi Saab, S., E. Hajj, B. Abi Salloum and E. Rahme, 2008. Seasonal and altitudinal variations on adaptation, growth and testicular activity of Baladi goats with vertical transhumance in Eastern Mediterranean. Lebanese Science Journal, 9(1): 99-111.

Crepali, P., M. Corti and M. Cicogna, 1999. Factors affecting milk production and prolificacy of Alpine in Lombardy (Italy). Small Ruminant Research 32: 83-88.

Darcan, N. and O. Güney, 2008. Alleviation of climatic stress of dairy goats in Mediterranean climate. Small Ruminant Research 74: 212-215.

Devendra, C., 1990. Comparative aspects of digestive physiology and nutrition in goats and sheep. In: Devendra, C. and E. Imazumi (ed.), Ruminant Nutrition and Physiology in Asia, pp. 45-60.

Khouri F., 1997. The encyclopedia of Goat Breeds in the Arab Countries. Conservation of Biodiverity and Environments in the Arab Countries. The Arab Center for the Studies of Arid Zones and Dry Lands (ACSAD/AS/P158/1996), Damascus, Syrian Arab Republic, 498 p.

King, J.M., 1983. Livestock water needs in pastoral Africa in relation to climate and forage. Res. Report N° 7. Int. Lives Center Africa (ILCA), Addis Ababa, Ethiopia.

Ministry of Agriculture, 2000. Resultats globaux du recensement agricole. FAO projet 'assistance au recensement agricole' (UTF/LEB/016). 122 p.

Mourad, M., 2001. Estimation of repeatability of milk yield and reproductive traits of Alpine goats under an intensive system of production in Egypt. Small Ruminant Research 42: 1-4.

NRC, 1981. Nutrient requirements of goats. National Academy Press, Washington, DC, 91 p.

Silva, E., M.A. Galina, J.M. Palma and J. Valencia, 1998. Reproductive performance of Alpine dairy goats in a semi-arid environment of Mexico under a continuous breeding system. Small Ruminant Research 27: 79-84.

Sleiman F.T. and S. Abi Saab, 1995. Influence of environment on respiration, heart rate and body temperature of filial crosses compared to local Awassi. Small Ruminant Research 16: 49-53.

Managing the breeding and calving seasons in the suckler cow herd

J. Álvarez-Rodríguez, I. Casasús, R. Revilla, M. Blanco and A. Sanz
CITA, Gobierno de Aragón, Av. Montañana 930, 50059 Zaragoza, Spain

Abstract

This study analysed the effects of shortening the breeding period from 85-90 to 63-66 days on herd fertility. The data set included 1,063 calving events comprising two breeds (Parda de Montaña, PA, and Pirenaica, PI), two calving seasons (autumn and spring) and 8 years (1999 to 2006). Overall actual pregnancy rate over a breeding period of 85-90 days was moderate in both seasons, but slightly higher in autumn than in spring (68% vs. 52%; $P<0.10$). Nevertheless, pregnancy rate within the first 63-66 days was similar in both seasons (56% vs. 46%; $P>0.10$). The day 65 after initiation of the calving period 86% and 92% of total calving had occurred in autumn and spring, respectively, regardless the season and year ($P>0.10$). Short breeding periods may allow for concentrated calving seasons, resulting in greater calf homogeneity at weaning and a more efficient labour organisation.

Keywords: beef cattle, extensive systems, management, reproduction, performance

Introduction

Permanent breeding turns into farming disorders because it increases the cost of supplementary feeding and it does not allow obtaining homogeneous lots of calves, thereby decreasing gross margin compared to concentrate breeding seasons (Daza and Martín, 2007).

Many Spanish mountain areas are characterized by marked forage seasonality, particularly from October to March, due to harsh climate constraints. Planning breeding and calving seasons may allow a more efficient use of forage resources and lower labour out of seasonal working peaks, but it draws higher control on physiological variables which can determine reproductive success in both genders.

The re-establishment of female ovarian function after previous calving is a limiting factor determining the success of concentrate breeding after male introduction and during defined mating periods. Meta-analyses from circulating progesterone during the post-partum period of 549 cows (11-year data: 1987-1998) showed that the interval to first ovulation after calving (IFO) was shorter in autumn than in spring (33 vs. 66 days), in response to a greater recovery of body reserves in cows calving at the end of summer grazing compared to cows calving at the end of winter housing (2.75 vs. 2.35, on a 1 to 5 scale) (Sanz *et al.*, 2004). In the previous work, cows in moderate condition which suckled *ad libitum* had longer IFO than those which were managed in restricted suckling twice-daily.

The objective of this study was to evaluate the effects of shortening the breeding period from 85-90 days to 63-66 days on herd fertility, by maintaining the same date of bull introduction into the herd and the same mating chances per cow (strata of 21-day cycles).

Material and methods

Animal management and study site

Data for this study consisted of the calving data and performances observed on an experimental suckling cattle herd in 'La Garcipollera' Research Station (North-eastern Spain, 42°37'N, 0°30'W, 920 to 2,200 m above sea level). The mean temperature and annual rainfall during the studied period (1999-2006) was 10.2 °C and 1,059 mm, respectively. There were two years (2004 and 2005) with annual precipitation below 800 mm.

Spring-calving cows calved from February 15 to May 15. They were housed from mid-December until June 15, that is, during late pregnancy and the first months of lactation. During the summer the dams and their calves grazed on the high mountain pastures, until weaning on September 15 (on average 6-month old calves). Then, the non-lactating, pregnant cows grazed on forest pastures until December 15. The average length of their grazing season was 183 days. Autumn-calving cows calved from September 15 to December 15. They were housed throughout lactation, from the start of the calving season until approximately April 15, the date of turnout depending on weather conditions. Their calves were weaned by mid-March (on average 4.5 months of age). The dry, pregnant cows grazed for about two months (April 15 to June 15) on forest pastures. During the summer (June 15 to September 15), they grazed on high mountain ranges with the spring-calving herd. The grazing season of the autumn-calving cows averaged 158 days.

In both seasons cows were bred by natural service to bulls of their same breeds during a 3-month breeding season, April 15 to July 15 for spring-calving cows and December 15 to March 15 for autumn-calving ones. Ninety days after first calving (average 50-55 days postpartum of the group) bulls were introduced into the herd. Sires were randomly assigned among cows, which were allotted according to physiological status and breed. Cows from both seasons calved with a body condition ranging between 2 and 3 (1 to 5 scale). The feeding level met between 75 to 100% of their energy requirements. Nearly 95% of the cows were kept under restricted suckling twice-daily for 30 min (08:00 and 16:00).

Data collection and analysis

Data were collected during the breeding and following calving notebooks. The data set included 1,063 calving events of multiparous cows from two breeds (PA and PI), two calving seasons (autumn and spring) and 8 years (1999-2006). The breeding notebook was used to record cow identification, bull assigned to each lot of cows and dates of breeding interval. The calving notebook was used to record the date of parturition events within 24 h after delivery, calf sex and the weight of dam and young at this moment. Also, cow was weighed again one week after parturition to reflect calving weight as the mean of both recordings.

The pregnancy rate (actual and stratified) was calculated as the number of calvings in the following year period divided by the number of cows bred (lactating and dry). Each year, we discarded 5% of cows that were culled from the herd prior to complete their production cycle. The simulated pregnancy rate was estimated from calving events occurred within the first 63-66 days (on a 21-22 days stratum) of the next calving season.

Data from pregnancy rate and female to male ratio were submitted to analysis of variance accounting for fixed effects of season and year as sources of variation. Weights of cows and calves at calving were analysed considering the effects of breed, calving season, year, calf sex, and their second degree interactions.

Results and discussion

The female per male ratio was similar across breeding seasons and years (average 15±1.3 females per male; P>0.05). There were no differences among years in overall pregnancy rates (P>0.10). The pregnancy rate was moderate in both calving seasons, but slightly higher in autumn than in spring (68% vs. 52±5.5%; P=0.10). Nevertheless, pregnancy rate within the first 63-66 days was similar in both seasons (56% vs. 46±4.7%; P>0.10).

From the beginning until day 43 of the calving season 63% and 71±2.9% of the total of calving events had occurred in the autumn and spring, respectively (P=0.09), irrespectively of year (P>0.10). On day 65 after initiation of the calving period 86% and 92±1.9% of the cows had calved in autumn and spring, respectively (P=0.07), without any influence of year (P>0.10). When parturition events within period were stratified per 21-day cycles, the number of calving during the second cycle (22-43 days) tended to be lower in autumn than in spring (28% vs. 39±3.1%; P=0.06), while the opposite occurred in the fourth cycle (66 days-end) (14% vs. 8±1.8%; P=0.07) (Figure 1).

As quoted earlier, the results of the period 22-43 days could be a consequence of delayed IFO in spring compared to autumn season due to lower body reserves at calving in the former group (Sanz *et al.*, 2004) and thus a greater proportion of anoestrus cows at mating. Regarding differences after day 66 of the calving season, they could be attributed to the different mating management on that period in each season, since in autumn the herd was kept indoors while in spring late breeding was completed outdoors, on large mountain areas with scattered shadows and drinking points which led to the formation of sub-groups without males and did not facilitate mating. Therefore, shortening the breeding period may have greater carry over effects in autumn than in spring, involving a loss of 11% of the calving events. However, the reproductive efficiency may be decreased only by 9%. On its favour, the annual combination of two differentiated breeding and calving periods with a 63-66 day length may allow conducting systematic pregnancy diagnosis 45 days after bull removal and subsequent early detection of non pregnant cows. This plan may accelerate the inclusion of these cows in the following breeding period, especially in the transition between autumn and spring calving seasons, where mating is initiated around two months after removing bulls from the previous season. Also, shortening the breeding season in spring may avoid bull transport from mountain ranges in summer, thereby improving attention on mating and ectoparasites.

The weights of cows and calves after delivery are shown in Table 1. Only 0.5% of twins were born within the studied period. There were no interactions among the evaluated fixed effects (P>0.10). For those cows that conceived, weight at calving was greater in autumn than in spring (598±4 vs. 576±3 kg, respectively; P<0.001), and lower in PA than in PI breed (574±2 vs. 600±5 kg; respectively; P<0.001). Calf sex did not affect weight of dams at calving (P>0.10). In contrast, calf weight at birth was lower in autumn than in spring (41.1±0.4 vs. 42.6±0.3 kg, respectively; P<0.01), and higher in PA than in PI (43.0±0.2 vs. 40.8±0.4 kg, respectively, P<0.001). Male calves were heavier at birth than females (43.5±0.3 vs. 40.3±0.3 kg, respectively; P<0.001). Year affected cow and calf weights: weight of dams at calving was greater during the first three years than in the rest (596±6 vs. 578±6 kg, in 1999-2002 and 2003-2006, respectively; P<0.05), but this variation was not linked with the observed year effect on calf weight at birth (higher in 1999 and 2003 than in the rest: 44.0±0.6 vs. 41.3±0.6; P<0.05). Earlier studies on the same herd pointed out that both breeds had similar weight

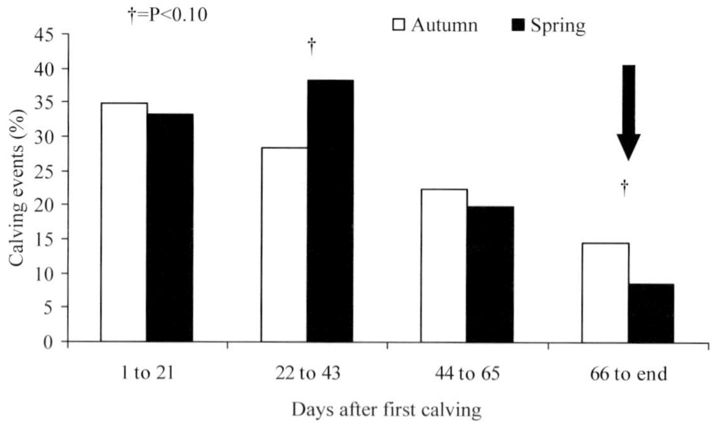

Figure 1. Calving distribution over autumn and spring calving seasons (period 1999-2006).

Table 1. LS Means ± standard error of cow and calf performance (period 1999-2006).

	Autumn		Spring	
	PA	PI	PA	PI
Cow weight at calving (kg)	582±3	615±8	566±4	585±5
Calf weight at birth (kg)	42.2±0.3	40.2±0.8	43.9±0.4	41.4±0.5

PA: Parda de Montaña, PI: Pirenaica.

at calving in the spring season (Casasús *et al.*, 2002), but herein we observed that PA breed might be directed towards a lighter frame-size with lower fatness degree in both calving seasons. This trait may be mediated by lower gains during grazing and possibly by a greater nutrient transfer to the foetus which may enhance calf birth weight in this breed compared to PI one.

In the present analysis, PA calves were lighter in autumn than in spring. Conversely, Villalba *et al.* (2000) reported that PA calves were heavier in autumn than in spring This difference might be due to the different nutrient sources used during the last third of pregnancy (forest and mountain pastures in autumn-calving cows vs. total mixed rations in spring-calving cows), which may have favoured the nutrient transfer to the fetus to the detriment of maternal reserves of spring-calving cows.

In addition, the inter-year effect on animal weight might be mediated by the variation in the analysed individuals throughout time and by the dependence of meteorological conditions, which generate years with high forage availability and quality and others with scarcity and short vegetative periods.

Overall results suggest that it is extremely difficult to maintain good reproductive efficiency in extensive cattle systems which meet 40% of their annual energy requirements from pasture (Casasús *et al.*, 2002). Seasonal differences on pregnancy rate may be coped with calf management practices: autumn-calving cows could raise their young *ad libitum*, whereas spring-calving cows should have restricted access to calves once-daily for 30-45 min so that to improve ciclicity at mating. Short breeding periods (around three 21-day cycles) may allow for concentrated calving seasons, resulting in greater calf homogeneity at weaning and a more efficient labour organisation.

References

Casasús, I., A. Sanz, D. Villalba, R. Ferrer and R. Revilla, 2002. Factors affecting animal performance during the grazing season in a mountain cattle production system. J. Anim. Sci. 80: 1638-1651.

Daza, A., and R. Martín, 2007. Estimación de la carga ganadera económicamente óptima en fincas de ganado vacuno de carne del ecosistema de la dehesa. ITEA Vol. Extra 28 (I): 294-296.

Sanz A., A. Bernués, D. Villalba, I. Casasús and R. Revilla, 2004. Influence of management and nutrition on postpartum interval in Brown Swiss and Pirenaica cows. Liv. Prod. Sci. 86: 179-191.

Villalba D., I. Casasús, A. Sanz, J. Estany and R. Revilla, 2000. Pre-weaning growth curves in Brown Swiss and Pirenaica calves with emphasis on individual variability. J. Anim. Sci. 78: 1132-114.

Improvement of the reproductive efficiency for the development of the Sardinian sheep breeding

V. Carcangiu, M.C. Mura, G.M. Vacca, M.L. Dettori, M. Pazzola, S. Luridiana and P.P. Bini
Dipartimento di Biologia Animale, Università degli Studi di Sassari, Via Vienna 2, 07100 Sassari, Italy

Abstract

Reproductive activity of the sheep bred at temperate latitude is conditioned by photoperiodic trend. Melatonin haematic levels are considered the biological informer of photoperiodic trend. The action of melatonin on reproductive activity is carried on through receptors present at hypothalamic level. Polymorphism of receptor MT1 gene in several sheep breeds resulted involved in the regulation of reproductive seasonality. Then, the study of this gene even in the Sarda breed sheep can give the possibility of a better rationalization of the reproductive activity. With this purpose, blood samples from 1000 sheep from different farms were taken. Age, sex and lambing date of the animals were recorded during at least 3 years. The DNA obtained from each blood sample was used for the typication of the MT1 receptor gene. Finally, the obtained genetic and reproductive data were subjected to statistical analysis to show the probable link between them. The confirmation of a relationship between genetic variants and reproductive activity more or less connected to photoperiod can supply a new approach both to selection and management of Sarda breed sheep. In fact, the animals less bound to photoperiod could be used for the deseasoning of lambing and consequently of productions. On the contrary, animals with a reproductive activity deeply linked with photoperiod could be directed to obtain spring lambing. All that, undoubtedly, can produce a great development of the Sardinian ovine sector according to the trend of product demands of market, bound to the tourist vocation of the island.

Keywords: ewe, melatonin, seasonal reproduction, MT1 receptor gene

Introduction

In the Mediterranean Basin animal production from sheep breeding is mainly influenced by seasonality of reproduction (Zuniga *et al.*, 2002), since at temperate latitudes, between the end of January and the beginning of April, sheep are in anoestrus period (Chemineau *et al.*, 2005). Photoperiod is the main factor which affects the sexual activity in the sheep (Malpaux *et al.*, 2001). The light signal perceived by the retina is transmitted to the pineal gland through bundles of neurons within the central nervous system, where it is transformed into a chemical signal, the melatonin. Melatonin is secreted mainly during the dark hours and its circadian and circannual fluctuation informs the body about the photoperiod (Simonneaux and Ribelayga, 2003) as high melatonin blood concentration for several hours affects GnRH secretion and supports reproductive activity (Malpaux *et al.*, 1998). It has been documented that administration of exogenous melatonin in the form of subcutaneous implants improves reproductive efficiency in European dairy and meat sheep breeds (Abecia *et al.*, 2006). However, during the middle anoestrus and among the dairy breeds, exogenous melatonin is not able to exert a sufficient stimulus (Abecia *et al.*, 2007). The attention of researchers has recently been focused on the study of melatonin receptors and their involvement in regulating seasonality of reproduction (Messer *et al.*, 1997). Later, the polymorphism of melatonin MT1gene receptor and, in several breeds, the link between some allele variants and seasonality of reproduction wereshown (Pellettier *et al.*, 2000; Notter and Cockett, 2005). The aim of this research was to study the polymorphism of MT1 gene receptor in the Sarda sheep breed and the possible correlation with the reproductive activity, also to plan future genetic selection schemes and obtain sheep production all over the year.

Material and methods

The research was carried out in 20 sheep farms located in Sardinia, with similar management practices and facilities. In each farm, 50 adult sheep (45 ewes and 5 rams) with an average age of 3.5±1.5 years were selected. For each ewe, mating and lambing date and number of newborn lambs were recorded in the three years period before this research had started. Rams were introduced in the flocks in the second week of April of each year. A blood sample was taken from each sheep and DNA was extracted using a commercial kit (Purgene, Gentra, Minneapolis, Minnesota, USA). After concentration and purity were assessed, 100-150 ng of DNA was amplified using the primers described by Messer *et al.* (1997) and the method by Pelletier *et al.* (2000). Amplified DNA was digested using the two endonucleases MnlI and RsaI (New England Biolabs, Beverly, MA, USA) in order to detect any possible polymorphism. Later, two samples of each different allele variant were sequenced. Data was analysed using Genepop to study the equilibrium of Hardy-Weinberg and χ^2-test (Minitab statistical software) to estimate the relationship between genotype and lambing period.

Results

The presence of a cleavage site at position 612 in both the allele, after endonuclease MnlI, indicated genotype +/+, while the absence genotype -/-. This polymorphism in the nucleotide sequence was caused by a G (allele +) or an A (allele -) at position 612. After endonuclease RsaI, the presence of a cleavage site at position 606 in both the allele, pointed out genotype C/C, while the absence genotype T/T. This polymorphism was caused by a C to T substitution. Each substitution was confirmed by nucleotide sequencing. Genotype frequency was 70% for +/+ and 12% for -/-, 55% for CC and 20% for T/T. Allele frequency was 0,79 for + and 0.21 for -, and 0.68 for allele C and 0.32 for T (Table 1). The examined population was in Hardy-Weinberg disequilibrium both for MnlI and RsaI, because of a heterozygote deficiency. 80% of ewes carrying genotype +/+ lambed in the period September-December and 20% in January-April (P<0.001) (Table 2). The majority of ewes (65%) carrying the genotype C/C lambed in September-December (P<0,001) while T/T in January-April (55%). Lambing of heterozygote ewes were equally recorded in the two period September-December and January-April. Prolificacy of the ewes was not affected by the genotype.

Discussion

The high frequency of allele + and C recorded in our study are similar to those of some sheep breeds reared in other countries and breeding methods, and to those breeds which shows a relationship between genotype and reproduction (Notter and Cockett, 2005). The link between the polymorphism of MT1 gene receptor and the reproductive activity, as it was evidenced from our trial, can pave the way to new progress in sheep farming management, given that ewes of +/+ and C/C genotype had an earlier restarting of reproduction than the other genotypes. On the contrary, the effect of genotype on reproductive activity was not shown for ewes carrying allele + or C in heterozygote state, as it

Table 1. Allele and genotype frequency in the Sarda sheep breed (n=1000).

Allele		Genotype		
+	-	+/+	+/-	-/-
0.79	0.21	70%	18.0%	12.0%
C	T	C/C	C/T	T/T
0.68	0.32	55%	25%	20%

Table 2. Distribution of lambing period in the Sarda female sheep breed according to genotype (n=900).

Lambing period	Genotype			
	+/+	-/-	CC	TT
September-December	504	24	322	81
January-April	126	84	173	99

χ^2 test $P<0.001$.

is also described by Hernandez *et al.* (2005). In our study, rams were introduced into the flock in April, in order to obtain the lambing period in the month of September and advance the beginning of lactation, also considering the availability of the pasture. Indeed, sheep breeding in Sardinia is strongly limited by the high ambient temperature registered during the summer, which causes the desiccation of natural vegetation and irrigated forage lands are necessary for animal feeding (Carcangiu *et al.*, 2005). Advancement of lambing is a necessary achievement in order to have a longer lactation. As a consequence, the knowledge of ewes genotype of the MT1 gene could permit to obtain a better management of reproduction. On the farms, the animals which are less seasonal in breeding activity should be used in the spring mating season, while seasonal animals in the summer-fall mating season. Breeding methods which do not consider animals' genotype is able to preserve animal biodiversity, which has been endangered since animals carrying 'useful' genes have been preferred for reproductive schemes. On the farms with irrigated forage lands, we can hypothesize, also by considering the genotype of animals, a better management of reproductive activity and a continuous production of suckling lambs, fresh and soft cheeses and ricotta. The availability of these milk products during summer, when the influx of tourists in Sardinia reaches its peak, is a starting point to improve market strategies and economic value of local productions. Moreover, animal welfare could be improved as pharmacological treatment are not necessary when natural reproduction based on genotyping MT1 gene is used. In conclusion, exploitation of animals with different polymorphism at MT1 gene is an innovation in the field of reproduction management for the Sardinian sheep breeding in order to reach an improvement of breeding methods and farm income.

References

Abecia J.A., I. Palacıın, F. Forcada and J.A. Valares, 2006. The effect of melatonin treatment on the ovarian response of ewes to the ram effect. Domest. Anim. Endocrinol. 31, 52-62.

Abecia J.A., J.A. Valares, F. Forcada, I. Palacıın, S. Martıın, and A. Martino, 2007. The effect of melatonin on the reproductive performance of three sheep breeds in Spain. Small Rum. Res. 69, 10-16.

Chemineau P., B. Malpaux, J.C. Thiéry, C. Viguié, H. Morello, L. Zarazaga and J. Pelletier, 1995. The control of seasonality: a challenge to small ruminant breeding. In: Enne, G., G.F. Greppi and A. Lauria, (eds.), Reproduction and Animal Breeding. Advances and strategy, Elsevier, Amsterdam, The Netherlands, 225-250.

Hernandez X., L. Bodin, D. Chesneau, D. Guillaume, P. Chemineau, B. Malpaux and M. Migaud, 2005. Relationship between MT1 melatonin receptor genepolymorphism and seasonal physiological responses in Île-de-France ewes Reprod. Nutr. Dev. 45, 151-162.

Malpaux B., A.D. Daveau, F.O. Maurice-Mandon, G. Duarte and P. Chemineau, 1998. Evidence That Melatonin Acts in the Premammillary Hypothalamic Area to Control Reproduction in the Ewe: Presence of Binding Sites and Stimulation of Luteinizing Hormone Secretion by *in Situ* Microimplant Delivery. Endocrinology 139, 1508-1516.

Malpaux B., M. Migaud, H. Tricoire and P. Chemineau, 2001. Biology of Mammalian Photoperiodism and the Critical Role of the Pineal Gland and Melatonin. J. Biol. Rhythms 16, 336-347.

Messer L.A., L. Wang, C.K. Tuggle, M. Yerle, P. Chardon, D. Pomp, J.E. Womack, W. Barendse, Crawford A.M., D.R. Notter and M.F. Rothschild, 1997. Mapping of the melatonin receptor 1a (MTNR1A) gene in pigs, sheep, and cattle. Mamm. Genome 8, 368-370.

Notter D.R. and N.E. Cockett, 2005. Opportunities for detection and use of QTL influencing seasonal reproduction in sheep: a review. Genet. Sel. Evol. 37 (Suppl. 1) S39-S53.

Pellettier J., L. Bodin, E. Hanocq, B. Malpoux, J. Teyssier, J. Timonier, and P. Chemineau, 2000. Association between expression of reproductive seasonality and alleles of the gene Mel_{1a} receptor in the ewe. Biol. Reprod. 62, 1096-1101.

Simonneaux V. and C. Ribelayga, 2003, Generation of the Melatonin Endocrine Message in Mammals: A Review of the Complex Regulation of Melatonin Synthesis by Norepinephrine, Peptides, and Other Pineal Transmitters. Pharmacol Rev 55: 325-395.

Zúñiga O., F. Forcada and J.A. Abecia, 2002. The effect of melatonin implants on the response to the male effect and on the subsequent cyclicity of Rasa Aragonesa ewes implanted in April. Anim. Reprod. Sci. 72, 165-164.

Part 2.
Innovations improving the efficiency of livestock systems in the Mediterranean area

B. Feeding systems, nutrition, health

Use of stochastic models to simulate long-term dynamics of mountain cattle herds under low-labour availability scenarios

D. Villalba[1], G. Ripoll[2], R. Ruiz[3] and A. Bernués[2]
[1]Departament de Producció Animal, Universitat de Lleida, Lleida, Spain; dvillalba@prodan.udl.es
[2]Centro de Investigación y Tecnología Agroalimentaria de Aragón, Zaragoza, Spain
[3]NEIKER-Tecnalia, Vitoria, Spain

Abstract

Mathematical models are useful tools to evaluate innovations or new management strategies in livestock farming systems because they allow advanced simulation of the consequences of their implementation, taking into account the existing relations between different components of the farm production system. In recent years, cattle farming systems in the Spanish Central Pyrenees have evolved towards extensification and lower dependence on inputs, linked to a reduction in the availability of labour dedicated to farming activities. A stochastic dynamic model (Nodriza) was designed to simulate long-term herd dynamics, taking into account the effects of nutrition and management applied at a batch level to animal reproduction performance and herd dynamics. Long-term effects of complex scenarios of nutrition and management on mountain cattle farming systems can be simulated and evaluated in terms of technical and economic performance, as well as the risk involved (derived from stochastic simulation of animals, variability within the herd and between years). The traditional management strategy observed in the study area was compared to an alternative low-labour management system. Their impacts on fertility, calf production and economic performance were analyzed. Preliminary results showed that, under certain circumstances, the low-labour scenario could be as good as the traditional one in terms of technical and economical performance, and also implied lower levels of risk under uncertain climate conditions.

Keywords: technical evaluation, economic evaluation, labour, environment

Introduction

Farming systems are very dynamic and change constantly depending on biophysical and socio-economic drivers. In general terms, the diversity of farming systems and the changes observed over time can be interpreted as attempts by farmers to adapt to three groups of factors (Manrique *et al.*, 1999): (1) the general socio-economic environment, which is predominantly influenced by agricultural policies; (2) specific local/regional factors that determine production potential, access to inputs and outputs markets, etc; and (3) the internal characteristics of the household, which can be structural, economic and sociological.

In recent years, beef cattle farming systems in the Spanish Central Pyrenees have changed substantially, evolving towards ecological and economic extensification (greater use of natural resources and lower dependence on inputs), this being partially explained by a reduction in labour dedicated to farming activities (García-Martínez *et al.*, 2008). These authors conclude that labour and household characteristics play a central role in explaining the evolution of farming systems in mountain areas.

On the other hand, due to the central role they play in the conservation of natural resources and mountain landscape in Europe (McDonald *et al.*, 2000), agricultural policies currently recognize their multiple productive, environmental and societal functions (Gibon, 2005). These functions are closely related, as the relationships between technical and economic factors, socio-economic and policy environments, the physical environment, and social factors are multiple and complex.

Simulation models are useful tools to evaluate technical innovation or new management strategies in livestock farming systems because they allow enable advance analysis of the consequences of their implementation, taking into account the relations between the different components mentioned above.

A stochastic dynamic model was designed to simulate long-term herd dynamics, taking into account the effects of nutrition and management applied at a batch level to animal reproduction performance and herd dynamics. The long-term effects of complex nutrition and management scenarios on mountain cattle farming systems can be simulated and evaluated in terms of technical and economic performance, as well as the risk involved (derived from stochastic simulation of animals, variability within the herd and between years) and the environmental implications.

In this paper, the structure and functionality of the software *Nodriza* for the simulation of beef cattle herds is presented. The traditional management strategies observed in a mountainous study area were compared to alternative more extensive/ low-labour management systems under diverse environmental conditions. The implications for economic performance and labour requirements are analyzed and discussed.

Material and methods

Overview of the model

The herd dynamics simulator (Nodriza[3]) is based on a stochastic and dynamic animal sub-model that represents interactions between management, nutrition and reproduction.

The animal sub-model was parameterised and validated under Spanish mountain conditions (Villalba *et al.*, 2006), for Parda de Montaña cows, one of the breeds widely used in the Spanish Pyrenees. From the initial cow characteristics, three basic compartments were defined: cow total fat, cow empty body weight and calf weight. These compartments were updated on a daily basis using a routine of metabolic energy (available from feed intake plus fat mobilisation from body reserves) partition into maintenance, gestation, lactation, and, if possible, for the deposition body reserves. To this end, the equations proposed by AFRC (1990), INRA (1978), Agabriel and Petit (1987) and Chilliard *et al.* (1987), were used. For a detailed description of the animal sub-model see Villalba *et al.* (2006). The body condition score (BCS) and live weight (LW) were obtained from the cow fat and cow empty body weight compartments. The BCS at calving and changes in LW after calving were used to calculate a postpartum anoestrus (PPA) for every cow, according to the results obtained by Sanz *et al.* (2004). The PPA is the primary variable to link nutrition with the reproductive performance of the cow, and hence with herd dynamics through the specific reproductive management practices.

The herd dynamics sub-model represents the main changes in the state experienced by each animal throughout the productive cycle (sexual cyclicity, mating, pregnancy, calving, lactation, culling and mortality). After calving, and once the APP is over (depending on the nutrition status of the animal, as stated in the animal sub-model), any adult cow will begin showing the average cycle length of 21 days typical of the cattle species. After each cycle, the cow may become pregnant provided there are bulls in the herd or the cow is artificially inseminated. If mating succeeds, the cow will enter the compartment of pregnant animals until calving or abortion (either as a consequence of sub-nutrition or of any other hazard). Otherwise, it will remain non-pregnant and will enter a new 21-day cycle. Other events in herd dynamics are also simulated by the model, like weaning (at a fixed date or age), culling and replacement.

[3] NODRIZA simulation software is available (in Spanish) at authors e-mail.

Nodriza software

The equations of the model were programmed in Visual Basic 2005 express edition® in order to facilitate its use. The Nodriza software designed had different visual interfaces where different scenarios can be defined.

Figure 1 shows the main interface of Nodriza software. The user can change the model parameters by changing the breed characteristics, the parameters defined as stochastic for the animal sub-model, and the productive-reproductive parameters that drive the animal sub-model and its relation with the herd sub-model.

From the main interface, three steps should be followed to run a simulation. First, the definition of natural resources (pastures) and available feed. In the natural resources interface the quality (in terms of metabolic energy) and daily intake per cow of pastures (grassland and forest pastures) should be introduced.

The second step refers to herd management and the feeding plan. Figure 2 shows the feeding plan interface, where the year can be divided into periods of different feeding management for lactating and non-lactating cows. Each period, which is defined by the corresponding initial and final dates, can include the use of natural resources and/or indoor supplementary feed for cows and calves. The existing climate conditions for the period can also be defined at this level. There are two options for the management of calves while they are kept indoors: either free access to the cow (*ad libitum* suckling), or restricted access (restricted suckling) when calves are kept apart from the cows and only have access to suckling twice a day, following the typical management in the Pyrenees.

To complete the second step, the management interface must be completed. This interface (Figure 3) includes the initial state of the herd: number of cows, percentage of pregnant cows, starting date and length of the calving season; key management dates such as age or date of weaning of calves, mating period, replacement and culling dates (including different management possibilities); number

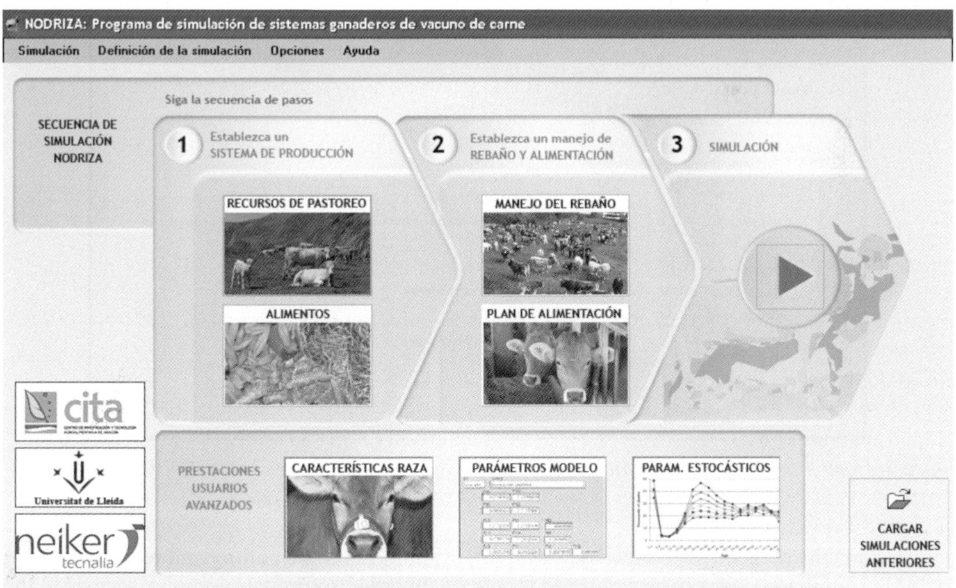

Figure 1. Main interface of Nodriza software.

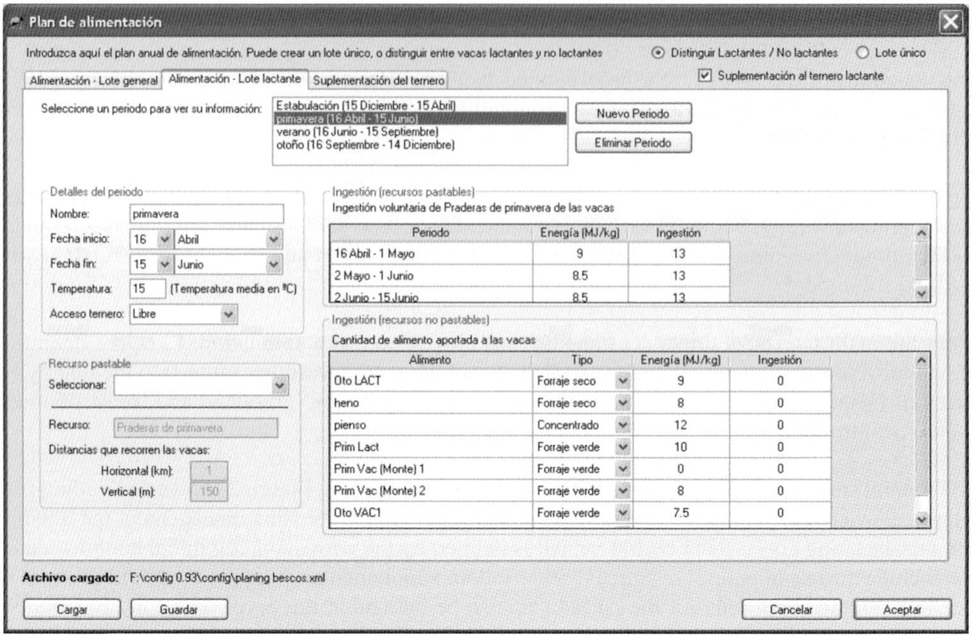

Figure 2. Feeding plan interface.

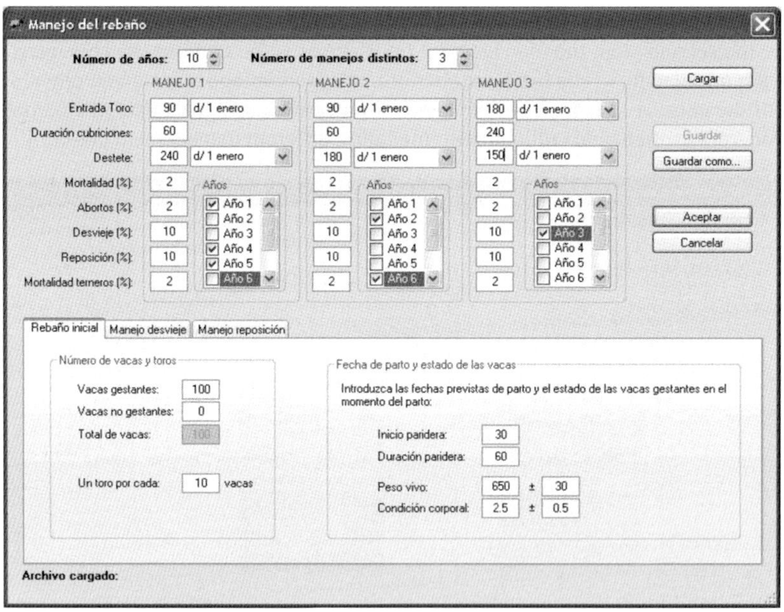

Figure 3. Herd management interface.

of bulls available; percentage of cow and calf mortality, abortions and culling and replacement rates. Finally, the number of years to be simulated is entered. In order to simulate complex scenarios, three different reproductive management systems can be entered in different simulation years.

The third simulation step allows the choice of the type of outputs (data files) and the interface output (Figure 4), where graphics representing changes in body weight, BCS, and calving distribution of different types of animals can be drawn at specific dates for one or several years of simulation.

Simulated strategies

To illustrate the functionality of the model, four scenarios were simulated for a 100-cow herd of 650±15 kg (mean±s.d.) of LW and BCS of 2.5±0.1. The period of time simulated was 15 years. It was assumed that the percentages of mortality, abortions, culling, replacement and calf mortality were 2, 1, 10, 12 and 2 respectively. Every calf was removed from the simulation (sold) at weaning.

The feeding management strategies simulated were based on the natural resources available in the Spanish Central Pyrenees (Casasús *et al.*, 2002). During summer, the herds usually graze in supra-forest mountain pastures that are normally located apart from the farm. In spring and autumn, the animals graze in valley grasslands or forest pastures depending on their physiological state (lactating or non-lactating), although animals can be supplemented with fodder and/or concentrates, especially during late autumn or early spring. In winter, the animals are normally housed and fed with on-farm fodder (collected during spring or summer) and/or purchased feed (either fodder or concentrates). However in some farms, it has become increasingly common to keep and supplement animals outdoors even during winter. In our scenario, in winter lactating cows received 2 kg of concentrate (10 MJ EM/kg MS) and 11 kg of hay (8 MJ EM/kg MS) per day. The remaining cows (dried-off, pregnant, etc.) received 11 kg of hay per day. Calves were not supplemented and had free access to their dams. The metabolic energy available (MEA) for lactating and non-lactating cows is presented in Figure 5.

In two scenarios, the quantity and quality of pastures varied in each simulation year in order to model between-year variability due to climatic variation. A real 15-year (1993-2007) rainfall dataset (April-September) was used to simulate variations in MEA in spring and summer pastures. Standardized differences in yearly rainfall from mean rainfall were transformed into a percentage (with 10 units

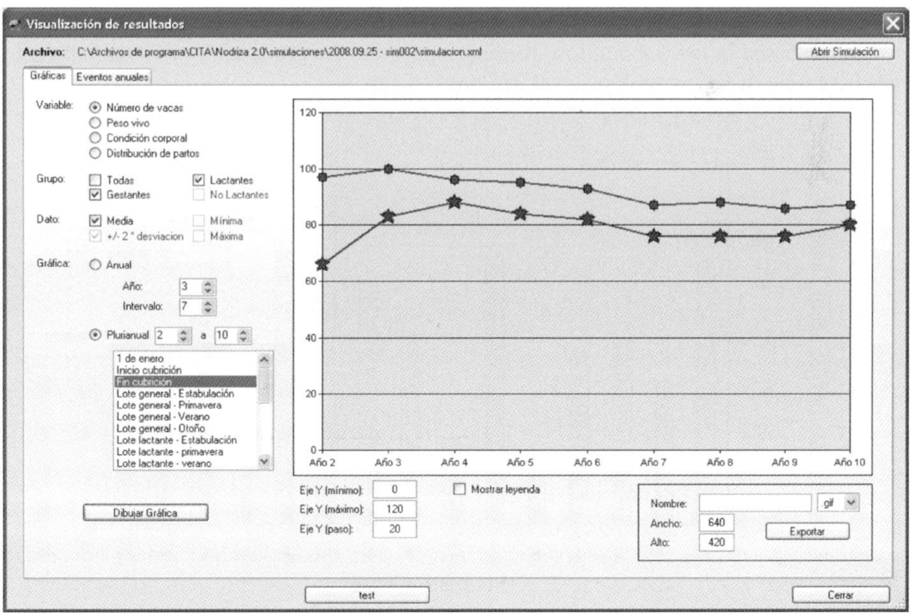

Figure 4. Graphical outputs of Nodriza software.

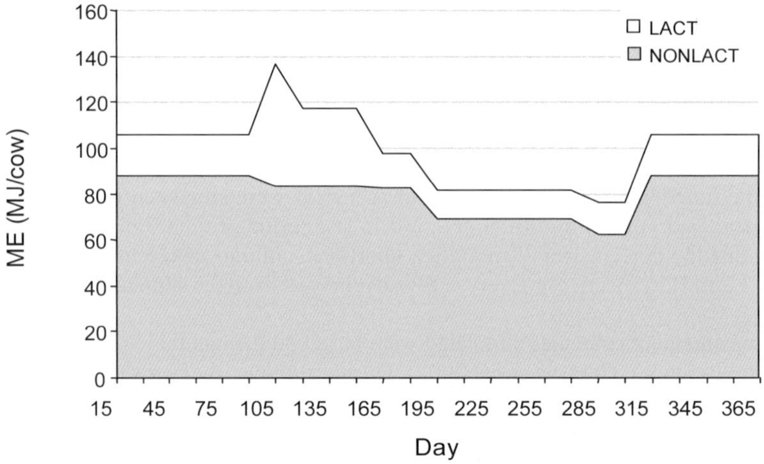

Figure 5. Metabolic energy available.

of standard deviation) in MEA (i.e. year 1993 had 548 mm of rainfall from April to September, 116 mm more than the 1987-2007 mean (0.8 standardized units), so simulated MEA from April to September would increase by 8%). Figure 6 shows the percentage of ME available each year in relation to the average situation.

Simulated management strategies are defined as follows (Figure 7). Winter calving (WC), with a mating period of three months through spring and summer to calve in winter and spring. Mating took place in low-valley pastures, allowing better control of reproduction compared to the summer mating strategy traditionally used for spring calving. Calves were weaned at the end of the summer grazing period at a mean age of 180 days.

Two calvings in three years (2C3Y) is a hypothetical management system corresponding to greater extensification based in low-labour, low-feeding inputs and the production of fewer calves. In this extensive scenario, weaning took place at 275 days of age.

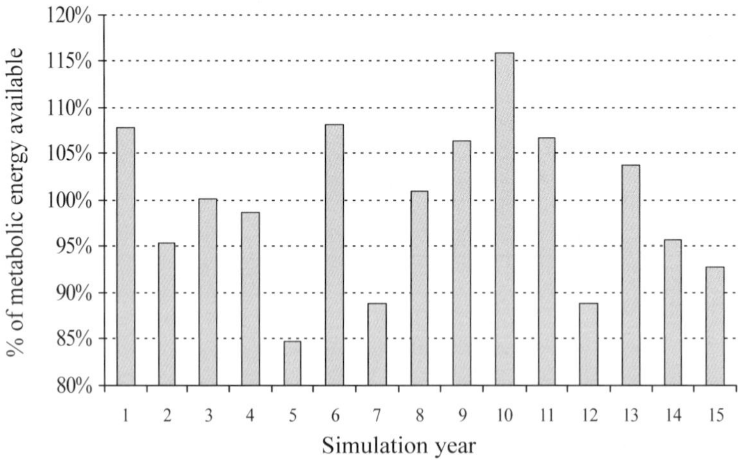

Figure 6. Percentage of metabolic energy available each year.

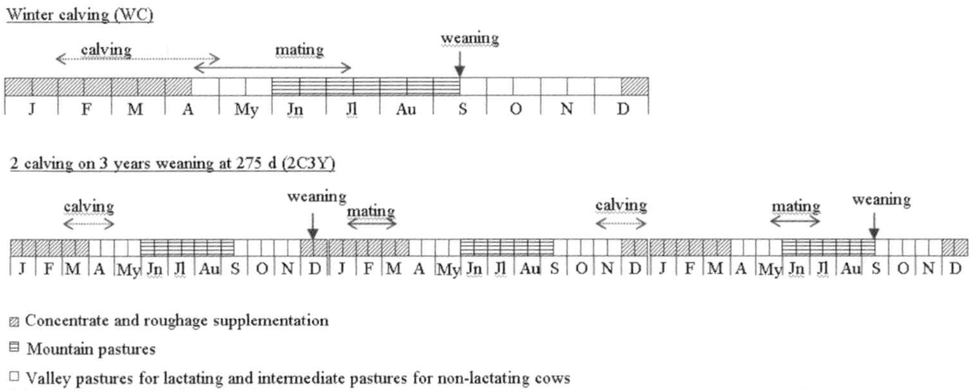

Figure 7. Reproductive and feeding management of different management strategies.

Finally, the four scenarios simulated were winter calving with and without MEA between-year variability (WC_0 and WC_10); and two calvings in three years with and without MEA variability (2C3Y_0 and 2C3Y_10). So the scenarios differed in the length of the indoor feeding period, energy intake and reproductive practices. Means and between-year standard deviations were used to evaluate: (1) technical aspects: fertility, BCS, and calf weaning weight; (2) economic aspects: the economic margin at weaning and labour costs; and (3) environmental aspects: use of natural resources.

The economic margin was calculated as income from the sale of calves at weaning (assuming 2.5 €/kg LW), minus the cost of the winter feed (2007 prices: 0.1 €/kg hay and 0.25 €/kg concentrate), minus labour costs. Labour costs were assigned to: 1) grazing management (1 hour per day of grazing); (2) in-door feeding (2 hours per day when lactating cows are present and 1 hour if all cows are non-lactating); and (3) calving supervision (1 hour per day during the calving season plus 0.1 hours per 3 calvings).

Results and discussion

The stochastic simulation was performed at the animal level, but outputs were compiled at the herd level. The model thus enabled evaluation of: (1) average results (cow LW, BCS and calf LW), (2) variability of these results, and (3) distribution of the most important events (calving, weaning).

Figure 8 shows the fertility for the 15 years of the simulation for all scenarios. Fertility for WC_0 was relatively stable from year 6 onwards at around 80% (WC_0 fertility mean was 0.81). WC_10 showed more variation in fertility, as in years 9 to 11 it reached values higher than 95%. The high value of fertility in year 12 (96%) with low MEA (Figure 6) reflects the capacity of the model to simulate the dynamic effect of nutrition. A one-year simulation of nutrition in year 12 would probably result in lower fertility rates, but in this long-term simulation the previous years (with high feed availability) allowed cows to increase body reserves and thus to tolerate low levels of nutrition. However, the effect of this low nutrition was felt the following year, because at the start of the mating season cows displayed a very low BCS, which could not be compensated for by a good level of nutrition during the mating season (see Figure 9) and, consequently yielded a low fertility rate.

The use of the 2C3Y management system, (in which the use of pasture was similar to WC), implied an extensification of reproductive management and calving concentration, which allowed for a reduction in labour costs. In addition, at certain times in this 3-year management system, the entire herd can use medium altitude pastures because all animals had low nutritional requirements (dry

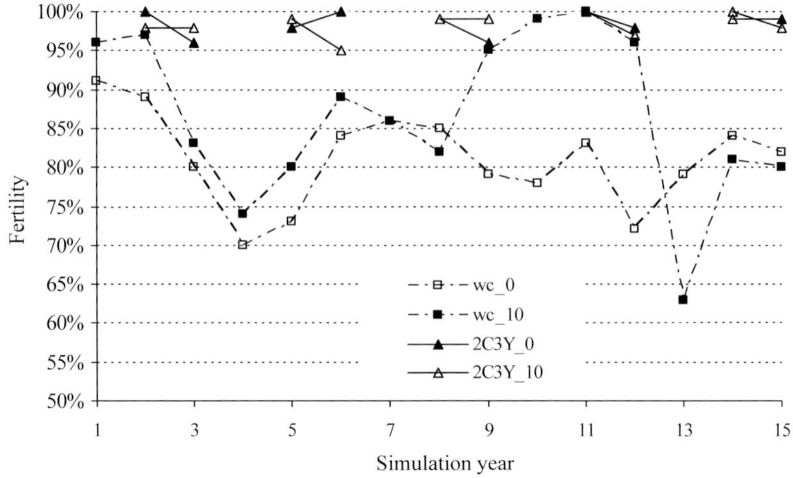

Figure 8. Herd fertility (number of pregnant cows at the end of the mating season).

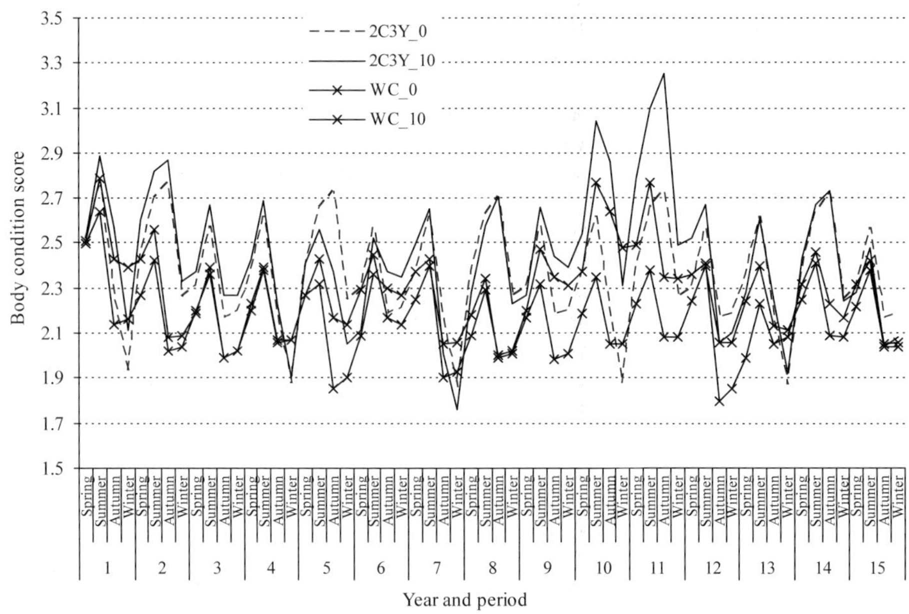

Figure 9. Body condition score at the start of spring, summer, autumn and winter periods.

or early pregnancy). The use of these transition zones, especially forest pastures with low nutritive value, is very important in terms of the conservation of biodiversity and prevention of forest fires (Luoto *et al.,* 2003), but only animals with low requirements are able to benefit without incurring sub-nutrition problems (Ferrer-Cazcarra *et al.,* 1998). In this scenario, simulations rendered very low BCS at the onset of the mating season in some years (Figure 9). Nevertheless, the time between the calving season and mating (11 months) assured the cyclicity at the male's entry, independently (to a certain degree) of the body reserves of the cows. Thus, although the mating season is very short, the fertility rate is high and constant (Figure 8). The reproductive extensification of the 2C3Y strategy allowed late weaning at 275 days, and the simulation shows that this late weaning does not impair reproductive performance. The late weaning option is only possible under the feeding conditions simulated with this strategy. When late weaning at 275 days was simulated for WC, fertility decreased more than 10 points (data not shown).

Some technical and economic indicators are presented in Table 1. The number of calves weaned annually per 100 cows in 2C3Y was well below (difference: 15 calves) the number in WC. Nevertheless, it was compensated for by the higher weaning weight in 2C3Y, producing similar incomes from sales of calves. Calves could be weaned at at 275 days of age was because this did not impair reproductive efficiency (see above) and because milk production predicted at 275 days of lactation was enough to maintain the growth of the calf. This fact could be questioned, but the model assumed that calf consumed only milk, while at ages over 100 days, calves can compensate for the decrease in milk production with pasture intake (INRA, 2007).

In the 2C3Y strategy, labour costs were 15% lower than in the traditional WC strategy. Basically, this reduction is a consequence of the shorter calving season and the reduction in labour needed for feeding, since in one out of every three years, no lactating cows are kept indoors. Other benefits derived from extensification were difficult to quantify in economic terms, but should be taken into consideration, such as off-farm labour opportunities, etc. As stated previously, the 2C3Y strategy represents the possibility of using low quality pastures for longer periods of time, implying lower labour management costs because cows are non-lactating, and with implications in terms of environmental management, which could have economic consequences in the current PAC context (agro-environmental measures).

The stability of outputs should also be taken into consideration when assessing the level of risk. For example, with the WC strategy, weaning weight was similar with and without climatic variability, but the between-year standard deviation (SDy) was higher with the WC_10 strategy (16.3 kg) than with WC_0 (6.7 kg). In contrast, the 2C3Y scenario was less affected by this indicator, with similar SDy in 2C3Y_10 (17.6 kg) and 2C3Y (17.8 kg).

Table 1. Average annual technical and economic indicators during the 15-year simulation period.

	Scenario					
	WC_0		WC_10	2C3Y_0		2C3Y_10
Weaned calves[1]	79		84	64		64
Calving period (d)	90		90	40		40
Weight at weaning (kg)	213		210	279		282
Total labour (h)		517			441	
Grazing management (h)		174			170	
Feeding indoor (h)		240			200	
Calving supervision (h)		103			71	
Incomes from sales of calves (€)	42,270		44,247	44,921		45,381
Cost of feed until weaning (€)	16,163		16,350	14,960		14,960
Labour costs until weaning (€)	4,528		4,535	3,860		3,860
Economic margin at weaning (€)	21,579		23,362	26,101		26,561

[1] Per 100 cows per year.

WC_0: winter calving without MEA between-year variability; WC_10: winter calving with MEA between-year variability; 2C3Y_0: two calvings in three years without MEA variability; 2C3Y_10: two calvings in three years with MEA variability.

Conclusions

The dynamic stochastic model (Nodriza) presented in this paper enabled simulation of the long-term effects of complex scenarios considering not only herd dynamics and interactions between nutrition, management and reproduction, but also the use of natural resources.

With regard to the scenarios tested, our results show that the alternative low-labour/extensive strategy is not only feasible, but even more advantageous than traditional management, under our study conditions. Nevertheless, simulations should be used to explore new management strategies and to analyze the consequences of their implementation under diverse scenarios, rather than being used as conclusive tests for the choice of the optimal strategy.

Acknowledgements

This study was funded by INIA (Spanish Ministry for Education and Science) - FEDER, Refs. RTA03-029-C2 and TRT2006-0004-C02.

References

AFRC, 1990. AFRC Technical Committee on responses to nutrients, Report Number 5, Nutritive Requirements of Ruminant Animals: Energy. Nutr. Abst. Rev. (Series B) 60(10): 729-804.

Agabriel, J. and M. Petit, 1987. Recommandations alimentaires pour les vaches allaitantes. Bull. Tech. C.R.Z.V. Theix, I.N.R.A. 70: 153-166.

Casasús, I., Sanz, A., Villalba, D., Ferrer, R. and R. Revilla, 2002. Factors affecting animal performance during the grazing season in a mountain cattle production system. J. Anim. Sci. 80: 1638-1651.

Chilliard, Y., Remond, V., Agabriel, J., Robellin, J. and R. Verité, 1987. Variations du contenu digestif et des réserves corporelles au cours du cycle gestation-lactation. Bull. Tech. C.R.Z.V. Theix, I.N.R.A. 70: 117-131.

García-Martínez, A., Bernués, A., Riedel, J.L. and A. Olaizola, 2008. Recent evolution of suckler cow farming systems in the Spanish Central Pyrenees. In, 2nd Seminar of the Scientific-Professional Network on Mediterranean Livestock Farming 'Mediterranean livestock production: Uncertainties and opportunities' Options Méditerranéennes A-78, Zaragoza, pp. 97-102.

Ferrer-Cazcarra, R., Casasus, I., Sanz, A., Villalba, D., and R. Revilla, 1998. Extensification of beef cattle production systems in the spanish pyrenees: Comparison of performances under indoors vs grazing conditions. Proceedings of European Workshop on Extensification on Animal Performance, Carcass Composition and Product Quality, 205-215.

Gibon, A. 2995. Managing grassland for production, the environment and landscape. Challenges at the farm and the landscape level. Liv. Prod. Sci. 96 (1): 11-31.

Garcia, F., Agabriel, J. And D. Micol, 2007. Alimentation des bovins en croissance et à l'engrais. Alimentation des Bovins, Ovins et Caprins, pp. 89-120.

INRA, 1978. Alimentation des Ruminants. R.Jarrige (ed). INRA Publications.Versailles.

Luoto, M., Pykälä, J., and M. Kuussaari, 2003. Decline of landscape-scale habitat and species diversity after the end of cattle grazing. Journal for Nature Conservation, 11(3): 171-178.

MacDonald, D., J.R. Crabtree, G. Wiesinger, T. Dax, N. Stamou, P. Fleury, J.G. Lazpita and A. Gibon, 2000. Agricultural abandonment in mountain areas of Europe: environmental consequences and policy response. Journal of Environmental Management 59, 47-69.

Manrique, E., A. Olaizola, A. Bernués, M.T. Maza and A. Sáez, 1999. Economic diversity of farming systems and possibilities for structural adjustment in mountain livestock farms. Options Mediterranéennes 27: 81-94.

Sanz, A., A. Bernués, D. Villalba, I. Casasús and R. Revilla, 2004. Influence of management and nutrition on postpartum interval in Brown Swiss and Pirenaica cows. Livestock Production Science 86, 179-191.

Villalba, D., I. Casasus, A. Sanz, A. Bernues, I. Estany and R. Revilla, 2006. Stochastic simulation of mountain beef cattle systems. Agricultural Systems, 89, 414-434.

Innovations in small ruminants feeding systems in arid Mediterranean areas

A. Nefzaoui[1], H. Ben Salem[2] and M. El Mourid[1]
[1]ICARDA, North Africa Program, 1 Rue des Oliviers, El Menzah V, 2037 Tunis, Tunisia
[2]INRAT, Laboratoire des Productions Animales et Fourragères, rue Hédi Karray, 2049 Ariana, Tunisia

Abstract

Small ruminant production is the main source of income of farmers living in the arid and semiarid Mediterranean regions. However, the development of this sector is constrained by numerous environmental, institutional, social and economical factors. These areas are characterized by low and erratic rainfall resulting in water scarcity leading to a low fodder potential. Furthermore, native rangelands are degrading due to overgrazing, high stocking rates and mismanagement. Options to improve small ruminant-based production systems include (1) innovative technologies targeting the increase of feed resources availability, improving diets' quality and alleviation of feeding cost, (2) organization of local institutions for better adoption of these technologies and to protect the main natural resources (rangelands and water), (3) participatory approaches involving all stakeholders concerned with the improvement of farmer's income and livelihoods and the better use of natural resources. This paper reports on updated knowledge and technologies available to improve the productivty and sustainabilty of production systems in the Mediterranean arid and semi-arid regions and discusses the success stories as well as failures of a set of technologies targeting the promotion of small ruminant production in the Mediterranean basin.

Keywords: small ruminants, feeding systems, technological innovations

Introduction

There is a contrasting situation of the production systems in the Mediterranean area. Indeed, while arid and semi-arid regions are suffering from feed and water scarcity, they are home to large flocks composed of a mixture of different animal species. Small ruminant (SR) systems are largely dominant in dryland areas in South and East Mediterranean Basin and in mountainous areas in Europe (Alary *et al.*, 2002).

SR production is an important component of the agricultural sector in most of arid Mediterranean countries where they contribute significantly to human diet. During the last fifty years the region has been facing a substantial increase in SR populations driven by more demands for animal products, primarily meat and milk. Moreover, rangeland areas and productivity decreased significantly. Rangeland contribution to meeting feed needs decreased from 80% in the sixties to 15-30% currently. Adhesion of North Africa and Near East countries to world trade agreements put them at a comparative disadvantage for SR production as not being competitive at a global level. Investment in agricultural sector and particularly in low rainfall areas has been very low. Climate change exacerbated this unfavorable environment and led to more water scarcity and poverty, resulting in an increased risk and vulnerability of herders (Nefzaoui *et al.*, 2008).

Because of the importance of small ruminants in arid Mediterranean basin, this contribution will focus on sheep and goat production in harsh environments.

Small ruminant production systems in arid Mediterranean area

The main features of sheep and goats production systems relate to the way small ruminants are raised. The major criteria adopted to distinguish the various systems are based on feed resources used and the nature of flock movement during the annual cycle.

Constraints to sheep production in arid and semi-arid regions of the Mediterranean basin are mainly (1) dramatic feed deficits, (2) shortage of drinking water, (3) mismanagement and husbandry, (4) poor control of animal health, (5) weak interaction between researchers, farmers, extension workers and policy makers, (6) poor organization of herders, and (7) unfavorable policies and legislations.

Main production systems

- Nomadic system (pastoral): Traditionally the nomadic system used to be the common system for small ruminants in the driest areas of North Africa and Near East (NENA) countries. However, the number of the truly nomadic flocks presently is quite low. In this system, flocks are in continuous movement in search for grazing.
- Transhumant system: This is the predominant production system in the NENA region because the majority of flocks are located within the limited rainfall marginal areas. In summer and autumn, the flocks move to the cultivated areas to graze cereal stubble and the residues of irrigated crops. The flocks come back to their home base once the winter rains trigger first growth of the annual pasture in the steppe. The transhumant flocks move over greater distances and to a larger number of locations during the annual feeding cycle than in the semi-sedentary system.
- Semi-sedentary system: In this system, the flocks are kept in cultivated areas during the winter season. However, in February they must leave the cultivated areas for grazing areas in the steppe to avoid crops' damage. Animals are moved back into cultivated areas immediately after crop harvest to graze cereal stubbles. By the end of summer, the flocks are moved back to their villages.
- Sedentary system: In the sedentary system, the small flocks are kept in villages throughout the year, and they depend on grazing communal lands, weedy fallows, roads and irrigated canals, and on cereal stubbles and irrigated crop residues. Animals receive some hand-fed supplements in almost all seasons.
- Oasis system: This is a specific system characterized by intensive irrigated agriculture, the production of alfalfa following cereal cropping, and production of D'man breed sheep in Morocco and goats in Tunisia. The feed calendar in this system includes; fresh alfalfa during March-September; and alfalfa hay, straw and waste dates palm in October-February.
- Peri-urban system: This system is more common in the peri-urban areas, where hundreds of lambs are confined in a yard or barn for fattening. Traditionally barley is the main diet's ingredient.

Small ruminants production systems challenges

Small ruminant production systems in these areas are facing serious challenges to their sustainability (Alary and El Mourid, 2007) deriving from: (1) climatic constraints represented by the low and erratic rainfall and the high incidence of droughts, affecting the productivity of rangeland ecosystems and the livelihoods of the population; (2) the desertification spiral which accelerated during the droughts of the 1980s and the late 1990s, coupled with changes in livestock and range management practices; (3) technical constraints underpinned by shortages in improved technologies to restore the ecological integrity, function and services of the degraded rangeland ecosystems, as well as the absence of monitoring and early warning systems; (4) socio economic limitations, including the high poverty and vulnerability rates of the population which are exacerbated by unstable feed and animal market conditions and limited diversification of income sources; and, (5) institutional obstacles linked mainly to continued cross-lawful inefficiencies on issues dealing with land use, coupled with the inadequate capacity of local institutions for land use control and management, and weaknesses in the system of incentives for adoption of improved land management practices, and in the drought mitigation approaches (Nefzaoui *et al.*, 2008; Alary and El Mourid, 2007).

Changes in small ruminant production systems

Two critical trends prevail in the current production context:

- *The first trend* involves the expansion of market demand for livestock products leading to opportunities for productivity and income improvement.
- *The second trend* involves a crisis in the feed supply reflecting water scarcity, exacerbated by the progressive decline of rangelands' productivity due to overgrazing, cultivation encroachment, or the disruption of institutional arrangements for resource utilization. Moreover we face a very low ratio of cultivated forages in the cropping systems.

The pastoral and agropastoral societies went through deep mutation during the last decades which includes (Bourbouze, 2000):

- Dismantlement of traditional organizations (informal institutions/community-based organizations).
- Privatization of communal rangelands and the development of barley and tree cropping.
- Regression of animal mobility with the sedentarization of the population. Only poor herders remain full transhumant.
- Increasing demand for livestock products leading to an increasing pressure on rangelands and subsequent degradation.
- Increasing reliance on supplemental feed.
- Mechanization (water and feed transport) that modified the management of rangeland.
- Inequality between poor and rich herders (less opportunity to purchase feed, drought mitigation policies favor pastoralists with large flocks).

In the mid-20th century, the mobility pattern of the pastoralists was perfectly matched to accessibility and availability of forage and water. With the mechanization of water transportation and the reliance on supplemental feed, animals can be kept continuously on the range, which disturbs the natural balance and intensifies the degradation process (Sidahmed, 1996; Nefzaoui *et al.*, 2002; Nefzaoui *et al.*, 2004). Mechanization profoundly modified rangelands' management in the steppes of the NENA. Water, supplements and other services are brought by trucks to flocks. As a result, the family is settled close to cities to have access to education and health services, and only sheepherders move up with their flocks to target grazing areas (transhumance).

Production systems are intensifying and it is possible nowadays to find in the steppe a continuum between intensive fattening units that are developing in peri-urban areas and along the main transportation axes, mixed grazing-fattening systems and pure intensive systems where hand feeding is only used to provide feed supplements to animals. In addition, and when terms of trade conditions are favorable, herders in North Africa are switching from permanent livestock production to 'opportunistic livestock production'.

Off-farm income and immigration are playing an increasing role in pastoralists' economy, especially to young generation. The overall impact is not known but migrants are reinvesting their earnings in livestock production and hold onto their right to access (and cropping) even during an extended absence. This fact is causing difficulties to the overall community management of rangelands. On the other hand, off-farm labor represents a complementary activity to livestock production and a risk management strategy and may actually improve community homogeneity and cohesion.

Inequality between poor and rich herders has been accentuated during the last decades due to several factors. The poorest herders, i.e. those with the smallest flock size are the ones affected most severely by rangeland degradation since they have less opportunity to purchase feed and rely mostly upon free range resources. Second, the drought mitigation policies have favored pastoralists with large flocks over those with small flocks since they are more often organized into associations in order to benefit from these actions (Hazell *et al.,* 2001)

Increasing reliance on supplemental feeds and stagnation of cereals and small ruminants productivity

Stagnation of yields, reliance on supplemental feed and failure of crop/livestock integration

While annual per capita consumption of cereal grain is around 250 kg, the annual per capita production is less than 150 kg. With a total population of over 80 million, the Maghreb spends over 4 billion dollars per year on cereal imports to meet national needs. Average yield of cereal grain crops in the Maghreb countries has increased from 0.7 to 1.2 t/ha over the past 50 years. However, yield is still low, representing about half the world average. In fact, the Maghreb is among the poorest regions in water resources, and cereal yields are not only low, but also extremely variable as they primarily depend on rainfall, which itself is low and highly variable.

Animal production in the Maghreb is much linked to cereal production in the cereal-based systems. While human consumption of animal products has increased considerably over the past 50 years, forage production has not kept pace with demands, covering less than 10% of the feed needs, the rest being met through imports of grain or feed concentrate. The recent soaring of feed commodity prices led to a widening of the feed deficit in Maghreb countries, and a consequent shrinking of local meat and milk production. Moreover food legumes production has decreased drastically and most of the needs are covered through imports.

Research conducted in the Maghreb and similar regions shows an opportunity to improve, i.e. at least double the productivity of cereals, food legumes and forage crops, and therefore, reduce Maghreb countries' dependence on grain, meat and milk imports.

Specific technologies therefore need to be developed and adopted in cereal-based systems to increase yield and reduce production fluctuation, while safeguarding the natural resources for their conservation and sustainable use.

What are the main feed sources for small ruminants?

Rangeland and fallow grazing, standing barley (green or whole dry crop), cereal stubble, crop residues, cereal and legume straws, barley, wheat bran, olive cake, olive tree leaves and twigs, cactus and shrubs are the main feed resources for SRs in the NENA region. Their contribution varies greatly between years according to rainfall and drought occurrence. In any case, the contribution of rangelands to cover livestock feed requirements is dramatically decreasing (10-30%) while concentrate feeds (barley and bran) represent more than 50% of animal needs.

On the other hand, the integration of crops and livestock is rather a fading illusion. Indeed, forage crops areas remained constantly low during the last three decades and its contribution is less than 10% in average in animal needs.

This trend is becoming alarming with the recent increase of cereal prices, particularly barley. The future of SR production in the NENA region appears uncertain. The NENA countries will have henceforth to allocate increasing budget for SR production and import.

Strategies to cope with climate fluctuation: managing drought

Managing the production risk caused by the variability of feed availability is the central issue in the SR production system in the NENA region.

Drought has long been a significant factor in the NENA region, particularly for the herders in the vast grazing areas of the steppe. Unfortunately, the level of wealth accumulated in these agropastoral societies is inadequate to provide full protection against severe droughts, and the economic and human losses in drought periods can be severe. The problem has worsened with population growth, as more and more people seek to earn a livelihood from the meagre resources available in these areas. It may also have been aggravated by more frequent and prolonged droughts associated with global warming. The high cost of droughts and the increasing vulnerability of agropastoral societies have led many governments in the region to intervene with various forms of drought assistance. However, many of these interventions are encouraging farming practices that could increase both the extent of future drought losses and the dependence of local people on government assistance. They are also costly to governments and use resources that could otherwise be spent for development purposes (Hazell, 2007).

How herders traditionally manage droughts? Agropastoral societies have developed their own strategies for coping with drought. These strategies include (Hazell, 2007; Alary *et al.,* 2007):

- Mobile or transhumant grazing practices that reduce the risk of having insufficient forage in any location.
- Reciprocal grazing arrangements with more distant communities for access to their resources in drought years.
- Adjustment of flock sizes and stocking rates as the rainy reason unfolds, to best match available grazing resources.
- Keeping extra animals that can be easily sacrificed in drought conditions, either for food or cash.
- Investment in water availability: wells, cisterns, and water harvesting.
- Diversification of crops and livestock (agropastoralism), especially in proximity to settlements, and storage of surplus grains, straw and forage as a reserve in good rainfall years.
- Diversification among animal species (sheep, goats, cattle, camels, donkeys) and different breeds within species.
- Income's diversification into non-agricultural occupations, particularly seasonal migration for off-farm employment in urban areas.

However, recent infrastructural and demographic changes (e.g. urbanization) have made such knowledge ineffective but without the availability of alternatives. Although solutions to major SR constraints resulted in some easing of the pressure caused by human needs for SR products, the consumption-production gap increases for most of the countries and imports are therefore increasing into the NENA region both in terms of feed and animal products. This trend is becoming alarming with the recent surge of cereal prices, particularly barley. The future of SR production in the NENA region is not clear. The NENA countries will witness an increasing budget load for small ruminant production and import. These countries will have to face an increased pressure on rangelands which requires an innovative approach to their effective management and complementation with better use of local natural resources with an emphasis on water harvesting and better use of adapted indigenous plant species, such as cactus and fodder shrubs, and introduction of feed blocks using agricultural by-products and treated straw. More attention should be paid to devising efficient animal nutrition and health management. Better integration of cereals and livestock production should be pursued, with more attention to the value of straw and fallow replacement with adoption of adequate rotations and conservation agriculture techniques that would increase feed production while maintaining the natural resource base for sustainable development. (Nefzaoui *et al.,* 2008)

Technological innovations

Lack of adequate year-round feed resources is probably the most important factor contributing to low animal production in arid and semiarid regions of the Mediterranean basin. The recent leap in prices of concentrate feeds and the continuous increase of the price of oil at the international level is putting machinery-based agriculture in serious difficulty and is threatening the livestock sector. Therefore, there is an urgent need to develop appropriate strategies for better use of local feed resources and to identify technologies optimizing the potential use of these unconventional feedstuffs in livestock feeding. Some of these strategies and technologies developed and or used by Mediterranean countries are described and discussed below.

Technology innovation focused on problem-solving interventions to improve productivity. National Agricultural Research Systems (NARS) in collaboration with the International Centre of Agricultural Research in the Dry Areas (ICARDA) worked (1) to develop new low cost-efficient diets using feed block technology as an integrating vehicle, suitable for both intensive and semi-intensive systems to make use of available by-products and crop residues; (2) to make a better use of shrubs and fodder trees; and (3) improve straw quality through treatment or selection; (4) to help disseminating research achievement through participatory research and development using community-based approaches. Figure 1 summarises the problematic and researchable issues of small ruminant production in the WANA region.

Conceptually ICARDA and NARS attempted to simultaneously conduct adaptive research and outscaling reaching more farmers in the environments of development projects.

Cultivated roughages

Different strategies have been decided by some Mediterranean governments to promote forage sector. Despite efforts made to achieve such objective, forage production in the arid and semi arid regions is low. Breeding programmes in the Mediterranean countries targeting selection of water

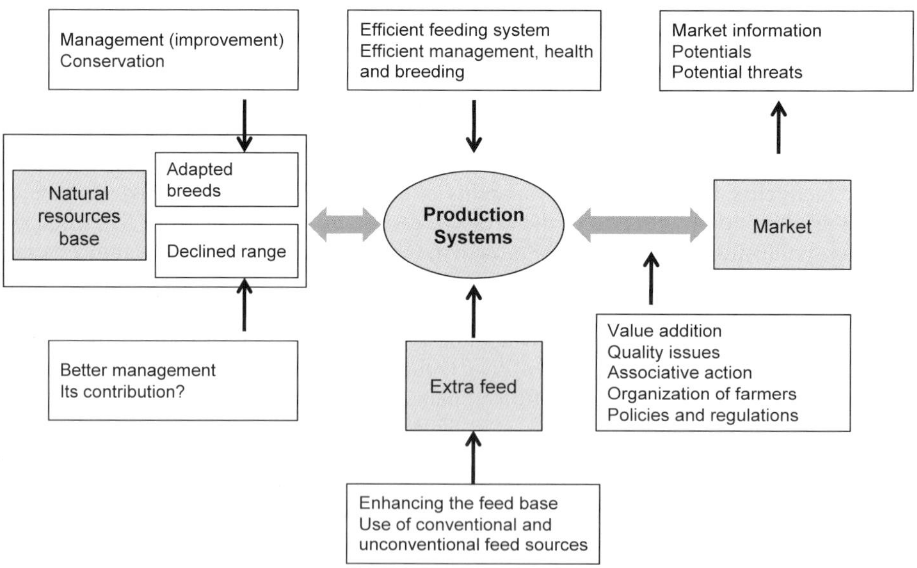

Figure 1. Problematic and researchable issue of small ruminants in the arid West Asia and North Africa Region (WANA) (Iniguez, 2008).

scarcity and high temperature tolerant forage cultivars are lacking. Often forage species and cultivars adapted to humid conditions are cultivated in the dry areas although they suffer from heat and water stress. Important quantities of dried forages moved from humid zones towards dry areas in the form of baled hay and or straw for livestock feeding. In Tunisia, transport fees of these feed resources from north to the centre and south is subsided. The need to produce adapted forages for feeding ruminants was recognised. Large-scale production of quality seeds should be undertaken to support fodder production. Extension of appropriate species and cultivars of various forages and legumes for specific agroclimatic and field situations is required to enhance livestock sector. More research is needed on the screening of species and cultivars, agronomy, nutritional evaluation, utilization, and large-scale propagation under commercial conditions. These efforts are the joint responsibility of crop and animal scientists.

Few improved cultivars of vetch, medics and perennial grasses adapted to dry conditions have been developed but most of them are not used due to the absence of seed enterprises. Promotion of fodder production in these dry areas is possible only through an appropriate integration of forage species into an agro-pastoral or agro-sylvopastoral systems.

Fibrous agricultural residues

Fibrous agricultural residues (straws, stubbles, sorghum stover, maize stalks, etc.) represent a valuable potential source of ruminant feed. However, most of these materials are characterised by extensive lignification of the cellulose fibres and by low levels of protein and minerals. In consequence, the intake and digestibility of these residues are not sufficient to sustain high levels of production. For convenience, we will focus on cereal straws as they are available in huge quantities in almost all Mediterranean countries. Treating straws for animal feeding has been a subject of considerable research, mainly in the eighties. Numerous review papers and original articles on straw treatment have been published by Mediterranean laboratories (e.g. Chenost and Kayouli, 1997; Nefzaoui *et al.*, 1999). There is no doubt that straw treatment (e.g. physical, chemical) has nutritional benefits, but the adoption rate of this technology at the farm level has been low. Ammonia treatment has attracted attention as a method of improving the nutritive value of cereal straws. To avoid the risk of intoxication and handling, development agencies, and specialized enterprises provided anhydrous ammonia and ensure that the treatment is risk-free and efficient. However, due to the high cost of this technology, especially to small holders, anhydrous ammonia was replaced by urea which under appropriate conditions could have the same effect on straw as anhydrous ammonia. Even though, the adoption of urea treatment is low and quite abandoned in some Mediterranean countries. Producers, mainly smallholders, are relying much more on alternative feed supplements.

Alternative feed resources

Agroindustrial by-products (AGIBs)

The most AGIBPs generated from food industry in the Mediterranean area are olive cakes, molasses, tomato pulp, fruit pulps such as grape marc and citrus pulp. Recycling, reprocessing and utilization of any AGIBPs offer the possibility of alleviating the current limited feed resources and to reduce feeding cost. Mediterranean AGIBPs are of four main categories:

- energy-rich feeds from citrus pulp and molasses;
- nitrogen-rich feeds from tomato pulp and brewery waste;
- by-products from cereal milling (e.g. wheat bran, wheat flour meal);
- secondary compounds-rich feeds (e.g. grape marc).

Although, high quantities of some AGIBPs are produced in the Mediterranean basin the utilisation of these feed resources in livestock feeding is still limited. This situation is likely due to proximity of the AGIBP to livestock flock (transportation and storage needs), alternative uses and the relative opportunity costs, the nutritive value of the new feed, and the managerial capabilities of the farmer. Some technologies have been developed to overcome this situation, thus to enhance the utilisation of these unconventional feed resources in ruminant feeding. Promising technologies are described and discussed below.

AGIBPs ensiling. For those farmers having transportation facilities and their flocks are raised in proximity of a specific food industry manufactory (e.g. olive oil extraction, fruit juice extraction, etc.) appropriate ensiling is a promising technique for better use of AGIBPs in livestock feeding. Even though the adoption rate of this technique in the Mediterranean area is still limited. Ensiled citrus pulp and tomato pulp are incorporated in the diet of sheep and dairy cattle in Italy and Tunisia, respectively.

Feed blocks (FBs). Feed blocks manufactured by the cold process are made from a mixture of one or more AGIBP (e.g. olive cake, tomato pulp, etc.), binder (e.g. quicklime, cement and clay), water and common salt, as well as urea with or without molasses. They should be air-dried until hardness and compactness criteria are met, and then offered to stall-fed or grazing ruminants on low quality diets. The technique of FB making is well described in the literature (e.g. Ben Salem and Nefzaoui, 2003; Ben Salem *et al.*, 2005a). Some variations in the blocks have been the incorporation of polyethylene glycol as a tannin-inactivating agent, which has increased the utilization of tanniniferous browse foliage in ruminant feeding (Ben Salem *et al.*, 2007). Medicated blocks containing anthelmintic agents and tannins to control internal parasites have been used in Australia and Ethiopia. Mineral enriched FBs (e.g. phosphorus, copper, etc.) were distributed to animals to mitigate their deficiency and improve reproduction in ruminants. Benefits from the integration of FBs in the diet of sheep and goats are reflected by data compiled in Table 1. It is clear that depending on the formula, FBs can partially or totally replace concentrate feeds, thus alleviates feeding costs without detrimental effects on livestock performances. Although it is ancient (Ben Salem and Nefzaoui, 2003), easy to make, to transport and to handle as well as their economical benefits only few Mediterranean countries (Algeria, Morocco, Syria and Tunisia) have tested and adopted FB technology. But it seems that this technology was adopted in Tunisia in the thirties to maintain animals during the drought periods (Cordier, 1947). Compared to the other Mediterranean countries, FB technology is much more developed and adopted in Tunisia. Indeed, Tunisian scientists have developed and evaluated numerous formulas of FBs on stall-fed and grazing sheep and goats. Some of these formulas have been transferred to farmers who shared the evaluation process of this technology and its economical benefit.

The leap in concentrate feed prices at the international incited Tunisian government to encourage the identification and use of alternative feed resources, like FBs, in livestock feeding. Some development agencies (e.g. The Pasture and Livestock Agency (OEP)) adopted this technology and are making efforts to transfer it among target farms. Recently a machine was manufactured to accelerate the manufacturing process of FBs, thus to produce high amounts of FBs in a day and to reduce the labour cost, thus the price of these feed supplements. Under the framework of bilateral project, Tunisia assisted Spain in developing formulas of FBs as a way for a better use of the two stage olive cake in livestock feeding. Promising results have been obtained with the incorporation of these FBs in the diet of dairy goats (Molina-Alcaide, unpublished data). In practice, although the interesting results obtained by some research laboratories in the Mediterranean area with the use of a wide range of FBs, the adoption of this technology is still low. There is an urgent need to devise strategies for popularizing the usage of FBs in Mediterranean countries and facilitate the exchange the experience of some of these countries in FBs making and utilization with inexperienced countries.

AGIBPs-based pellets. Conserving AGIBPs in the form pellets is another promising option. Nefzaoui and Ben Salem (unpublished data) have developed and determined the nutritive value of olive cake based pellets. The formula was inspired from ingredient composition of FB. These pellets were composed of olive cake, wheat bran, rapeseed meal, wheat flour residue, salt and minerals. Urea

Table 1. Compiled data on the potential use of feed blocks as alternative feeds supplements for sheep and goats in the Mediterranean area (Ben Salem et al.*, 2005a).*

Basal diet	Supplement	Animals	Growth rate (g/day)	Feeding cost variation	Country
Stubble grazing	Concentrate (250 g/d)	Lambs	95		Algeria
Stubble grazing	Conc. (150 g/d) + FB1	Lambs	136	-81%	Algeria
Wheat straw *ad lib*	Conc. (500 g/d)	Lambs	63		Tunisia
Wheat straw *ad lib*	Conc. (125 g/d) + FB2	Lambs	66	-11%	Tunisia
Straw (310 g/d)	Conc. (800 g/d)	Lambs	120		Lebanon
Straw (310 g/d)	Conc. (300 g/d) + FB3	Lambs	110	-5%	Lebanon
Acacia leaves	FB4	Lambs	14		Tunisia
Acacia leaves	FB5 enriched with PEG	Lambs	61		Tunisia
Rangeland grazing	Conc. (300 g/d)	Kids	25		Tunisia
Rangeland grazing	FB4	Kids	40		Tunisia

FB1: wheat bran (10%), olive cake (40%), poultry litter (25%), bentonite (20%), salt (5%); FB2: wheat bran (25%), wheat flour (15%), olive cake (30%), rapeseed meal (10%), urea (4%), quicklime (8%), salt (5%), minerals (1%); FB3: wheat bran (28%), ground barley (10%), molasses (44%), sesames hull (5%), white cement (5%), minerals (3%), urea (5%); FB4: wheat bran (28%), olive cake (38%), wheat flour (11%), quicklime (12%), salt (5%), minerals (1%), urea (5%); FB5: wheat bran (23%), olive cake (31.2%), wheat flour (9%), quicklime (9.9%), salt (4.1%), minerals (0.8%), urea (4.1%), PEG (18%).

was removed to avoid any risk of intoxication with excessive ammonia in the rumen. The *ad libitum* intake of these pellets by sheep averaged 2.5 kg/day. The cost of these pellets was about half of the price of Lucerne pellets which are imported and subsidised. In contrast to FBs, mechanisation is necessary to make pellets. In any case, AGIBPs-based pellets seem an interesting technology to reduce the use of conventional feed resources and to satisfy feed demands in the dry Mediterranean areas. Formulation should continue and evaluation of these pellets on different animal species should be encouraged, then popularizing this technology should be scheduled.

Fodder shrubs and trees (FST) in the smallholders farming systems

Trees and shrubs are part of the Mediterranean environment. They are present in most natural grazing lands of the Mediterranean region. FST are integral parts of smallholder farming systems in the Mediterranean area. Shrubs are made up of different fractions of diverse quality. Some species are high in essential nutrients but low in anti-nutritive factors (e.g. *Morus alba*), some others are low in nutrients but high in secondary compounds (e.g. *Pistacia lentiscus*) while some shrubs are high in both nutrients and secondary compounds (e.g. *Acacia cyanophylla*, *Atriplex* spp.). This is an ecological factor adapted to withstand grazing and to provide ground for selective grazing.

In arid and semi-arid Mediterranean rangelands where forage species available are of poor quality, FST could be used as feed supplements to increase animal intake of native resources. They can be used also to defer grazing after the autumn/winter opening rains, so that more production could be obtained. With proper treatment and management, FST could constitute a greater proportion of livestock diets. The type of shrubs and trees used as fodder by smallholder farmers varies from country to country. But, these species are in general drought and or salt-tolerant. Saltbushes are planted in the dry zones of numerous Mediterranean countries such as Tunisia and Morocco (*Atriplex nummularia*) and Syria (*Atriplex halimus* and *Salsola vermiculata*).

There are many advantages of promoting fodder shrubs and trees, because of their wider adaptability to harsh agro-climatic conditions and ability to produce for a longer period. As trees require little care after the establishment, the cost of production will be low.

(a1) Cactus, a welcomed shrub species by farmers. Multipurpose shrubs and trees not requiring high inputs are generally welcomed by farmers. Preferred species are those which can withstand harsh conditions prevailing in the dry areas and improve farmers' income. Unfortunately, the list of these shrubs and trees is too short in the Mediterranean basin. Cactus is a promising shrub species and its plantation is welcomed mainly by farmers from the southern Mediterranean region. In Tunisia for example, most farmers opted for growing cactus in their wastelands. This preference is not only based on the profitability, but also on market demand. Cactus plantations are available and expending at high speed in North Africa but also in many countries in Africa (Ethiopia, Mauritania, South Africa, etc.), Asia (India, Iran, Jordan, Pakistan, etc.) and America (Argentina, Brazil, Mexico, etc.). The popularity of cactus lies in its low demand for water, its high water use efficiency, tolerance to high temperature, rapid growth, high biomass, and the high content of its cladodes in water and energy. Additionally, it represents a source of income through selling its fruits and cladodes. Benefits from the integration of cactus cladodes in the diet of sheep and goats are well documented (see for example the review by Ben Salem and Smith, 2008). Literature data suggest that cactus could be incorporated of sheep and goats to improve feed intake and digestibility, thus to increase growth rate and to decrease and even to stop beverage water consumption. But, one should bear that such positive effects could be obtained only when appropriate nitrogen supplementation and adjunction of fibrous source are emphasized (Nefzaoui and Ben Salem, 2002).

(a2) Benefits from secondary compounds containing shrubs. Tannins and saponins are the most widely occurring secondary compounds in plants. They can have both beneficial and adverse effects depending upon the nature and the amount an animal consumes. Tannins and saponins could be used advantageously to improve livestock performance and product quality.

Tannins, rich feedstuffs could be integrated in the diet to improve feed efficiency:
- Reduce ruminal degradation of dietary protein, thus to increase protein and amino acid flow in the intestine and consequently increase growth rate of young animals and milk production in dairy ewes and cattle. Only few studies have been performed in the Mediterranean area. In Tunisia, lambs on oaten hay receiving small amount of air-dried leaves of *Acacia cyanophylla* (100 g/d, source of tannins) and soybean meal (200 g/d, source of protein) grew more than those receiving the same diet without acacia. This is the result of the interaction between acacia condensed tannins and soybean meal proteins (Ben Salem *et al.*, 2005). This technology, i.e. association of small amounts of CT to dietary protein, is simple, cost-effective and can promote livestock sector in the Mediterranean area. In practice, SRs could be allowed to graze in Acacia plantation (CT-source) during a short time then placed for longer period in a grass dominated pasture (protein source, e.g. Lucerne).
- Reduce protozoa count in the rumen, thus improve bacterial activity, and impair methanogenesis. Therefore, they improve feed efficiency in ruminants.

There is a recent interest in many countries to use dietary condensed tannins as a biologic solution to control gastrointestinal parasites (GIP) in ruminants. Veterinary drugs commonly used to reduce faecal egg count are expensive. Moreover, there are alarming calls by specialists for the development of anthelmintics resistance in worm populations, particularly in small ruminants. Tannins in some plant species proved efficient in controlling parasites According to Hoste (2005), tannins might interfere directly with the biology of various nematode stages and they could indirectly improve the host nutrition by protecting the diet proteins from ruminal degradation and this could modulate worm biology. Recent studies in France and Tunisia showed that repeated distribution of sainfoin hay (Hoste *et al.*, 2006) and *Acacia cyanophylla* foliage (Akkari *et al.*, 2008a,b) reduced faecal egg counts in kids and lambs, respectively.

Saponins, glycosides of aglycone linked to one or more sugar chains have a promising activity to increase nutrient absorption and to reduce proptozoa and methane production in the rumen. Therefore, like tannins, saponins could be used to improve diet efficiency, thus to ameliorate livestock performances. In contrast to tannins, studies, mainly in the Mediterranean area, on positive effects of saponins in ruminant feeding are scarce. Based on a recent study in Tunisia, it seems that the administration of small amounts (30-40 g/day) of fenugreek (*Trigonella foenum-graecum* L.) seeds, which are relatively high in saponins, increased lamb growth and milk production in dairy ewes (H. Ben Salem, unpublished data). Further investigations are in progress to identify other promising saponin-containing plants.

Overall the interesting results about the advantageous use of tanniniferous and/or saponin containing forages to improve feed efficiency and to control GIP, thus to improve productive and reproductive performances of ruminants should encourage the establishment of practical options for agronomical applications of plants containing these natural secondary compounds in grazing systems. These options offer also solutions to reduce the use of chemicals in livestock feeding and health control.

(a3) Alley-cropping technique. It consists of cultivating herbaceous vegetation (graminae and or legumes) between rows of tree or shrub species (Acacia, Atriplex). Among the reasons of the low adoption of shrubs planting is the technical design of plantation, mismanagement, and competition for land often dedicated to cereal crops. Alley cropping technique overcomes some of these disadvantages because it (1) improves soil; (2) increases crop yield; (3) reduces weeds and (4) improves animal performance. Properly managed, alley-cropping allows diversification to benefit from several markets. It also promotes sustainability in both crop and livestock production. Benefits from cactus-barley cropping system were evaluated in Tunisia (Alary *et al.*, 2006, 2007; Shideed *et al.*, 2007). Compared to barley alone, the total biomass (straw plus grain) of barley cultivated between the rows of spineless cactus increased from 4.24 to 6.65 tonnes/ha and the grain from 0.82 to 2.32 tonnes/ha. These results are due to the change of the micro-environment created by alley-cropping with cactus, which creates a beneficial 'wind break' role reducing water loss and increasing soil moisture. The barley crop stimulated an increase in the number of cactus cladodes and fruits, while the cactus increased the amount of root material contributing to the soil organic matter. The alley-cropping system with *Atriplex nummularia* proved efficient in the semi arid regions of Morocco (annual rainfall 200-350 mm). Barley was cropped (seeding rate 160 kg/ha) between atriplex (333 plants/ha) rows. Compared to farmers' mono-cropping system, dry matter consumable biomass yield of atriplex was significantly high in the alley-cropping system. The later system was more profitable than mono-cropping. Indeed, Laamari *et al.* (2005) determined the net benefit from atripex planting and barley-atriplex alley cropping over 15 years. The cumulative net benefit was 732.18 $/ha and 3,342.53 $/ha, respectively. The economic and agronomic assessment of alley cropping shows that this technology is economically profitable. Therefore, it should be extended on a large scale in the agro-pastoral areas of the Mediterranean basin.

(a4) Shrub mixing technique. Most Mediterranean fodder shrubs and trees are either low in essential nutrients (energy and/or digestible nitrogen) or high in some secondary compounds (e.g. saponins, tannins, oxalates). These characteristics explain the low nutritive value of these fodder resources and the low performances of animals. For example, *Acacia cyanophylla* foliage is high in CT but low in digestible nitrogen. *Atriplex* spp. are low in energy and true protein although they contain high levels of crude protein, fibre and oxalates. Cactus cladodes are considered an energy source and high in water but they are low in nitrogen and fibre. Moreover, they are remarkably high in oxalates. Wealth information on the complementary nutritional role of these three shrub species and the benefits from shrub mixed diets for ruminants, mainly sheep and goats, reported in the literature

(Ben Salem *et al.*, 2000, 2002, 2004, 2005b). This technique permits to balance the diet for nutrients and to dilute the adverse effects of secondary compounds and the excess of minerals, like salt. The association cactus-atriplex is a typical example of the benefits from shrub mixing. The high salinity and the low energy content of atriplex foliage could be overcome by cactus. Some examples of the effects of shrub mixed diets on sheep and goats performances are reported in Table 2. In summary, diversification of shrub plantations should be encouraged to promote livestock production in the dry areas of the Mediterranean basin.

Land tenure and institutional arrangements

The resilience of SRs' production system is also function of social issues and social uncertainty related, for example, to family events and to property rights at the community level. More generally, the social, institutional and policy changes could induce important changes in rangeland-based systems.

Decision-makers and all research and development partners are increasingly aware that 'the heart of the rangeland sustainable management' is linked to institutional issue. Indeed, in the past the situation of rangelands was relatively better not only because population pressure and demand for meat were lower, but also because the management of rangelands was more strictly controlled by traditional institutions (jmaas in Morocco, Myaad in Tunisia) that enjoyed effective power. Numerous policy and institutional reforms have been carried out in several countries of the WANA. In most cases, policy and institutional reforms weakened pastoral institutions. These institutional reforms can be classified into three main approaches: state appropriation of rangeland resources, strengthening customary tribal claims, and privatization with titling (Ngaido and McCarthy, 2004).

Recent experience of communal rangeland management by Agricultural Development Grouping (GDA) in South Tunisia is quite promising. The GDAs are built up on socio-territorial units that correspond to the traditional tribe boundaries. They are fully participating in the design and the implementation of their integrated development. The approach used involves the real participation of agropastoral communities, in a new bottom-up mode, for the establishment of community development plan (CDP) that reflects the real issues and priority needs of the community. This is developed through the joint inputs of all stakeholders including community members, agricultural specialists, extension services, local administration and state representatives. Best-bet technologies for technical, institutional and policy issues are jointly identified for implementation, monitoring and evaluation. The community is represented by a formal community-based organization (CBO),

Table 2. Compiled data on the effects of shrub mixed diets on sheep and goat growth.

Basal diet[1]	Supplement[2]	Animal	Daily gain (g)	References
Acacia (417 g/d)	Atriplex (345 g/d); Barley (280 g/d)	Lambs	54	Ben Salem *et al.* (2002)
Cactus (437 g/d)	Atriplex (310 g/d); Acacia (265 g/d)	Lambs	28	Ben Salem *et al.* (2002)
Cactus (499 g/d)	Straw (207 g/d); Atriplex (356 g/d)	Lambs	81	Ben Salem *et al.* (2004)
Atriplex grazing	Cactus (290 g/d)	Lambs	20	Ben Salem *et al.* (2005)
Native shrubland grazing	Cactus (100 g/d); Atriplex (100 g/d)	Kids	60	Ben Salem *et al.* (2000)

[1] Acacia: *Acacia cyanophylla*; Cactus: *Opuntia ficus indica* f. *inermis* (cladodes); Atriplex: *Atriplex nummularia*.
[2] Values between parentheses are daily dry matter intakes.

directly elected by community members and fully recognized by government authorities as their equal partner for implementation of all actions set out in the jointly developed CDP. This includes such crucial issues as management of communal pasture and rangelands (20,000 ha of collective rangelands are put under rest and fully controlled by the communities), as well as the procurement of funds and necessary inputs and facilities, and the independent and transparent contact with all stakeholders and similar CBOs in the WANA region for exchange of relevant information and experiences (Nefzaoui *et al.*, 2007).

Conclusion and recommendations

Although solutions to major SR production constraints succeeded in alleviating the pressure caused by human needs for SR products, the consumption-production gap increases for most of the countries and imports are therefore increasing into the NENA region both in terms of feed and animal products. This trend is becoming alarming with the recent surge in prices of cereals, particularly barley. The future of SR production in the NENA region appears uncertain. The NENA countries will have to allocate increasing budget for SR production and import. These countries will have to face an increased pressure on rangelands which require an innovative approach to their effective management and complementation with better use of local natural resources. A set of technologies developed and or used in the Mediterranean could help alleviating feed shortage and feeding cost. Some of these technologies could ameliorate SR production and farmers' income. Therefore, they are worth to transfer at the farm level and to encourage their adoption. Strengths and weaknesses of technologies reported in this paper are summarized in Table 3.

Research conducted by NARS and ICARDA over the last decade under the framework of the Mashreq/Maghreb project (Crop-livestock integration, community development and improving livelihoods of agropastoral communities in eight countries of WANA), facilitated the development and evaluation of integrated technical, institutional and policy options to help building coping strategies:

1. Help livestock keepers building strong institutions that can facilitate both collective and individual adaptation and response to climate change and other external pressures, both short and long term; and ensure the participation of the livestock keepers in devising coping and risk management approaches to climate change through awareness building, collective action.
2. The increased pressure on rangelands will require an innovative approach to their sustainable improvement and management including institutional solutions for access to communal/collective ranges.
3. Better use of local natural resources with an emphasis on water harvesting and better use of adapted indigenous plant species, such as cactus and fodder shrubs, and introduction of AGIBPs-based feed blocks and treated straw.
4. More attention should be paid to devising efficient animal nutrition and health management options.
5. Better integration of cereals and livestock production should be pursued, with more attention to the value of straw and fallow replacement with adoption of adequate rotations and conservation agriculture techniques that would increase feed production while maintaining the natural resource base for sustainable development.
6. Use of biotechnology in breeding programs as a potentially effective tool to select drought tolerant forages and cereals and to sustain biodiversity (livestock and feed sources) as measures to cope with climate uncertainties.
7. Improved livestock capacity to cope with climate change through the identification and improvement of local breeds adapted to the local feed resources and tolerant to heat/cold stress.
8. Develop early warning systems based on local knowledge, livelihood strategies, and modern tools to forecast information on biophysical, economical, and markets environment to agropastoral communities.

Table 3. Advantages and limits of new technologies to promote livestock production in the dry areas of the Mediterranean basin.

Technologies	Strengths
Forages	• Dietetic quality of milk and meat of animals on forage-based diets • Reduce the use of concentrate feed
Straw treatment	• Nutritive value improved • Possibility of incorporating larger quantities of straw in livestock diets
AGIBPs ensiling	• Possibility of integrating perishable AGIBPs in livestock feeding • Reduce the use of some expensive feeds
Feed blocks	• Easy to make, to transport, to use • Solution to integrate AGIBPs in livestock feeding • Farmer can manage feed resources depending on year condition (rainfall) • Reduce the use of expensive feed supplements
AGIBPs pelleting	• Reduce the use of expensive feeds • Replace imported pellets • Formulation according to nutrient requirements of animals
Cactus plantations	• Withstand drought conditions • Rapid growth and high biomass • Source of water for animals in dry areas • Source of income • Reduce the use of expensive feed sources (barley, etc.)
Other shrub plantation	• Bank of evergreen forage resource • Multipurpose use
Use of tanniniferous shrubs	• Availability, easily used by smallholders • Avoid the use of chemicals and additives
Use of natural saponins	• Availability, easily used by small holders • Enhance rumen fermentation

Weaknesses	Opportunities
• Cannot grow under the rain-fed conditions of dry areas • Machinery demanding • Farmer cannot do the treatment himself • Reagent cost is high • Supplementation of treated straw is needed (energy, protein and minerals)	• More research focussing on selection of drought tolerant cultivars is needed • Not ready for transfer to farmers • Local institutions (associations, GDAs, etc.)
• Transport facilities needed • Tractor needed to press the material	• This technology could be adopted by small farmers when small silos (plastic bags) are considered
• Risk of intoxication with urea • The provision of some ingredients is not easy for small farmers (urea, minerals)	• This technology is ready for transfer to farmers in all Mediterranean countries • Local institutions (Associations, GDAs) should be encouraged to establish a mechanised feed block unit • Scientists and technicians should assist target producers and farmers in the manufacturing process and utilisation by animals
• Specific and expensive machines are needed	• Encourage youth holder of a qualification to invest in a project of pellets manufacturing and commercialisation
• Animals are not allowed to the above ground herbaceous vegetation in high density cactus plantation • Cut and carry is the only way to use cactus as feed • Cladodes should be chopped	• Ready to transfer to farmers for fruit and forage production • Experienced countries (e.g. Italy, Tunisia) could help technology demanding countries
• Need irrigation during the first two years of plantation • Should be regularly pruned • High cost of establishment of shrub plantation • Proper grazing management of shrub plantations which depends on plant species and cultivars, animal breeds and the local and socio-economic conditions	• Assisting farmers in utilisation mode of fodder shrubs is needed
• High levels have adverse effects on livestock performance	• Grazing systems based on the association of tannin-rich plants and protein sources could be transferred • Scientists and technicians should help farmers formulating appropriate tannin-containing diets
• High levels could result in negative effects	• Not ready for transfer • More research is needed to identify appropriate levels and sources of saponins to use

Table 3. Continued.

Technologies	Strengths
Alley-cropping	• Fits with farmers preferences • Increases crop yield • Soil improvement and conservation • Weed control • Improves animal productivity
Shrub mixed diets	• Complementary role between shrubs • Low quality shrubs could be integrated in the diet • Fodder bank available year-round at the farm
Rest technique	• Avoid irreversible degradation of the rangeland • Increase vegetation cover and biomass • Biodiversity (flora and fauna)

References

Akkari H., H. Ben Salem, M. Gharbi, S. Abidi and M.A. Darghouth, 2008a. Feeding Acacia cyanophylla Lindl. foliage to Barbarine lambs with or without PEG: Effect on the excretion of gastro-intestinal nematode eggs. Animal Feed Science and Technology 147: 182-192.

Akkari H., M.A. Darghouth and H. Ben Salem, 2008b. Preliminary investigations of the anti-nematode activity of Acacia cyanophylla Lindl.: Excretion of gastrointestinal nematode eggs in lambs browsing A. cyanophylla Lindl. with and without PEG or grazing native grass. Small Ruminant Research 74: 78-83.

Alary, V., M. El-Mourid, P. Lecomte, A. Nefzaoui, A. Waterhouse, I.A. Wright and A., Gibon, 2002. Assessment of livestock and farming systems in harsh environements – Approaches adopted by farmers through management practices. In: Annual meeting of the European Association For Animal production, Cairo, Egypt, September 1-4.

Alary, V., A. Nefzaoui, H. Ben Salem and S. Chouki, 2006. Impact assessment of the adoption of cactus pear cropping in livestock-cereal systems in arid and semi-arid Tunisia. The case of Zoghmar, central Tunisia. In: C. Mondragon Jacobo, G. Aranda Osorio, W.B. Phippen (eds.). Proc. Vth Int. Congress on cactus pear and cochineal, Acta Hort. 728, 279-288.

Alary, V., A. Nefzaoui and M. Ben Jemaa, 2007. Promoting the adoption of natural resource management technology in arid and semi arid areas: Modelling the impact of spineless cactus in alley cropping in Central Tunisia. Agricultural systems 94 (2007): 573-585.

Alary, V. and M. El Mourid, 2007: Changement réel et changement induit; Décalage ou perpétuelle recherche pour les zones arides d'Afrique du Nord. Cahiers Agricultures vol.16, No.4,juillet-aout 2007: 330-337.

Alary, V., A. Nefzaoui and M. El Mourid, 2007. How risk influences the adoption of new technologies by farmers in low rainfall areas of North Africa? In: El-Beltagy, A., M.C. Saxena and Tao Wang (eds). Human and Nature – Working together for sustainable Development of drylands. Proceedings of the Eighth International Conference on Development of Drylands, 25-28 February 2006, Beijing, China. ICARDA, Aleppo, Syria. pp: 802-810.

Ben Salem, H., A. Nefzaoui and L. Ben Salem, 2000. Supplementing range goats in central Tunisia with feed blocks or a mixture of *Opuntia ficus indica* f. *inermis* and *Atriplex nummularia* L. Effects on behavioural activities and growth. In: Proceedings of the 7th International Conference on Goats, Tours, France. pp. 988-989.

Ben Salem, H., A. Nefzaoui and L. Ben Salem, 2002. *Opuntia ficus indica* f. *inermis* and *Atriplex nummularia* L. Two complementary fodder shrubs for sheep and goats. Acta Horticultura 581, 333-341.

Ben Salem, H. and A. Nefzaoui, 2003. Feed blocks as alternative supplements for sheep and goats. Small Rumin. Res. 49, 275-288.

Ben Salem, H., A. Nefzaoui and L. Ben Salem, 2004. Spineless cactus (*Opuntia ficus indica* f. *inermis*) and oldman saltbush (*Atriplex nummularia* L.) as alternative supplements for growing Barbarine lambs given straw-based diets. Small Rum. Res. 51, 65-73.

Weaknesses	Opportunities
• High labour input required to prune trees	• Extension activities are needed for wider adoption of this technology • Scientists should continue the evaluation of other combinations (shrub species and forage and or cereal species
• Great variability in the growth rate of shrub species	• Adopted by farmers but scientists should make some adjustments of diets composition
• Rangelands users should agree and respect the period and conditions of rest technique • Need to find other area where animals can graze	• Ready for transfer provided that local institutions accept to be involved

Ben Salem, H., N. Al-Jawhari, M.A. Daba, A. Chriyaa, S. Hajj Hassan, M.L. Dehimi, M.Y. Masri, 2005a. Feed block technology in West Asia and North Africa. ICARDA, 111 p.

Ben Salem, H., H. Abdouli, A. Nefzaoui, A. El-Mastouri and L. Ben Salem, 2005b. Nutritive value, behaviour and growth of Barbarine lambs fed on oldan saltbush (*Atriplex nummularia* L.) and supplemented or not with barley grains or spineless cactus (*Opuntia ficus indica* f. *inermis*) pads. Small Rum. Res; 59, 229-238.

Ben Salem, H., A. Nefzaoui and H.P.S. Makkar, 2007. Feed supplementation blocks for increased utilization of tanniniferous foliages by ruminants. In: H.P.S. Makkar, M. Sanchez, A.W. Speedy (Eds.): Feed supplementation blocks. FAO Animal Production and Health paper 164, FAO Rome. 185-205.

Ben Salem, H. and T. Smith, 2008. Feeding strategies to increase small ruminant production in dry environments. Small Rumin. Res. 77, 174-194.

Bourbouze, A., 2000. Pastoralisme au Maghreb: la révolution silencieuse. *Fourrages. Revue française pour la production fourragère.* 161: 3-21.

Chenost, M. and C. Kayouli, 1997. Utilisation des fourrages grossiers en régions chaudes. Etude FAO Production et Santé Animales 135, FAO Rome, 226 p.

Cordiez, G., 1947. De la composition de quelques produits fourragers tunisiens et de leur valeur pour l'alimentation du mouton. Ann. Service Botan. Agron. Tunisie 20, 25-108.

Hazell, P., P. Oram and N. Chaherli, 2001. Managing livestock in drought-prone areas of the Middle East and North Africa: Policy Issues. In: H. Lofgren (ed.). Food and agriculture in the Middle East. Research in Middle East Economics, Vol.5. New York: Elsevier Science.

Hazell, P.; 2007. Managing drought risks in the low-rainfall areas of the Middle East and North Africa. In: 'Perinstrup-Anderson and Cheng, eds. Food policy for developing countries: the role of Government in the global food system'. Cornell university, Ithaca, New York.

Hoste H., J.F. Torres-Acosta, V. Polini, A. Aguilar-Caballero, E. Etter, Y. Lefrileux, C. Chartier and Broqua, 2005. Interactions between nutrition and gastrointestinal infestions with parasitic nematodos in goats. Small Rumin. Res. 60, 141-151.

Hoste H., F. Jackson, S. Athanasiadou, S. Thamsborg and S.O. Hoskin, 2006. The effects of tannin rich plants on parasitic nematodes in ruminants. Trends Parasitol. 22, 253-261.

Iniguez, L., 2008. Future trends in small ruminant production in the NENA region. In:' Expert Consultation 'Small Ruminant Research Strategy for the Non-Tropical Dry Areas in the Near East North Africa Region', 9-11 March 2008, Cairo, Egypt.

Laamari A., M. Boughlala and A. Chriyaa, 2005. Adoption and impact studies in Morocco. In: H. Shideed Kamil and M. El Mourid (eds). Adoption and impact assessment of improved technologies in crop and livestock production systems in the WANA region. The development of integrated Crop/Livestock Production in Low Rainfall Areas of Mashreq and Maghreb Regions (Mashreq/Maghreb Project).ICARDA, Aleppo, Syria, viii + 160 pp. En. pp. 107-118.

Nefzaoui A., A. Chriyaa and M.Y. El-Masri, 1999. A review of the research in North Africa on cereal straw use in animal feeding. ICARDA, 100 p.

Nefzaoui, A., 2002. Rangeland management options and individual and community strategies of agropastoralists in Central and Southern Tunisia. International conference on policy and institutional options For the management of rangelands in dry areas. CAPRi Working paper N° 23; pp.14-16.

Nefzaoui A. and H. Ben Salem, 2002. Forage, fodder, and animal nutrition. Chapter 12. In: P.S. Nobel (ed.), Cacti, biology and uses. University of California Press, 280 pp.

Nefzaoui A., 2004. Rangeland improvement and management options in the arid environment of Central and South Tunisia. In: Ben Salem, H., Nefzaoui A., Morand-Fehr P. (eds.). Nutrition and feeding strategies of sheep and goats under harsh climates. Zaragoza: CIHEAM-IAMZ, 2004. Options Méditerranéennes: Série A. Séminaires Méditerranéens; n. 59, p. 15-25.

Nefzaoui, A., M. El Mourid, V. Alary, T. Ngaido and K. El Harizi, 2007. Empowering rural communities for better management of desert collective rangelands- from concept to implementation. In: El-Beltagy, A., M.C. Saxena and T. Wang (eds). Human and Nature – Working together for sustainable Development of drylands. Proceedings of the Eighth International Conference on Development of Drylands, 25-28 February 2006, Beijing, China. ICARDA, Aleppo, Syria. pp: 620-632.

Nefzaoui, A., M. El Mourid, H. Ketata and A. Sidahmed, 2008. Future trends in small ruminant production in the NENA region. In:' Expert Consultation 'Small Ruminant Research Strategy for the Non-Tropical Dry Areas in the Near East North Africa Region', 9 – 11 March 2008, Cairo, Egypt.

Ngaido, T. and N. McCarthy, 2004. Collective action and property rights for sustainable development. Institutional Options for Managing Rangelands. Vision 2020 for Food, Agriculture, and Environment. Focus 11, February, 2004.

Sidahmed, A., 1996. Restocking destitute nomads: post-disaster rehabilitation. IFAD Technical Advisory Division, Staff working paper N° 27. www.ifad.org/lrkm.

Fattening Cinta Senese pigs with chestnuts: effect on meat and fat characteristics

F. Sirtori, C. Pugliese, S. Parenti, S. D'Adorante and O. Franci
Dipartimento di Scienze Zootecniche, Università di Firenze, Via delle Cascine 5, 50144 Firenze, Italy

Abstract

The aim of this trial was to identify the effect of feeding pigs chestnut on some meat and fat characteristics. Thirty-three Cinta Senese pigs were reared indoors and allotted into three groups. One group was fed only commercial feed (0-CH); the other two groups were fed chestnut for one (1-CH) and three (3-CH) months before slaughtering. The chestnut diet contained a minimal supplement of bran. Pigs were slaughtered at 130 kg of live weight, on average. Regarding carcass characteristics, the only parameter affected by the diet was pH_{24} where the 1-CH group showed the lowest value. Also in sample joint dissection, a rare effect of diet was found; the only significant difference among groups was in intermuscular fat content, which was highest in the 1-CH group. There was a strong effect of diet on the colour parameters of *Longissimus lumborum:* the chestnut diet led to more coloured meat, due to higher L*, a* and b* values in the 1-CH and 3-CH groups. A negative effect of chestnut on meat quality traits was found: the 3-CH group showed the highest values of drip loss, cooking loss and WB on raw meat. As for fatty acid composition, the animals fed chestnut throughout the fattening period showed the highest level of unsaturation due to the highest percentage of mono-unsaturated fatty acid (MUFA) and polyunsaturated fatty acids (PUFA). The 1-CH group was in an intermediate position for all unsaturated fatty acids detected.

Keywords: Cinta Senese, chestnut, meat quality, fatty acid

Introduction

Cinta Senese is a local swine breed reared mainly in Tuscany. Among the Italian autochthonous swine breeds, Cinta Senese has a good chance of being developed thanks to the high number of animals and the strong linkage to traditional production methods and to the territory. At present, the breed is one of the more interesting examples of a well conducted safeguard operation of autochthonous germoplasm and it is a witness to the link among rustic breeds, territory and typical products. Grazing in woods is practiced to a varying extent for Cinta Senese pig and this type of rearing allows feed resources to be exploited that are otherwise unused.

Thanks to extensive scientific literature on autochthonous pigs (Lopez-Bote, 1998; Coutron-Gambotti *et al.*, 1998; Pugliese *et al.*, 2004; Pugliese *et al.*, 2005) it is known that rearing pigs in woods, and consequently feeding on acorns, chestnuts etc., affects the physical, chemical and sensorial traits of fresh and processed meat products.

Nowadays, one of the main problems of rearing Cinta Senese is linked to the exploitation of woods which, in many cases, has been shown to be inadequate. Frequently the number of animals per hectare is very high and the period of pasture in the woods is longer than the time of effective availability of natural resources, causing irreparable damage to the soil. Thus it is very useful to know the minimal period of pasture in woods required to obtain a characterization of the product and, at the same time, to avoid damaging the forest. At the same time, it is possible to examine the hypothetical issue of feed traceability. In fact, nowadays the meat of Cinta Senese pigs is guaranteed by PDO (Protected Designation of Origin) and in the future it is probable that the products will have a commercial differentiation based on the difference in the rearing system. In this case the use of a technique of *a posteriori* traceability is inalienable since the documental traceability, even if useful, is not enough to guarantee the origin of the products. The aim of this trial was to study the effect of the use of chestnuts at various periods during fattening on the physical and chemical characteristics of meat and fat in Cinta Senese pigs.

Material and methods

Thirty-three Cinta Senese pigs (barrow and gilt) were reared indoors. During the fattening period (last three months before slaughter) animals were allotted into three groups balanced for sex. One group was fed a commercial mixture (0-CH) for the whole fattening period; the other two groups were fed chestnuts for one (1-CH) or three (3-CH) months before slaughter. The chestnut diet contained a minimal supplement of wheat bran (about 10%) to ensure intestinal functions. Average feed intake and diet characteristics are detailed in Table 1. The animals were slaughtered at an average of 130 kg of live weight. At 45 minutes (pH_{45}) and 24 hours (pH_{24}) post-slaughter, pH was measured in *Longissimus lumborum* muscle near the 10th lumbar vertebra. A sample joint (portion of loin from the 2nd to the 5th lumbar vertebra inclusive) was removed. The sample joint was dissected into lean, fat and bone. The lean portion was separated into *L. lumborum* (Ll), *Psoas major* (Pm) and 'other muscles' and subcutaneous fat in the outer and inner layers. Each portion was weighed. The following analyses were carried out on the Ll: (a) colour measurement (L*, a* and b*) with a Minolta Chromameter CR-200. Colour was analysed 24-h post-mortem on the cut surface after 1 h rest at 4 °C, according to the official method (Boccard *et al.,* 1981). The hue angle (H°= arctan b*/a*) and chroma ($C^*= ((a^*)^2+(b^*)^2)^{0.5}$) parameters were also calculated. (b) moisture was measured by lyophilising to constant weight intramuscular fat (IMF) as ether extract and protein contents (AOAC, 1990); (c) water-holding capacity (WHC) was determined by the following techniques: drip loss from slices stored horizontally for 48 h, expressed as a percentage of the initial weight of the slice; cooking loss by water bath, the meat samples were cooked until the temperature at the centre reached 75 °C; (d) shear force measured with an Instron 1011 apparatus on raw and cooked meat. All shear force measurements were made on a core 2 cm in diameter and replicated; (e) fatty acid profile of total lipids of adipose tissue, the inner and outer layer were measured separately. For the latter determination, fatty acids methyl esters were prepared by esterification in the presence of sulphuric acid and were analysed by gas chromatography using a DANI 86.10 apparatus equipped with a flame ionisation detector (FID). Fatty acids were separated in a capillary column coated with FFAP-TPA stationary phase (length: 30 mm, internal diameter: 0.32 mm, film thickness: 0.25 µm). The temperature of the column started at 160 °C and reached 220 °C, increasing by increments of 2 °C/min. The temperature of the detector was set at 260 °C. The individual methyl esters were identified by their retention time. Results are expressed as g/100 g of dry matter. The GLM procedure of SAS (2003) was used to analyse the data of the factorial experimental design. The model included the fixed effects of diet, sex and layer (for fatty acid composition) or muscle (for chemical traits of Ll and Pm).

Table 1. Daily feed intake and nutritive characteristics of the diets.

	Chestnut	Mixture
Feed intake (kg of d.m./d)	2.7	2.7
Diet composition		
Crude protein (% on d.m.)	5.54	18.9
Digestible energy (kcal/kg of DM)	3,029	2,958
Energy/protein	546.7	156.5
Fatty acids (g/100 g of DM)		
Saturated fatty acids (FSA)	0.39	0.26
MUFA	0.91	0.30
PUFA - n3	0.16	0.07
PUFA - n6	0.92	0.72

Results and discussion

Table 2 lists the characteristics and means of the carcass and of the sample joint. As regards the characteristics of the carcass, the only parameter affected by the diet was pH_{24} with the 1-CH group showing the lowest value. In the sample joint, a rare effect of diet was found; the only significant difference among groups was in intermuscular fat content, which was higher in pigs in the 1-CH group. In a study on the effect of substituting a diet of concentrate for chestnuts in Corsican pigs, Coutron-Gambotti *et al.* (1998), found no difference in the percentage of subcutaneous fat. In our previous work (Pugliese *et al.*, 2007) on Cinta Senese pigs pastured in chestnut woods or reared indoors on a commercial mixture, we obtained the same results in the sample joint dissection, but differences in the percentage of subcutaneous fat, which was higher in animals pastured in woods.

Table 3 shows the chemical composition of the *Longissimus lumborum* and *Psoas major* muscles. The means of the muscles are tabulated according to diet because of the significant diet*muscle interaction, especially for intramuscular fat content. Generally, the 3-CH group had the highest percentage of intramuscular fat, and Ll muscle showed lower moisture and higher IMF content than Pm; these differences between muscles increased with the length of chestnut feeding period. Due to the higher energy/protein ratio than in the commercial mixture (Tabe 1), it is possible that the chestnut diet led to higher adiposity during fattening, which occurred as intramuscular fat in late developing tissue; moreover some muscles appeared to be much more influenced than others.

As regards colour parameters (Table 4), the main differences were found in *L. lumborum* where a clear effect of the chestnut diet was observed. The 1-CH and 3-CH groups showed the highest value of L*, a* and b* and, consequently their meat was more coloured. These results are in contrast with previous findings (Pugliese *et al.*, 2007) which revealed differences in colour parameters not in Ll, but only in subcutaneous fat, where pigs fed chestnut exhibited the lowest values of a* and b*. As far as physical parameters are concerned (Table 4), the 3-CH group showed the lowest water holding

Table 2. Carcass characteristics and sample joint composition.

	Diet			RSD
	0-CH	1-CH	3-CH	
Slaughter weight (kg)	136.0	134.5	131.5	15.08
Fat thickness at UT (mm)	37.57	34.93	37.57	6.38
Fat thickness at GM (mm)	42.48	40.39	42.84	9.71
pH_{45}	6.44	6.44	6.45	0.22
pH_{24}	5.60^{a}	5.41^{b}	5.60^{a}	0.14
Sample joint weight (kg)	2.17	2.20	2.00	0.32
Sample joint composition (%)				
Longissimus lumborum	29.63	28.11	28.01	2.83
Psoas major	9.51	9.26	8.91	1.13
other lean	6.71	6.80	6.53	1.21
Subcutaneous fat				
outer layer	19.00	19.40	17.68	2.49
inner layer	23.98	22.98	24.70	4.03
Intermuscular fat	2.50^{b}	4.61^{a}	3.87^{ab}	1.77
Bone	8.67	8.85	10.29	2.11

Values within traits followed by different superscript letters are significantly ($P<0.05$) different.
RSD: residual standard deviation.

Table 3. Chemical composition of Longissimus lumborum *(Ll) and* Psoas major *(Pm) muscles (% on wet basis).*

Diet	0-CH		1-CH		3-CH		RSD
Muscle	Ll	Pm	Ll	Pm	Ll	Pm	
Moisture	72.25^{b}	73.02^{c}	73.27^{c}	74.60^{d}	69.88^{a}	71.79^{b}	0.84
Ether extract	3.34^{c}	2.46^{d}	4.03^{b}	2.58^{d}	6.10^{a}	3.41^{c}	0.82
Protein	23.03	23.19	21.51	21.20	22.66	23.00	1.01
Ash	1.15	1.28	1.08	1.21	1.14	1.28	0.13

Values within traits followed by different supersript letters are significantly (P<0.05) different.
RSD: residual standard deviation.

Table 4. Colour parameters and physical characteristics of Longissimus lumborum.

	Diets			RSD
	0-CH	1-CH	3-CH	
Colour of *L. lumborum*				
L*	47.02^{b}	49.32^{a}	50.17^{a}	2.16
a*	10.36^{b}	12.09^{a}	11.89^{a}	1.07
b*	3.02b	4.39^{a}	4.56^{a}	0.76
Chroma	10.80^{b}	12.86^{a}	12.74^{a}	1.14
Hue	0.285^{b}	0.350^{a}	0.366^{a}	0.06
Colour of outer fat				
L*	79.17	79.03	79.12	2.19
a*	4.01	3.84	3.42	1.97
b*	3.29^{ab}	3.74^{b}	3.05^{a}	0.68
Chroma	5.24	5.52	4.62	1.89
Hue	0.710^{b}	0.841^{a}	0.739^{ab}	0.15
Colour of inner fat				
L*	80.77	81.39	80.95	0.81
a*	2.45	2.31	2.50	0.69
b*	2.40	2.75	2.37	0.80
Chroma	3.46	3.60	3.46	0.98
Hue	0.772	0.870	0.769	0.14
Water holding capacity (WHC) (%)				
Drip loss	2.25^{b}	3.21^{b}	6.64^{a}	1.43
Cooking loss	22.58^{b}	26.37^{a}	27.13^{a}	3.26
Shear force (kg)				
on raw meat	7.97^{b}	8.71^{ab}	9.83^{a}	1.66
on cooked meat	13.86	12.76	12.59	3.03

Values within traits followed by different superscript letters are significantly (P<0.05) different.
RSD: residual standard deviation.

capacity and the highest value of shear force on raw meat. These results are in contrast with the negative correlation between IMF percentage and water losses and between IMF percentage and shear force observed by Wood *et al.* (2007).

Concerning fatty acid composition (Table 5), the animals fed the chestnut diet throughout the fattening period showed the highest level of unsaturation due to the highest percentage of MUFA, PUFA n-3 and PUFA n-6. These results reflect the fatty acid composition of the diet: the chestnuts have higher MUFA and PUFA contents than the commercial mixture. So, considering the daily feed intake of dry matter of 2.7 kg for all animals, it is possible to calculate the feed intake of MUFA, PUFA n-3 and PUFA n-6 for animals fed with chestnuts (1-CH and 3-CH) and for animals fed with the commercial mixture (0-CH): 24.57, 4.32, 24.84 vs. 8.1, 1.89, 19.44 g/d respectively. This average daily intake led to a deposition of fatty acid – especially PUFA n-3 and n-6 which are hexogen – which was high throughout the chestnut fattening period. Concerning the absolute value, the 1-CH group was in an intermediate position for all unsaturated fatty acids detected. Coutron-Gambotti *et al.* (1998) in the aforementioned study on Corsican pigs, also found the highest percentage of PUFA in adipose tissue of animals fed with chestnuts, while no effect of diet was found on MUFA content.

Conclusion

Substituting commercial concentrate for chestnuts significantly affected the quality of both meat and fat. Briefly, chestnut appeared to have a negative effect on physical characteristics (lower WHC and higher shear force) in spite of the higher intramuscular fat that it provides. Moreover the chestnut diet led to a more unsaturated level of adipose tissue especially for PUFA n-3 and n-6. The latter result could be considered to be positive from the dietetic point of view but negative from a technological one. So, considering that among Cinta Senese products that most appreciated by consumers are dry

Table 5. Chemical and fatty acid composition of subcutaneous adipose tissue (g/100 g of DM).

	Diets			RSD
	0-CH	1-CH	3-CH	
Moisture %	3.45^c	5.39^b	4.87^a	0.74
Lipids %	79.06	79.93	78.48	3.92
Fatty acid[1]				
C14:0	0.68^b	0.72^a	0.72^a	0.07
C16:0	13.17	13.52	13.69	1.28
C16:1	1.06^b	1.13^a	1.08^{ab}	0.14
C18:0	7.10	7.27	7.37	0.80
C18:1	25.52^b	26.99^a	28.11^a	2.50
C18:2	5.53^b	6.20^a	6.47^a	0.77
C18:3	0.24^c	0.32^b	0.41^a	0.05
C20:1	0.59^b	0.64^{ab}	0.69^a	0.10
C20:2	0.28^b	0.32^a	0.33^a	0.04
SFA	21.21	21.79	22.04	2.07
MUFA	27.29^b	28.90^a	30.02^a	2.65
PUFA-n3	0.24^c	0.32^b	0.41^a	0.05
PUFA-n6	5.86^b	6.60^a	6.91^a	0.82

Values within traits followed by different superscript letters are significantly ($P<0.05$) different.

[1] Other fatty acids detected: C12:0, C15:0,C17:0, C17:1, C20:0, C20:4, C20:3

RSD: residual standard deviation.

cured ham and cured lard, it appears that chestnuts are not advisable in the fattening of Cinta Senese pigs. Pasturing in the woods, included chestnut woods, is a main point in the rearing of Cinta Senese pigs because it is recognised by consumers as 'natural' and 'genuine', at least in their representation. So, in order to check the effects of chestnut on noticeable parameters, further investigations on aromatic and sensorial traits of Cinta Senese are needed.

Acknowledgements

We would like to thank 'Cassa di Risparmio di Firenze' for providing financial support for the SAS licence and the managerial and technical staff of the *Corpo Forestale dello Stato* (ex-ASFD administration) of Siena. This study was supported by a grant from PRIN-MIUR prot. 2005077118_001.

References

AOAC, 1990. Official methods of analysis of the association of official analytical chemist (15th ed.) No. 960.39. Washington: Association of the Office of Analytical Chemists.

Boccard, R., L. Buchter, E. Casteels, E. Cosentino, E. Dransfield, D.E. Hood, R.L. Joseph, D.B. Macdougall, D.N. Rhodes, L. Schon, B.J. Tinbergen and C. Touraille, 1981. Procedures for measuring meat quality characteristics in beef production experiments report of a working group in the commission of the European Communities (CEC) beef production research programme. Livest. Prod. Sci. 385-397.

Coutron-Gambotti, C., G. Gandemer and F. Casabianca, 1998. Effects of substituting a concentrated diet for chestnuts on the lipid traits of muscle and adipose tissues in Corsican an Corsican × Large White pigs reared in sylvo-pastoral system in Corsica. Meat Sci. 50: 163-174.

Lopez-Bote, C.J., 1998. Sustained utilization of the Iberian pig breed. Meat Sci. 49 (suppl. 1): S17-S27.

Pugliese C., G. Calagna, V. Chiofalo, V.M. Moretti, S. Margiotta, O. Franci and G. Gandini, 2004. Comparison of the performances of Nero Siciliano pigs reared indoors and outdoors. 2. Joints composition, meat and fat traits. Meat Sci. 68: 523-528.

Pugliese C., R. Bozzi, G. Campodoni, A. Acciaioli, O. Franci and G. Gandini, 2005. Performance of Cinta Senese pigs reared outdoors and indoors. 1 Meat and subcutaneous fat characteristics. Meat Sci. 69: 459-464.

Pugliese, C., L. Pianaccioli, F. Sirtori, A. Acciaioli and O. Franci, 2007. Effect of pasture on chestnut wood on fat and meat quality of Cinta Senese. Option Méditerranéennes, 76: 263-267.

SAS, 2003. SAS/STAT software, release 9.1. SAS Institute Inc., Cary, NC.

Wood, J.D., M. Enser, A.V. Fisher, G.R. Nute, P.R. Sheard, R.I. Richardson, S.I. Hughes and F.M. Whittington, 2007. Fat deposition, fatty acid composition and meat quality: A review. Meat Sci. 78:342-358.

Importance of Mediterranean forest products as food resource of domestic herbivores: the case of oak acorn

A. Gasmi-Boubaker[1], R. Mosquera Losada[2], H. Abdouli[3] and A. Rigueiro[2]
[1]Institut National Agronomique, Tunis, Tunisie
[2]Escuela Polytecnica Superior, Universidade de Santiago de Compostela, Lugo, Spain
[3]Ecole Supérieure d'Agriculture de Mateur, Tunisie

Abstract

Many Mediterranean trees and shrubs constitute real fodder reserves for livestock and wildlife. Because of their resistance to heat and frequent harvesting, browse species are the major feed resources especially under harsh conditions. In the Mediterranean area, the fruits of oak (acorns) are used in extensive feeding of cattle, goat, sheep and pig. Proximate analyses have revealed the chemical composition of cork oak acorn to be similar to that of cereals (e.g. barley). Acorn is rich in starch (56.6-59.9% DM) but poor in protein (less than 10% DM) content. Another particularity of acorns is their high level of tannins (4-15.9% DM) that are found particularly in the hulls. In goats, voluntary intake of acorn containing concentrate ranged from 763-1,091 g DM/day without any signs of physiological problems. However, there were significant decrease in apparent digestibility coefficients for dry matter, organic matter and crude protein. N retention was positive in kids offered acorns and live weight gain reached 43 g/day. Inclusion of acorn in the diet of rabbit decreased the in vivo digestibility by 10.8%, 11.5% and 17.5%, respectively, for DM, organic matter and crude protein, but no significant effect was observed on average daily gain, carcass yield and feed conversion ratio. These results suggest that acorns could replace partially conventional feedstuffs as either forages or concentrate feeds without affecting livestock performance but should reduce the feeding cost.

Keywords: acorn, broiler, goat, sheep, rabbit

Introduction

Animal production is one of the most important activities of global agriculture. It plays a significant economic role in most developing countries and it is essential for the food security of their rural population. However, feed shortages, both quantitative and qualitative, limit livestock productivity. Supplementary feeding could resolve feed scarcity through commercial feedstuffs (maize, barley, soya bean meal...). However, the high cost of these feed resources puts pressure on the national balance of payments and limits their use by many smallholders. During the last years, new scientific research has focused on valorization of alternative feed resources, including leaves and fruits of shrubs and trees such as oak fruits (acorn). Acorns are not a novel feedstuffs but their usage in human and animal nutrition has a long history especially in the Mediterranean region. In some of the large oak forests in Spain and Portugal, pigs are still turned loose in oak groves in the autumn, to fill and fatten themselves on acorns.

In Tunisia, oak trees occupy an area of 90,000 ha and produce an average yield of between 190 and 700 kg acorns/ha/year (Kayouli and Buldgen, 2001). These fruits are valuable feed for wildlife, birds and domestic animals (goat, sheep, cow) and the small holders recognize their fodder potential. Several results will be summarised here.

Nutrient content of acorn from differents species

Acorns are rich in nutrients. Percentages vary from species to species (Table 1), but all acorns contain large amounts of carbohydrates (56.6-59.9% dry matter, DM) and fats, as well as the minerals calcium, phosphorus and potassium (Kayouli and Buldgen, 2001). However, acorns have low contents of crude

Table 1. Chemical composition of acorn from different species.

	DM	OM	Ash	CP	CF	NDF	ADF	Fat	References
Quercus aegilops	53.0	97.6	2.4	3.9	23.6	51.6	35.1	2.5	Aljassim *et al.* (1998)
Quercus coccifera	53.7	97.2	2.8	2.9	25.3	49.8	26.4	2.3	Aljassim *et al.* (1998)
Quercus coccifera	97.3	97.8	2.2	11.1	5.96	-	-	2.2	Nowar *et al.* (1994)
Quercus ilex	59.3	98.4	1.6	4.8	25.2	-	-	12.1	Nieto *et al.* (2002)
Quercus ilex	63.5	98.2	1.8	5.6	-	-	-	11.3	Lopez-Carrasco *et al.* (2004)
Quercus suber	71.55	97.15	2.85	7.95	11.85	33.1	11.4	-	Gasmi-Boubaker *et al.* (2007b)
Barley	86.0	97.4	2.6	10.8	5.3	20.6	6.8	1.7	

DM: dry matter, OM: organic matter, CP: crude protein, CF: crude fat, NDF: neutral ditergent fibre, ADF: acid detergent fibre.

protein (CP) and should therefore be fed together with another source of nitrogen. Acorns also contain bitter tannins, the amount varying with the species (4-15.9% DM) and they are found mainly in the hulls. These phenolic compounds are known to impair feed utilization, growth and animal health. The tannin in bitter acorns of some oaks can cause poisoning in livestock if fed in high percentage.

Digestibily of diets including acorn

The results from nylon bag technique studies show that 79.7% (Kayouli and Buldgen) to 82.9% dry matter of acorn are potentially degradable in the rumen with a slow rate (0.031 h^{-1}) due to the high fibre content (Moujahed *et al.*, 2005). In a previous paper, authors reported from their *in vivo* digestibility study that acorns decreased digestibility coefficient of dietary constituent in lambs fed on the 50% acorn (El Jassim *et al.*,1998) and goats fed on 75% acorn (Gasmi-Boubaker *et al.*, 2007) in the diets. Nitrogen retention was also lower in acorn diets. Also, research work with rabbits (Gasmi-Boubaker *et al.*, 2007b) has shown that inclusion of acorn in the diet decreased by 10.8%, 11.5% and 17.5%, respectively, for DM, OM and CP digestibility (Table 2). In fact, protein digestibility as

Table 2. Digestibily (%) of diets including acorn.

	Digestibility (%)			References
	DM	OM	CP	
Sheep				
Control diet (25-50% barley)	78.8	81	74	Al-jassim *et al.* (1998)
Diet including acorn (25-50%)	71.7-66	73.7-68.3	69-63.3	
Goats				
Control diet (75% barley)	69.7	77.2	77.2	Gasmi-Boubaker *et al.* (2007a)
Diet including acorn (75%)	57.2	53.9	53.9	
Rabbits				
Control diet (25% barley)	65	66	77.3	Gasmi-Boubaker *et al.* (2007b)
Diet including acorn (25%)	57.9	58.5	63.8	

DM: dry matter, OM: organic matter, CP: crude protein.

well as the digestibility of other nutrients was reported to be depressed by the presence of tannins through inhibition of digestive enzymes.

Growth performance

Acorns are very palatable to domestic animals (pigs, cow, goat, sheep, rabbit..). Previous *in vivo* results (Table 3) have suggested that replacing barley grains or corn by varying levels of oak acorn decreased average daily gain (Al Jassim *et al.*, 1998; Gasmi-Boubaker *et al.*, 2007a). Daily weight gain of lambs fed the control diet or 25% acorn diet was similar (186 and 189 g) but higher (P<0.01) than lambs on the 50% acorn diet (142 g). Feed conversion ratio (kg DM/kg live weight gain) was lower (P<0.01) for the 50% acorn diet. Nevertheless, in a study conducted to evaluate the effect of inclusion of acorn on growth performance and carcass characteristics in fattening rabbits, Boubaer *et al.* (2007b) indicated that acorn could be used at a level of 25% as a replacement of barley in concentrate feed. The final live weight, carcass yield and daily weight gain from weaning through slaughter did not show any significant difference between rabbits fed the diet containing barley or cork oak acorn and no signs of health problems were observed. Moreover, Nowar *et al.* (1994) found that feed efficiency and protein efficiency ratio were improved with the presence of oak acorn at a level of 20% in the diet and no differences in carcass traits and internal organs were recorded with acorn or corn diets. In the opposite, Bouderoua and Selselt Attou (2003) reported that diets containing 60% acorn decreased growth (-29%) of chickens but also an abdominal fat deposition. Moreover, the proportion of linoleic acid was significantly higher in the chickens fed oak acorn-based diets (23%). This means that acorn could modify the quantity and the quality of lipids stored in animals. While Al Jassim *et al.* (1998) suggested that substitution of acorns for barley at a maximum level of 25% would be economically advantageous.

Conclusion

Acorns can be partially substituted for barley and corn to meet energetic requirements of domestic animals such as sheep, goat, rabbit and chiken. Their low price should reduce the feeding cost. Consequently, the existing local acorn market could be greatly expanded and provide new income for rural people.

Table 3. Performance of animals fed diets including acorn.

	Average daily gain (g/d)	References
Sheep		
Control diet (25-50% barley)	186	Al Jassim *et al.* (1998)
Diet including acorn (25%)	189	
Diet including acorn (50%)	144	
Goats		
Control diet (75% barley)	80	Gasmi-Boubaker *et al.* (2007a)
Diet including acorn (75%)	43	
Rabbits		
Control diet (20% corn)	25.6	Nowar et *al.* (1994)
Diet including acorn (20%)	24.9	
Control diet (25% barley)	23	Gasmi-Boubaker *et al.* (2007b)
Diet including acorn (25%)	26	

References

Al Jassim, R.A.M., K.I. Ereifiji, R.A. Shibi and A. Abudabos, 1998: Utilisation of concentrate diets containing acorns (Quercus aegilops and Quercus coccifera) and urea by growing Awassi lambs. Small Ruminant Research, 29: 289-293.

Ben Mustapha, C., 2004: Potentialités nutritionnelles des glands de chêne Kermès: Utilisation pour l'alimentation des petits ruminants. Mémoire de diplôme d'études approfondies. ESA Mateur.

Bouderoua, K and G. Seleselt-Attou; 2003. Fatty acid composition of abdominal adipose tissue in broilers fed green-oak (Quercus ilex), cork oak acorn (Quercus suber. L) based diets. Anim. Res. 52:377-382.

Gasmi-Boubaker, A., H. Abdouli, Hajer Khelil, Rabia Mouhbi and L. Tayachi., 2007a. Nutritional value of cork oak acorn (Quercus suber L.) as an energy source for growing goats. Asian Journal of Animal and Veterinary Advances, 2(1):32-37.

Gasmi-Boubaker, A., H. Abdouli, M.R. Mosquera Losada, L. Tayachi, M. Mansouri and I. Zaidi, 2007b. Cork oak (Quercus suber L.) acorn as a substitute for barley in the diet of rabbits: effect on in vivo digestibility, growth and carcass characteristics. Journal of animal and veterinary advances 6 (10): 1219-1222.

Kayouli, C. and A. Buldgen, 2001. Elevage durable dans les petites exploitations du Nord Ouest de la Tunisie; F.U. S, Gembloux, Belgique. 198 p.

Lopez-Carrasco, C., A. Daza, A. Rey and C. Lopez-Bote, 2004. Efectos de las heladas y los carpofagos (*Curculio sp.*) sobr la calidad de bellotas en una dehesa de Castilla-La Mancha. In: Garcia-Criado et al. (eds.)XLIV Reunion Cientifica de la SEEP, Salamanca, pp: 427-432.

Moujahed, M., C. Ben Mustapha and C. Kayouli, 2005. Effect of maturity stage on chemical composition, in sacco degradation and in vitro fermentation of acorns (*Quercus coccifera L.*). Options Méditerranéennes, Série A, N°.67.

Nieto, R., M. Ribera, M.A. Garcia and J.F. Aguilera, 2002. Amino acid availability and energy value acorn in the Iberian pig. Livest.Prod.Sci. 77: 227-239.

Nowar, M., K. Al-Shawabkeh and H. Nissour, 1994. Evaluatin of Oak Acorn (Quercus coccifera) as Untraditinal Energy Feedstuff for complete Substitution of Corn Grains in Fattening Rabbit Ration. In: Rabbit Production in Hot Climates. Baselga, M. and I.F.M. Marai (Ed), Zaragoza: CIHEAM-IAMZ International Conference f Rabbit Production in Hot Climates, Cairo (Egypt), pp: 177-182.

Finishing Cinta Senese pigs at pasture: fatty acids composition of cured lard

L. Giuliotti[1], J. Goracci[1], N. Benvenuti[1], F. Sirtori[2], C. Pugliese[2] and O. Franci[2]
[1]Dipartimento Produzioni Animali, Viale delle Piagge, 2, 56124 Pisa, Italy
[2]Dipartimento di Scienze Zootecniche, Via delle Cascine 5, 50144 Firenze, Italy

Abstract

Extensive farming system increasingly involves swine husbandry based primarily on wood pasture; this practice could cause problems linked with woods exploitation and needs substantial skills about sustainable woods management. For many years, grass pasture represented a valid option in Italian and foreign regions. The utilisation of grass could represent a valid alternative to achieve good performances in finishing pigs with effects on fatty acids composition. For these reasons we tried to evaluate the effects of grazing on fatty acids (FA) composition and malonaldehyde (MDA) content of cured lard, a typical Tuscan product. 16 Cinta Senese finishing pigs were divided into two groups: 'Concentrate' (C) fed a common diet utilised in the farm (2.6 kg/head/d) and 'Pasture' (P) fed on grass pasture (lucerne, oat, lolium and clover) with a daily integration of 1.4 kg/head of the same concentrate used by 'C'. Slaughtering happened at 150 kg of live weight. FA and MDA content were determined on lard cured for 6 months, according to dry seasoning technique. FA composition did not show significant differences between the two groups revealing a substantial homogeneity in fatty acids composition. More in detail linolenic acid, that is particularly influenced by the diet, did not show higher values in the group fed pasture (0.59 vs. 0.64 respectively for C and P). MDA content was similar in the two groups (1.46 vs. 1.53 mg/kg) with higher values than those found in other investigations conducted on Cinta Senese pigs.

Keywords: pig, Cinta Senese, grass pasture, fatty acids, cured lard

Introduction

The quality of a product has been defined as 'the sum of the properties and characteristics of a product or of a service which confer on it the ability to satisfy implicit or explicit requirements' (Edwards, 2005); in the last years quality expectation of consumers about animal food derivates point toward the respect of animal welfare that commonly match with the outdoor production system (Stern and Andrens, 2003). In swine enterprise there is an increase in the opportunity of extensive farming systems with pigs reared outdoor fed primarily on woods pasture. There are well known examples of extensive Iberian pig production systems with animals kept on acorn and grass pasture during the finishing period (Soto *et al.*, 2008; Daza *et al.*, 2007) whose product achieves good appraisal by consumers and reaches higher prices in the markets (Rey *et al.*, 2006). The utilisation of grass in the diet of finishing pigs results in noticeable effects on fatty acids composition that, whether in adipose tissue or muscle, contribute importantly to various aspects of meat quality and are central to the nutritional value of meat (Wood *et al.*, 2008). This trial represents a part of a large project regarding performance traits, carcass and meat quality of Cinta Senese pigs reared outdoor on grass pasture; information on seasoned products are few (Sirtori *et al.*, 2005) thus the aim of the present study was to investigate on fatty acids profile and malonaldehyde content of cured lard of Cinta Senese pigs fattened on grass pasture.

Material and methods

The study was carried out in an organic Cinta Senese pig farm near Pisa. Fattening happened in large areas where rest and feeding spaces were provided. Sixteen animals were divided in two groups (each with 3 castrates and 5 females) homogeneous for body weight: 'concentrate group' – 'C' –

fed concentrate usually utilised in the farm (2.6 kg/pig/d) and 'pasture group' – 'P' – fed on grass pasture (6.2 and 7.3 q of live weight/ha at the beginning and at the end of the trial, respectively) with an integration based on the same concentrate (1.4 kg/pig/d). Concentrate consisted in barley (40%), corn (20%), *Vicia faba minor* (15%), wheat bran (15%) and wheat bran shorts (10%). Pasture sampling (Numata, 1982) resulted in lucerne, oat, lolium and clover. Details on the experimental design and results on performance in vita and meat quality are reported in previous paper (Giuliotti *et al.*, 2007). Pigs were slaughtered at 150 kg of live weight. On cured lard the following analysis were carried out: fatty acid profile (Morrison and Smith, 1964) and MDA content (Pikul *et al.*, 1983) for inner and outer layer separately. Statistic was performed with JMP (2002), considering the diet and layer effects.

Results and discussion

In the data of Table 1 the fatty acid profile is reported; values do not show any significant difference between the two groups.

Data regarding fatty acids on fresh lard (Giuliotti *et al.*, 2007) did not reveal significant differences between group fed grass pasture and control except for C18:3, higher in the latter one; the seasoning period likely did not affect fatty acid profile confirming the lack of differences between the two experimental groups. Other researches carried out on the same breed (Sirtori *et al.*, 2005), showed significant variation among groups fed different diets (chestnut, acorn and concentrate); in that case animals were reared exclusively on wood pasture.

Table 1. Effect of the diet on fatty acids (FA) composition (% of the total FA) and on and malonaldehyde (MDA) content.

Fatty acid	Control		Pasture		*P*
	Mean	Std Err	Mean	Std Err	
C12	0.04	0.004	0.05	0.005	0.059
C14	1.14	0.034	1.17	0.036	0.624
C16	23.25	0.383	23.92	0.400	0.241
C16:1	2.31	0.091	2.34	0.095	0.843
C17	0.22	0.021	0.17	0.022	0.164
C17:1	0.20	0.020	0.16	0.021	0.123
C18	10.76	0.497	11.68	0.519	0.218
C18:1	48.59	0.498	47.37	0.520	0.104
C18:2	11.48	0.459	11.07	0.480	0.536
C18:3	0.59	0.065	0.64	0.068	0.593
C20	0.11	0.012	0.14	0.012	0.092
C20:1	0.82	0.029	0.82	0.031	0.963
C20:2	0.37	0.024	0.35	0.026	0.759
C20:3	0.04	0.014	0.06	0.014	0.271
C20:4	0.06	0.009	0.06	0.009	0.866
SFA	35.53	0.871	37.13	0.909	0.217
MUFA	51.93	0.574	50.69	0.600	0.148
PUFA n-6	11.95	0.491	11.55	0.512	0.574
PUFA n-3	0.59	0.065	0.64	0.068	0.593
PUFA/SFA	0.36	0.022	0.33	0.023	0.437
n6/n3	22.98	2.230	20.04	2.329	0.373
MDA (mg/kg)	1.46	0.197	1.53	0.205	0.798

The level of MDA as indicator of lipid oxidation is similar in both groups; the values are higher than those reported by Sirtori in studies carried on Cinta Senese (Sirtori *et al.*, 2005), however they do not seem to modify the normal process of seasoning.

Regarding differences on the fatty acids composition between the two layers of subcutaneous fat (Table 2) it is possible to notice that the outer layer shows a lower percentage of C16, C18, C20 and total SFA ($P<0.001$) and higher levels ($P\leq0.01$) of C16:1, C18:1, C18:2, C20:2 MUFA, PUFA n-6 and n6/n3, according to Sirtori *et al.* (2005).

Conclusion

Fatty acids content of cured lard was not influenced by the diet: probably the amount of grass ingested in replacement of the concentrate was not able to modify the fatty acid profile. Hence Cinta Senese pigs can profitably utilize pasture even in fattening period.

Table 2. Fatty acids composition (% of the total fatty acids) of inner and outer layer.

Fatty acid	Inner		Outer		*P*
	Mean	Std Err	Mean	Std Err	
C12	0.04	0.005	0.05	0.005	0.131
C14	1.16	0.036	1.15	0.034	0.761
C16	24.47	0.313	22.75	0.299	0.001
C16:1	2.12	0.071	2.52	0.068	0.001
C17	0.18	0.023	0.21	0.022	0.284
C17:1	0.20	0.021	0.17	0.020	0.257
C18	12.55	0.353	9.97	0.338	0.001
C18:1	46.83	0.427	49.08	0.408	0.001
C18:2	10.47	0.416	12.04	0.399	0.013
C18:3	0.63	0.068	0.60	0.065	0.750
C20	0.15	0.012	0.11	0.011	0.023
C20:1	0.81	0.030	0.83	0.029	0.617
C20:2	0.31	0.020	0.41	0.019	0.001
C20:3	0.05	0.015	0.05	0.014	0.799
C20:4	0.05	0.009	0.07	0.008	0.105
SFA	38.54	0.656	34.24	0.628	0.000
MUFA	49.96	0.475	52.60	0.455	0.001
PUFA n-6	10.87	0.442	12.57	0.423	0.011
PUFA n-3	0.63	0.068	0.60	0.065	0.750
PUFA/SFA	0.30	0.019	0.39	0.018	0.003
n6/n3	18.33	2.163	24.56	2.071	0.050
MDA (mg/kg)	1.73	0.192	1.27	0.184	0.098

Acknowledgements

Work supported by ARSIA (Tuscany Region).

References

Daza A., A. Mateos, A.I. Rey, I. Ovejero and C.J. Lopez-Bote, 2007. Effects of duration of feeding on free-range conditions on production results and carcass and fat quality in Iberian pigs. Meat Sci., 76: 41-416.

Edwards S.A., 2005. Product quality attributes associated with outdoor pig production. Livest. Prod. Sci. 94: 5-14.

JMP, 2002. J.M.P. User's Guide ver. 5.0, S.A.S. Institute Inc., Ed. Cary (NC), USA.

Giuliotti L., J. Goracci, M.N. Benvenuti, A. Acciaioli and G. Campodoni, 2007. Effect of pasture on meat and fat quality in Cinta Senese pigs. Proceeding of the 6th Intern. Symposium on the Mediterranean Pig. Oct. 11-13, 168-171 Capo d'Orlando (ME), Italy.

Morrison W.R. and L.M. Smith, 1964. Preparation of fatty acid methyl esters and dimethylacetals from lipids with boron fluoride methanol. J. Lipid Res. 5: 600-608.

Numata M., 1982. A methodology for the study of weed vegetation. In: W. Holzner and M. Numata (eds.). 'Biology and Ecology of Weeds', Junk Publ., The Hague, The Netherlands, 21-34.

Pikul J., D.E. Leszczynski and F.A. Kummerow, 1983. Elimination of sample autoxidation by butylated hydroxytoluene additions before thiobarbituric acid assay for malonaldehyde in fat from chicken meat. J. Agric. Food Chem. 31: 1338-1342.

Rey A.I., A. Daza, C. Lopez-Carrasco and C.J. Lopez-Bote, 2006. Feeding Iberian pigs with acorn and grass in either free-range or confinement affects the carcass characteristics and fatty acids and tocopherol accumulation in Longissimus dorsi muscle and backfat. Meat Sci. 73:66-74.

Sirtori F., 2005. Caratteristiche dei prodotti di suini Cinta Senese ingrassati in bosco. PhD Thesis, University of Florence.

Sirtori F., C. Pugliese, G. Meo Zillo, G. Campodoni, L. Pianaccioli and A. Bonelli, 2005. Characteristics of cured lard of Cinta Senese pig fed acorn and chestnut. Ital. J. Anim. Sci., 4 (suppl. 2): 467-469.

Soto E., L. Hoz, J.A. Ordonez, E. Hierro, B. Herranz, C.J. Lopez-Bote and M.I. Cambero, 2008. Impact of feeding ad rearing systems of Iberian pigs on volatile profile and sensory characteristics of dry-cured loin. Meat Sci. 79:666-676.

Stern S. and N. Andresen, 2003. Performance, site preferences, foraging and excretory behaviour in relation to feed allowance of growing pigs on pasture. Livest. Prod. Sci. 79:257-265.

Wood J.D., M.B. Enser, A.V. Fisher, G.R. Nute, P.R. Sheard, R.I. Richardson, S.I. Hughes and F.M. Whittington, 2008. Fat deposition, fatty acid composition and meat quality: a review. Meat Sci., 78: 343-358.

Effects of winter shearing on reproductive and lactation performances of the Latxa dairy ewe

A. Garcia-Rodriguez, I. Beltran de Heredia, J. Arranz, L.M. Oregui and R. Ruiz
NEIKER, A.B. - Granja Modelo de Arkaute, Apartado 46, 01080 Vitoria-Gasteiz, Spain

Abstract

A study was carried out to assess the effect of winter shearing on reproductive and lactation performance in the Latxa breed over a two year period. In each year two trials were performed: Trial 1 during late pregnancy and trial 2 during lactation. In trial 1 dry matter intake (DMI), body weight (BW), body condition score (BCS), litter size, and lambs′ birth weight (LBW) were recorded. In trial 2, BW, BCS, DMI, milk yield and milk quality were recorded. During late pregnancy shearing significantly increased grass silage and total DMI. Upon lactation, a tendency was found for increased grass silage and total DMI in shorn ewes. A tendency was found for increased LBW in shorn ewes, especially of twin borns. Shearing did not significantly affect lactation performance except for fat yields concomitant with a significant higher BCS mobilisation. In conclusion, shearing increased DMI with little effect upon productive performance.

Keywords: winter shearing, dairy ewe, milk production, milk quality, intake

Introduction

In the Latxa system, the lambing season mainly takes place during winter and ewes are milked twice a day after the suckling lambs are weaned. In the Basque country this milk production system is very important from a social point of view, as most of the milk produced is processed into high quality cheese.

Under southern Europe conditions the removal of the fleece is usually timed to coincide with favourable climatic conditions which occur in early summer, but recently winter shearing is also becoming a common practice. However, information concerning the influence of winter shearing on milk production and quality of dairy sheep is lacking. Therefore, the purpose of this study was to determine the effect of winter shearing on gestation, reproductive and lactation performance over a two year period.

Material and methods

The study was carried out during two different years (2003-04 and 2004-05) and in each of these years two different trials were performed: Trial 1 was performed during late pregnancy and trial 2 during lactation. Fifty six pregnant Latxa ewes in 2003-04 and 64 in 2004-05 were selected on day 109 of pregnancy in 2003-04 and 100 in 2004-05 from a larger flock of pregnant ewes. Animals were blocked into two homogeneous groups on the basis of body weight (BW) and previous years' milk potential. In both years ewes were shorn at the beginning of December leaving 15 to 30 mm of stubble. Ewes were managed as one mob under commercial farming conditions except for the periods in which they were under experimental conditions.

Experimental design

Trial 1

In both years, the unshorn and control lot (C) and the shorn group (S) were housed separately. Within each treatment, ewes were blocked into 4 subgroups on the basis of BW. The length of this trial was

13 d, consisting of 5 d for adaptation and 8 d for dry matter intake (DMI), BW and body condition score (BCS) monitoring. During this period each ewe was offered grass silage *ad libitum* and was supplemented with 340 g DM/d of a mixture of wheat and barley grain (1:1) in 2003-04 and with 340 g DM/d of a commercial concentrate in 2004-2005.

Trial 2

After the lactation peak, C and S groups were housed separately again and within each treatment group ewes were blocked into 3 subgroups on the basis of milk yield (MY). The length of this trial was 12 d, consisting of 5 d for adaptation and 7 d for BW, BCS, DMI, MY and milk quality determinations. During lactation ewes were milked twice a day, and during each milking half of the daily supplement allowance was offered, which consisted of 612 g DM/d of a commercial concentrate. In addition, each ewe was offered with 820 g DM/d of alfalfa hay and grass silage *ad libitum*.

Samplings, calculations and statistical analysis

After lambing, litter size and lamb birth weight (LBW) were recorded as soon as possible. Body weight and BCS were recorded at the beginning and at the end of trial 1 and 2. Concentrate (CI), alfalfa hay (AHI) and grass silage intake (GSI) were measured as the difference between the quantities offered and rejected per lot. Individual milk yield was recorded daily and its concentration of fat, protein and lactose were determined once from one successive morning and afternoon milk sample. Standard milk production (SMY) was calculated as described by Bocquier *et al.* (1993). Protein, fat and lactose concentrations in milk were measured with an infra-red analyser.

Data were analyzed separately for trial 1 and trial 2. In trial 1 data were analyzed as a factorial design (2 years × 2 treatments × 4 subgroups) using the PROC GLM procedure. Lamb birth weight was analyzed as a factorial design (2 years × 2 treatments) using the PROC GLM procedure. In trial 2, except for milk yield, data were analyzed as a factorial design (2 years × 2 treatments × 3 subgroups) using the PROC GLM procedure. Milk yield was analyzed (2 years × 2 treatments × 3 subgroups) using PROC MIXED for repeated measures. Multiple means comparison was analyzed using the Tukey's test and differences were declared at $P<0.05$.

Results

Shearing increased total DMI in the pregnancy trial, but these effects were found to depend on the year of experience: whereas in 2003-04 DMI was 9.5% higher in S sheep (1,173 vs. 1,284 g DM, $P<0.001$), in 2004-05 shearing caused a 20.6% increase (1,132 vs. 1,365, $P<0.001$).

Shearing caused a 5.8% increase in LBW (4,318 vs. 4,079, $P=0.07$), but whereas in 2003-04 no differences were found between S and C ewes (4,278 vs. 4,327 g, $P>0.10$), a 13.7% increase was found in 2004-05 (4,357 vs. 3,831 g, $P=0.09$). In addition this shearing effect on LBW was found to significantly depend on the litter size: no differences were found in LBW for single lambings (5,002 vs. 4,660 g, $P>0.05$), but a 14.4% increase was found for twin-born lambings in the S ewes (4,420 vs. 3,865 g, $P<0.05$).

Winter shearing had a consistent effect on intake during the lactation trial in both years. Shearing did not have a significant effect on AHI or CI but showed a 13.5% and 8.3% increase for GSI and DMI, respectively. Shearing did not significantly affect MY, milk protein or fat content, protein yields, SMY, but it showed a tendency for higher fat yields (Table 1). Similarly, shearing did not significantly affect body weight gain but differences were found in the BCS evolution. While C ewes had a positive BCS evolution S ewes lost BCS. However, while in 2003-04 shearing did not modify significantly BCS evolution, in 2004-05 it resulted in a BCS loss.

Table 1. Effects of shearing on the lactation performance and body reserves evolution.

	Treatment			Effects		
	Shorn (S)	Control (C)	Sem	Year	Treatment[1]	Year × treatment
Milk yield (ml d^{-1})	1,540	1,547		**	NS	NS
SMY (ml d^{-1})[2]	1,447	1,386	175	NS	NS	NS
Milk fat content						
g kg^{-1}	6.87	6.68	0.58	NS	NS	NS
g d^{-1}	107.50	100.80	15.00	NS	τ	NS
Milk protein content						
g kg^{-1}	4.96	5.05	0.28	NS	NS	**
g d^{-1}	77.50	77.30	10.60	NS	NS	NS
BWG (kg)[3]	0.56	1.12	1.98	***	NS	NS
Body condition score	-0.079	0.102	0.291	*	*	**

[1] τ: $P<0.1$; *: $P<0.05$; **: $P<0.01$; ***: $P<0.001$; NS: non significant.

[2] SMY: standard milk yield.

[3] BWG: body weight gain.

Discussion

Increases in LBW associated to winter shearing have generally been reported in either singleton (Kenyon *et al.,* 2002b) or twin-born lambs (Kenyon *et al.,* 2002a), but rarely in both (Kenyon *et al.,* 2002a). This result is of interest because multiple-born lambs are smaller than their singleton counterparts and are particularly susceptible to starvation-exposure. Thus, any technique that can increase the LBW has the potential to substantially increase the chances of survival of multiple-born lambs. In some studies with shorn wethers it has been observed that heat production increased linearly with a decrease in environmental temperature below 23 °C irrespective of the feeding level (Graham *et al.,* 1959). Outdoors average temperatures in those periods ranged from 3.8 to 5.0 °C, The additional energy necessary to compensate heat production should have been obtained from a higher intake of digestible energy by a rise in the DMI in the lactation trial and/or by reduction in the caloric stress in the gestation trial (Vipond *et al.,* 1987). In the present experiment whereas the observed 15% DMI increase in trial 1 in the shorn ewes is within the 12-43% range reported in the literature (Vipond *et al.,* 1987) the 8.3% higher DMI found in the shorn ewes in the lactation trial is below this range. As a consequence of such increase in DMI, it seems likely that the increased LBW observed after shearing was the result of metabolic adaptations to a cold environment.

The lower effect of shearing on DMI in the lactation trial could be related to two issues. First, trial 2 took place around 2.5 months after shearing, and by that time fleece growth is supposed to provide some protection against cold stress, reducing its effect and diminishing the demands for body temperature control during the lactation trial. Then, it is the fact that in the late pregnancy trial the novelty produced a strong hormonal and metabolic reaction in animals, which would not be such during lactation.

These metabolic adaptations to winter shearing could result in endocrine changes in ewes during lactation likely to increase the partition of nutrients towards milk production (Symonds *et al.,* 1990). It was apparent that the shorn ewes were deriving significantly more energy from body tissues than unshorn controls as assessed by a significant higher BCS mobilization which could explain the observed higher milk fat yields. Although a significant higher DMI intake was found in shorn ewes in both trials, MY and SMY were unchanged probably because a prolonged exposure of sheep to a cold environment resulted in a depression of whole tract apparent DM and GE digestibility in association with an increased passage of digesta through the reticulo-rumen (Kennedy, 1985).

The Latxa dairy production system implies other values than only the production of low cost and high quality food. Winter shearing would improve management practices leading to better care/assistance at lambing, easier access of sheep to milking parlours, cleaner conditions and better hygienic standards for milking. Therefore, a more attractive professional activity for the farmer could be achieved.

In conclusion, winter shearing in the dairy Latxa sheep production system improved the pregnancy and reproductive performance in terms of lamb birth weight but it did not improve total milk production, standard milk yield or milk quality parameters. The observed improvement in reproductive parameters could lead to maintain the Latxa dairy production system with the concomitant environmental and socio-economical benefits.

References

Bocquier, F., T. Barillet, P. Guillouet and M. Jacquin, 1993. Prévision de l'énergie du lait de brebis à partir de différents résultats d'analyses: Proposition de lait standard pour les brebis laitières. Ann. Zootech. 42, 57-66.

Graham, N.M., F.W. Wainman, K.L. Blaxter and D.G. Armstrong, 1959. Environmental temperature, energy metabolism and heat regulation in sheep. I. Energy metabolims in closely clipped sheep. J. Agric. Sci. (Camb) 52, 13-24.

Kennedy, P.M., 1985. Influences of cold exposure on digestion of organic matter, rates of passage of digesta in the gastrointestinal tract, and feeding and rumination behaviour in sheep given four forage diets in the chopped, or ground and pelleted form. Br. J. Nutr. 53, 159-173.

Kenyon, P.R., S.T. Morris and S.N. McCutcheon, 2002a. Does an increase in lamb birth weight through mid-pregnancy shearing necessarily mean an increase in lamb survival to weaning? Proc. NZ Soc. Anim. Prod 62, 53-56.

Kenyon, P.R., S.T. Morris, D.K. Revell and S.N. McCutcheon, 2002b. Maternal constraint and the birthweight response to mid-pregnancy shearing. Aus. J. Agric. Res. 53, 511-517.

Symonds, M.E., M.J. Bryant and M.A. Lomax, 1990. Metabolic adaptation during lactation in winter-shorn sheep. J. Agric. Sci. (Camb) 114, 201-205.

Vipond, J.E., M.E. King and D.M. Inglis, 1987. The effect of winter shearing of housed pregnant ewes on food intake and animal performance. Anim. Prod. 45, 211-221.

Effect of concentrate quantity and administration pattern on milk parameters and grazing time in a rationed dairy sheep grazing system

A. García-Rodriguez, R. Ruiz, L.M. Oregui and N. Mandaluniz
NEIKER-Tecnalia, P.O. Box 46, 01080 Vitoria-Gasteiz, Spain

Abstract

Dairy sheep production systems in the Basque country are based on pasture utilisation in partial time grazing during spring. The aim of this study was to evaluate the effect of concentrate quantity and its administration pattern on standardized dairy milk yield (DMYs) and quality (% fat and crude protein), and on grazing time (GT) and live weight (LW). The experiment was conducted over 4 weeks. Sheep were blocked and randomly assigned to 4 different treatments: (A) 500 g, offered half at each milking time, (B) 750 g, offered half at each milking time, (C) 1000 g half at each milking time and (D) 750 g, offered 250 g at morning milking and 500 g at evening milking. Ewes grazed during 4 hours continuously. Moreover, each ewe received 275 g DM of lucerne hay after evening milking. Data were analyzed as a factorial design. As for concentrate quantity, group B and C showed significantly higher DMYs and quality, while group A and B showed significantly longer GT and lower LW. As for concentrate administration pattern, group B and D showed similar DMYs, milk quality and LW, but B had significantly higher grazing time. In conclusion, in this rationed grazing system no productive advantage is achieved when offering more than 750 g/d of concentrate or by altering the concentrate administration pattern.

Keywords: dairy sheep, rationed grazing system, concentrate quantity, concentrate administration pattern

Introduction

Dairy sheep production systems in the Basque Country are based on pasture utilisation during an important part of the year. Herds range by continuous grazing on high pastures during summer and in partial time grazing during spring, where ewes spend a limited number of hours outdoors (Oregui *et al.,* 1997).

During this valley period forage resources are managed by shepherds who match herbage, forage and supplement availability with production requirements. This is achieved thorough a rationed grazing system, where ewes spend 3 to 7 hours grazing (García-Rodriguez *et al.,* 2005). As there is a lack of knowledge concerning the substitution or complementation of grass with concentrate under these conditions, the aim of this study was to evaluate the effect of concentrate quantity and its administration pattern on sheep production parameters (milk yield, milk quality and live weight) and grazing behaviour under this system with restricted access to pasture.

Material and methods

The experiment was conducted over 4 weeks during the spring of 2006 with 48 multiparous Latxa dairy ewes. Sheep were blocked into 4 homogeneous groups of 12 on the basis of the lactation number, days in milk, milk yield and body weight. Initial mean values were 2nd or 3rd lactation, 40-60 day of lactation, milk yield of 1.40-1.50 l day^{-1} and 57-62 kg of live weight.

Each block was randomly assigned to 4 different treatments referred to concentrate quantities (CQ) and administration patterns: (A) 500 g d^{-1} offered half at each milking time; (B) 750 g d^{-1} offered half at each milking time, and (C) 1000 g d^{-1} half at each milking time and (D) 750 g d^{-1} offered 250 at morning milking and 500 at evening milking. Moreover, after the evening milking each ewe was offered indoors 275 g DM of (alfalfa) lucerne hay. Referring to grazing, sward was kept between 6-8 cm, and each group accessed pasture after morning milking, during 4 consecutive hours.

Fat and protein concentration and milk yield were determined from one successive morning and afternoon milk sample taken each week. Milk samples were analysed for crude fat (CF) and protein (CP) contents with an infra-red analyser. Standardised dairy milk yield (DMYs) was calculated as described by Bocquier *et al.* (1993). Animal behaviour was recorded once a week every week. Observations were recorded continuously during the time animals had access to pasture. To measure grazing activity the number of ewes engaged in grazing was recorded every 5 minutes.

Data were analysed using the GLM procedures with concentrate level (CL), week (W) and their interaction (CQxW) as fixed effects. Two independent analysis were made, first considering CQ (comparing group A, B and C) and the second considering the administration pattern (AP) of the concentrate (group B vs. D).

Results and discussion

Average DMYs, fat and protein content were 1.10±0.28 l d^{-1}, 6.33%±0.81 and 4.97%±0.72. The week of lactation and quantity of concentrate significantly affected the lactation performance but the interaction of treatments was not significant. Whereas, the DMYs was significantly reduced with week of experience, non significant differences were found in terms of milk fat or protein contents. On the other hand, the concentrate quantity significantly increased DMYs with increasing concentrate quantities resulting in increases of DMYs (Figure 1). In terms of milk quality, increasing the quantity of offered concentrate resulted in a significant increase of the milk crude protein content. However, whereas non significant differences were found between groups B and C in terms of milk crude protein content, a significant lower milk crude fat content was found for the C group (Figure 2). As the quantity of concentrate offered increases there is more protein and energy available for microbial protein synthesis which could lead to the observed increases in milk crude protein contents (Wright *et al.*, 1998). This increase of energy intake obtained with the increase of the quantity of offered concentrate would also result in increases of the milk crude fat content, but above 750 g concentrate per day the forage intake could be limited as observed by the decreased time spent grazing (Figure 3). The lower forage to concentrate ratio of the diet could negatively affect the acetate synthesis in the rumen which is one of the most important milk fat synthesis precursor Van Soest (1994).

Average grazing time per hour spent on pasture was 48.9 min h^{-1} which is similar to that observed in previous experiences where ewes were allowed to pasture 4 h (Perojo *et al.*, 2005). Time spent

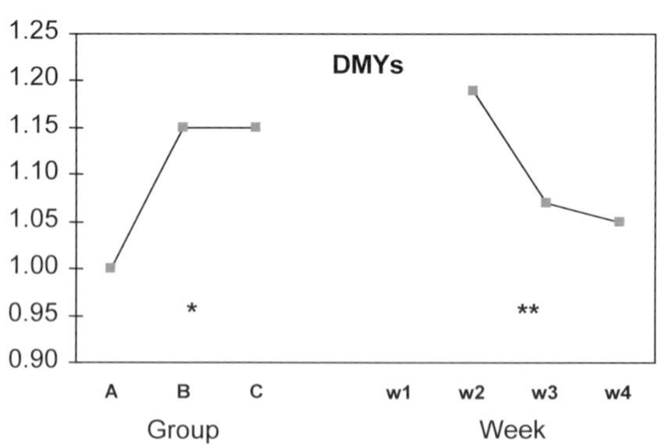

Figure 1. Effect of concentrate quantities (CQ) and week on dairy milk yields (DMYs) (l d^{-1}). *$P<0.05$; **$P<0.01$.

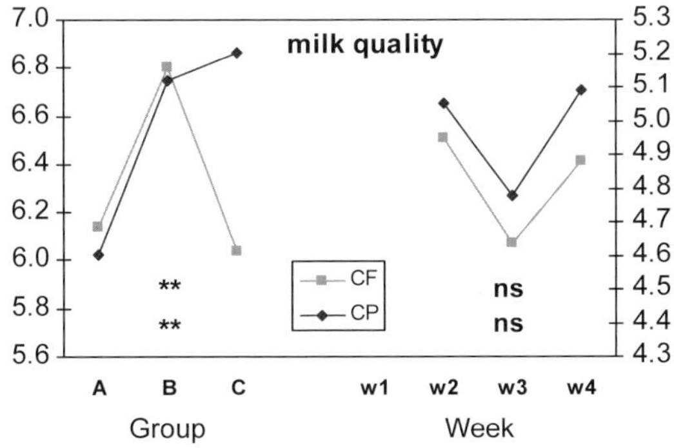

Figure 2. Effect of concentrate quantities (CQ) and week on milk quality (crude fat content (CF) and crude protein content (CP) in percentages). **P<0.01.

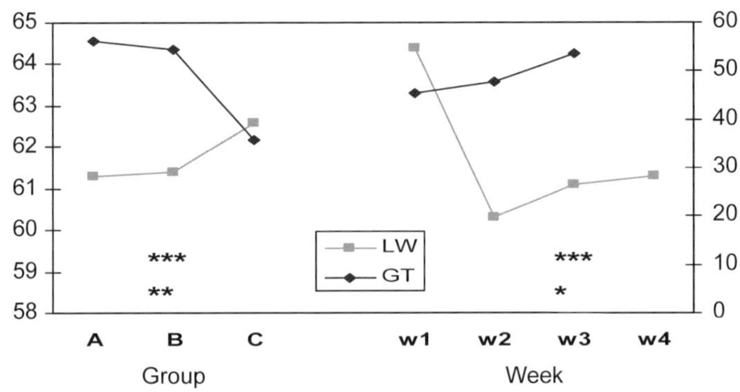

Figure 3. Effect of concentrate quantities (CQ) and week on live weight (LW) (kg) and grazing time (GT) (min h^{-1}). **P<0.01; ***P<0.001.

grazing per hour significantly increased with the week of experience so it could be hypothesised that in some way animals learn that they can only stay a limited number of hours on pasture and modify their grazing behaviour accordingly (Figure 3). Another possible explanation could be that the increase in the fibre content of pasture as the experience evolved resulted in a longer time spent grazing. The quantity of offered concentrate significantly affected the grazing behaviour by decreasing the time spent grazing. These results are in agreement with Ligios *et al.* (2002), who found that more than 750 g d^{-1} of concentrate in the diet caused a significant reduction of the grazing time probably due to a substitution of forage intake with concentrate. Although significant differences were observed in grazing time per hour between groups B and C, the latter group showed a higher LW probably because this group had a higher concentrate intake, and therefore a higher energy availability.

As for the concentrate distribution pattern is concerned, there were not significant differences between groups B and D in any of the measured parameters (Table 1). It was hypothesised that supplementing ewes with less concentrate before going out to pasture could increase pasture utilization by a higher intake due to a lower substitution rate. However, in the present experiment ewes were only allowed 4 hours on pasture which could have not been enough time to observe differences in grazing behaviour and lactation performance due to AP.

Table 1. Effects of concentrate administration pattern (AP) and, week (W) on lactation performance, grazing behaviour and live weight.

	Administration pattern		Effects[1]		
	B	D	AP	W	CQ×W
Lactation performance					
Standard milk yield (l day^{-1})	1.15	1.17	NS	NS	NS
Milk fat content (%)	6.80	6.34	NS	NS	NS
Milk protein content (%)	5.12	5.29	NS	NS	NS
Grazing time (min h^{-1})	54.6	48.0	NS	NS	NS
Live weight (kg)	60.5	60.5	NS	**	NS

[1] ***P*<0.01; NS: non significant.

With these experiments pasture is considered as part of the diet, and not only as a complement. This implies not only a reduction in feed costs of the system but also a more sustainable management of the flocks with the concomitant benefits from the environmental and animal health point of view.

Conclusions

In this grazing system increasing concentrate supplementation beyond 750 g per day did not improve lactation performance which could be due to its negative effect on grazing behaviour. In addition, a different concentrate administration pattern did not imply a productive performance improvement.

References

Garcia-Rodriguez, A., N. Mandaluniz and L.M. Oregui, 2005. Effect of time spent on pasture on milk production and quality in Latxa dairy sheep. XX Int. Grassland Congress: offered papers, pp.168.

Ligios, S., M. Sitzia, N. Fois, M. Decandia, G. Molle, P.P. Roggero and S. Casu, 2002. Effect de la disponibilite en herbe et de la structure du couvert herbace sur l'ingestion et la production de brebis au paturage. Opt. mediterraneennes, 42: 73-84.

Oregui L.M., J. Garro, M.S. Vicente and M.V. Bravo, 1997. Estudio del sistema de alimentación en las razas ovinas Latxa y Carranzana: utilización de los pastos comunales y suplementación en pesebre. Información técnica económica agraria, 93: 173-182.

Perojo, A., Garcia-Rodriguez, A., Arranz, J. and Oregui, L.M., 2005. Effects of time spent on pastures on milk yield, body reserves, herbage intake and grazing behaviour. Options Méditerranéennes A-67, 275-279.

Van Soest, P.J., 1994. Nutritional Ecology of the Ruminants. Cornell University Press, Ithaca, NY.

Wright, T.C., Moscardini, S., Luimes, P.H., Susmel, P. and McBride, B.W., 1998. Effects of Rumen-Undegradable Protein and Feed Intake on Nitrogen Balance and Milk Protein Production in Dairy Cows. J. Dairy Sci. 81, 784-793.

Relation between the distribution of time spent on pasture and protein content of the concentrate on milk yield and grazing behaviour

A. Garcia-Rodriguez, I. Goiri, R. Ruiz, N. Mandaluniz and L.M. Oregui
NEIKER, Granja Modelo de Arkaute, Apdo. 46, 01080 Vitoria-Gasteiz, Spain

Abstract

The objective of this study was to assess the influence of the crude protein (CP) content (130 vs. 190 g CP kg^{-1}) of the concentrate on milk yield and body reserves in a production system where ewes can spend a limited time on pasture. A total of 40 multiparous Latxa ewes were blocked into 4 groups on the basis of milk yield, number of days in lactation, body weight and condition score. Ewes either grazed during 4 hours continuously after the morning milking or grazed 2 hours continuously after the morning and evening milking. Ewes were assigned to one of the following experimental treatments: 2+2 hours high protein; 2+2 hours low protein; 4 hours high protein and 4 hours low protein. Ewes spending 2+2 hours grazed longer on a daily or per hour basis. Nevertheless, neither concentrate CP content nor time spent on pasture significantly affected milk yield or milk composition, but ewes with the high CP concentrate lost more weight. In conclusion, under this rationed grazing system, the inclusion of high CP supplementary concentrates resulted in very poor differences in terms of milk yield or milk composition but distribution of time spent grazing could affect the grazing behaviour.

Keywords: dairy sheep, rationed grazing, crude protein, milk yield, grazing behaviour

Introduction

In the Basque country, the Latxa dairy production system is very important from a social point of view as most of the milk produced is processed into high quality cheese. Moreover, this production system implies other values than only the production of low cost and high quality food. These are related to nature, landscape or social functions and are relatively invisible but latently present in the society. Although this production system is based on a high pasture utilisation, natural resources available to livestock are scarce. Shepherds manage these resources trying to match herbage, forage and supplement availability with production necessities (requirements). This is achieved by a rationed grazing system where ewes spend between 3 and 7 hours outdoors (Oregui *et al.*, 1997).

Well managed pastures may contain important crude protein (CP) levels which can be degraded rapidly and to a great extent. For lactating ruminants this may result in an effective rumen degradable protein (ERDP):fermentable metabolizable energy (FME) ratio exceeding microbial requirements and resulting in poor nitrogen use efficiencies. One method of improving animal performance and the efficiency of nitrogen utilisation to reach a more integrated production system is the use of supplements that optimise the rumen ERDP:FME ratio. Therefore, the objective of the current experiment was to investigate the response of lactating ewes under a rationed grazing system during the spring to variations in the CP content in a supplementary concentrate.

Material and methods

The experiment was conducted over 6 weeks during the spring of 2004. A total of 40 multiparous Latxa dairy ewes were blocked into 4 groups of 10 on the basis of milk yield, days in milk (DIM) and body weight (BW). Initial mean values for milk yield, BW and DIM were 1,530±530 ml, 57.3±9.6 kg and 45±5 d. In each group 6 ewes were selected and identified. Groups were randomly assigned to one of the following experimental treatments: (1) supplement containing 150 g CP kg^{-1} (LP) or (2) supplement containing 210 g CP kg^{-1} (Hp). Half of the daily allowance (526 g DM) of the supplements, which was formulated to be isoenergetic, was offered during each milking. Ewes were on pasture either 4 hours continuously (4 h) resulting in (4hHP) and (4hLP), or 4 hours divided in 2 hours after the morning milking and 2 hours after the evening milking, resulting in (2+2hHP) and (2+2hLP). When not on pasture groups were housed separately in a free stall barn. In the evening ewes were fed 255 g DM of alfalfa hay. Sward height was estimated twice a week using a swardstick and maintained between 6 and 8 cm by removing ewes from pasture. Anyway, none of the identified ewes were removed. Samples of herbage were collected by hand-plucking from each plot every two weeks. Milk yields were recorded three times a week. The concentration of fat and protein were determined from one successive morning and afternoon milk sample taken each week with an infra-red analyser. Standard milk yield (SMY) was calculated as described by Bocquier *et al.* (1993). BW was measured weekly.

Animal behaviour was recorded once a week every week. Observations were recorded continuously during the time animals had access to pasture. To measure grazing activity the number of ewes engaged in grazing was recorded every 5 minutes.

The effect of time spent on pasture and CP content of the supplementary concentrate on milk yield, milk composition and BW were subjected to an analysis for repeated measures taking the data used to make the experimental groups as covariates. Total time spent grazing per day and time spent grazing per hour were subjected to analysis of variance (GLM).

Alfalfa hay, herbage and concentrates were dried at 60 °C for 48 h to determine dry matter content and ground to pass a 1 mm screen. Ash content was measured by ignition of samples in a muffle furnace at 550 °C overnight. Crude protein was determined by Kjeldahl procedure.

Results

Animal behaviour

The time spent grazing was longer in the groups that had 2+2 h-access to pasture (223 vs. 207 min/d, $P<0.01$).In addition the 2+2 h groups used the time spent on pasture in a more efficient way spending 56 min/h grazing in comparison with the 52 min/h used by the 4 h groups ($P<0.05$).

In terms of grazing time per hour spent on pasture (Figure 1) both groups (4 h and 2+2 h) spent most of the first three hours grazing. However in the last hour the 2+2 h group maintained higher grazing time compared with the 4 h group ($P<0.01$).

On the other hand, the protein content of the concentrate did not have any influence on grazing animal behaviour either in terms of time spent grazing per hour (53.75 vs. 53.72 min/h, $P>0.05$) or in terms of total grazing time per day (215 vs. 214 min/d, $P>0.05$).

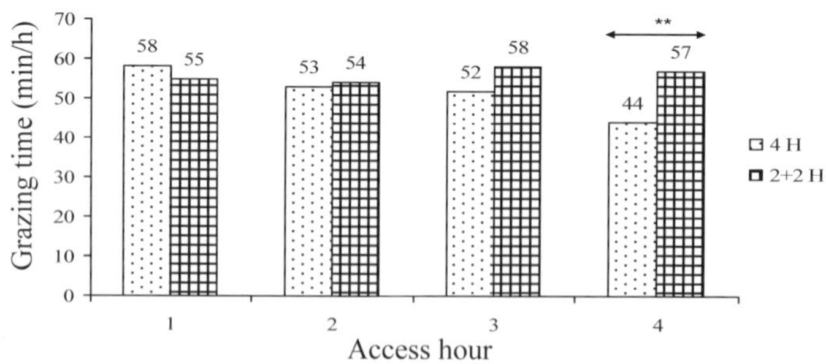

Figure 1. Grazing behaviour of ewes with different distribution of time spent on pasture. ** $P<0.01$.

Production data

Milk yields and overall performance for each group are presented in Table 1. The distribution of time spent on pasture (4 h vs. 2+2 h) did not affect milk yield, SMY, or milk quality measured in terms of protein and fat content and protein and fat yield.

Similarly, the protein content of the concentrate (Hp vs. LP) affected neither milk yield, SMY nor milk quality traits, but increased body weight loss (Table 1).

Discussion

Management of time spent on pasture affected grazing behaviour by increasing the time spent grazing per hour when ewes had 2+2 h access to pasture, especially in the last hour with the concomitant increase in total time spent grazing per day. Therefore, it could be hypothesised that in some way animals learn that they can only stay a limited number of hours on pasture and modify their grazing behaviour accordingly. This adaptation to grazing management has also been observed in previous experiences where ewes allowed to pasture 4 h had a higher time spent grazing per hour than those allowed to pasture 7 h (Garcia-Rodriguez and Oregui, 2003; Perojo *et al.*, 2005). However, this increase in total grazing time with the discontinuous grazing management (2+2 h) did no result in changes in the milk production or quality traits probably because the increase in total grazing time (7.7%) was not enough to affect in a significant way grass dry matter intake (Perojo *et al.*, 2005).

In agreement with previous results observed by Garcia-Rodriguez and Oregui, (2004) concentrate CP content did not affect milk yield or milk quality traits in ewes allowed to spend 4 h on pasture. In the conditions of the present trial although the ewes receiving the HP concentrate lost more weight than LP ewes, in general both groups lost weight and it could be hypothesised that both groups were in a negative energy balance. In this condition in which energy was likely to be the limiting factor for microbial protein synthesis, an increase in the diet CP content would not lead to any improvement in microbial synthesis and consequently in milk production (Wright *et al.*, 1998), and could even lead to a greater negative energy balance status (Nocek and Russell, 1988). Moreover, this CP excess in

Table 1. Milk yield and overall performance of ewes receiving concentrates with 150 or 210 g CP kg⁻¹, spending either 4 h or 2+ 2 h on pasture.

	Time		Protein			Significance[1]		
	4 h	2+2 h	LP[4]	HP[4]	Se[2]	Time	Protein	Interaction
Milk yield (ml d^{-1})	1,182	1,152	1,140	1,194	53.2	ns	ns	ns
SMY (ml d^{-1})[3]	1,080	1,070	1,036	1,113	46.6	ns	ns	ns
Fat								
g kg^{-1}	66.7	68.9	66.1	69.5	1.53	ns	ns	ns
g d^{-1}	77.2	78.8	74.5	81.5	3.59	ns	ns	ns
Protein								
g kg^{-1}	53.7	51.9	52.6	53.1	1.22	ns	ns	ns
g d^{-1}	62.5	59.1	59.1	62.5	2.51	ns	ns	ns
Body weight (kg)	57.6	57.5	58.4	56.6	0.55	ns	*	ns

[1] Ns: non significant; * $P<0.05$.

[2] Se: standard error of mean.

[3] SMY: milk production standardised to 5.0 MJ l^{-1} according to Bocquier *et al.* (1993).

[4] LP: low protein concentrate; HP: high protein concentrate.

the HP concentrate diet not only would result in a concentrate price increase but also in a higher N emission through urea and faeces to environment with the concomitant environmental and social detrimental effects.

Conclusions

Under this rationed grazing system, the inclusion of high CP supplementary concentrates result in very poor differences in terms of milk yield or milk composition but distribution of time spent grazing could affect the grazing behaviour.

References

Bocquier, F., T.Barillet, P. Guillouet and M. Jacquin, 1993. Prévision de l'énergie du lait de brebis à partir de différents résultats d'analyses: Proposition de lait standard pour les brebis laitières. Ann. Zootech. 42, 57-66.

Garcia-Rodriguez, A. and L.M. Oregui, 2003. Effects of time spent on pasture and the characteristics of the concentrate on milk yield, milk composition and body reserves in the Latxa dairy ewe. British Society of Animal Science, York, pp. 187.

Garcia-Rodriguez, A. and L.M. Oregui, 2004. Effect of time spent on pasture and protein content of the concentrate on milk yields and body reserves., In: Lúscher, A., Jeangros, B., Kessler, W., Huguenin, O., Lobsiger, M., Millar, N., Suter, D. (eds.), Land use systems in grassland dominated regions. Proceedings of the 20th General Meeting of the European Grassland Federation, vdf Hochschulverlag AG an der ETH Zurich, Luzern, Switzerland, pp. 662-664.

Nocek, J.E. and Russell, J.B., 1988. Protein and Energy as an Integrated System. Relationship of Ruminal Protein and Carbohydrate Availability to Microbial Synthesis and Milk Production. J. Dairy Sci. 71, 2070-2107.

Oregui, L.M., J. Garro, M.S. Vicente and M.V. Bravo, 1997. Estudio del sistema de alimentación en las razas ovinas Latxa y Carranzana: Utilización de los pastos comunales y suplementación en pesebre. Información Técnica Económica Agraria 93, 173-182.

Perojo, A., A. Garcia-Rodriguez, J. Arranz and L.M. Oregui, 2005. Effects of time spent on pastures on milk yield, body reserves, herbage intake and grazing behaviour. Opt. Méditerranéennes A-67, 275-279.

Wright, T.C., S. Moscardini, P.H. Luimes, P. Susmel and B.W. McBride, 1998. Effects of Rumen-Undegradable Protein and Feed Intake on Nitrogen Balance and Milk Protein Production in Dairy Cows. J. Dairy Sci. 81, 784-793.

Comparison of animal performance of domestic herbivores grazing on partially improved heath lands

L.M.M. Ferreira[1], R.Celaya[2], R. Benavides[2], U. García[2] and K. Osoro[2]
[1]CECAV- Departamento de Zootecnia, Universidade de Trás-os-Montes e Alto Douro, P.O. Box 1013, 5001-801 Vila Real, Portugal
[2]Servicio Regional de Investigación y Desarrollo Agroalimentario (SERIDA), P.O. Box 13, 33300 Villaviciosa, Asturias, Spain

Abstract

Animal performance (live-weight changes, LWC) of 5 mares, 5 cows, 35 goats and 35 ewes suckling their offspring, managed on natural shrub land with an area (24%) of improved pasture, was compared across the grazing season. Mothers' live weight (LW) was recorded in the beginning and thereafter monthly until the seasons' end. Their offspring were weighed at turn-out and at weaning. Grazing season was divided into four seasons: spring (May to July), summer (July to end of September), autumn (end of September to end of November) and winter (end of November to January). On the overall grazing season, animal species had a significant (P<0.01) effect on LWC (g/day/livestock unit-LU). While cows lost weight (-170 g/day/LU), mares, goats and ewes were able to increase their LW in 63, 76 and 64 g/day/LU, respectively. LWC differed significantly (P<0.001) between animal species in all periods. All four species had positive LWC in the spring (859, 842, 377 and 308 g/day/LU in cows, mares, ewes and goats, respectively) as a result of a higher improved pasture availability (6.8 cm). In the summer and autumn, when the sward height decreased (4.6 and 2.9 cm, respectively) cows, mares and goats lost weight (average of -189, -194 and -72 g/day/LU), especially in the summer, while ewes were able to increase their LW (129 g/day/LU). In the winter, cows and mares increased their weight loss rate (-1,878 and -604 g/day/LU) and ewes showed negative LWC (-290 g/day/LU). Contrastingly, goats were able to increase their LW (64 g/day/LU). Within the offspring, lambs showed the highest LW gains (1,083 g/day/LU), while kids the lowest ones (696 g/day/LU), but not significantly different from those observed in calves and foals. These presented similar LWC from turn-out to weaning (804 and 785 g/day/LU, respectively).

Keywords: herbivores; heath land

Introduction

Heath lands actually cover large areas in the Northwest of the Iberian Peninsula, mostly marginal and poor lands of acidic soils, constituting a poor resource for animal production due to its low nutritive value (Celaya *et al.*, 2007). Therefore, the development of sustainable livestock production systems in these areas (mothers with suckler offspring), require areas of improved pasture to meet the animals' nutritional requirements (Osoro *et al.*, 2008a). Due to different eco-physiological adaptations (mouth and dental anatomy, digestive strategy), domestic herbivores species show different abilities to exploit the available plant resources (Hofmann, 1989; Illius and Gordon, 1993). Consequently, different responses in performance will be observed, depending on the available vegetation whose quantity and nutritive value vary with season. It is important to know the performance of different livestock species when managed under the same changing conditions, to set the best management option in order to enhance the profitability and hence the sustainability of the production system. The aim of this work was to study the live weight changes of different herbivore species (cattle, equines, sheep, goats), managing nursing mothers with their growing offspring, under the same conditions (i.e. mixed grazing on the same plot) of heath land vegetation with 24% area of improved ryegrass-clover pasture.

Material and methods

The experiment was carried out on a hill experimental farm (900 to 1000 m a.s.l.) at the Carbayal Research Station, Sierra de San Isidro, western Asturias, Spain. Five crossbred mares (350±50.3 kg LW), 5 Asturiana de los Valles beef cows (479±60.3 kg LW), 35 Gallega and Lacha crossed ewes (41±2.7 kg LW) and 35 Cashmere does (37±2.4 kg LW) were managed together in a mixed herd under continuous grazing to have the same opportunities for diet selection in a plot (22.3 ha) established on natural shrubland (76%) dominated by short heather-gorse shrubland (*Erica umbellata, Eucalyptus cinerea, Calluna vulgaris* and *Ulex gallii*) and tall scrubland (*Erica australis* subsp. *aragonensis* and *Erica. arborea*), with an area (24%) of improved pasture (*Lolium perenne* cv 'Phoenix', *Lolium x hybridum* cv 'Dalita' and *Trifolium repens* cv 'Huia').

Mothers' live weight (LW) was recorded in the beginning of the grazing season (May) and thereafter monthly until the seasons' end (January). Their offspring were weighed at turn-out and at weaning. Grazing season was divided into four seasons: spring (May to July), summer (July to end of September), autumn (end of September to end of November) and winter (end of November to January). Grass availability was assessed by recording weekly the sward surface height at 100 random points on the improved pasture area using the HFRO swardstick (Barthram, 1986). Statistical analysis was performed using SPSS version 11.5 (SPSS Inc., 2002). Individual daily LW changes during the different periods of the grazing season were subjected to a General Linear Model ANOVA, including in the model the effects of animal species, grazing season and their interaction.

Results

Mean sward height in the improved pasture area steadily decreased as grazing season advanced, as consequence of consumption by grazer animals and senescence. It accounted for a mean of 6.6 cm during spring, lowering to 4.6 cm during spring, 3.0 cm during autumn and 2.0 cm during winter (Figure 1).

Table 1 shows the mothers daily LW changes of the four herbivores species across the grazing season. On the overall grazing season, animal species had a significant ($P<0.01$) effect on LWC (g/day/livestock unit-LU). While cows lost weight (-170 g/day/LU), mares, goats and ewes were able to increase their LW in 63, 76 and 64 g/day/LU, respectively.

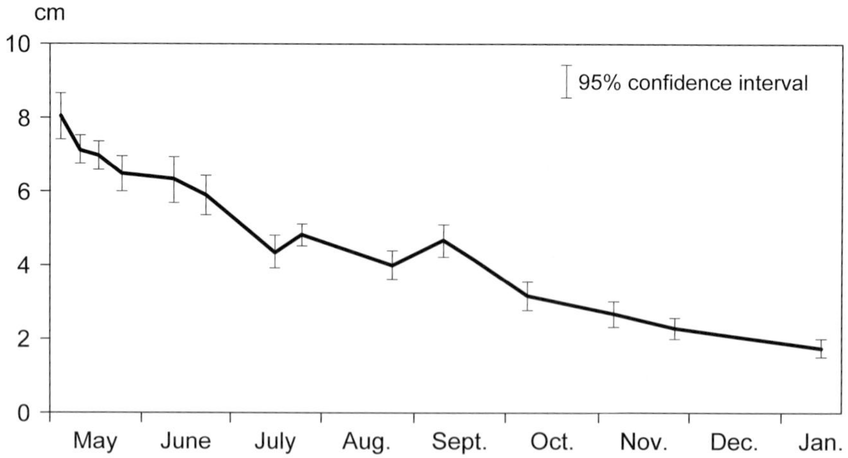

Figure 1. Sward height in the improved pasture area during the grazing season.

Table 1. Daily live-weight (LW) changes per livestock unit (LU) in suckler cows, mares, ewes and goats on partially improved heathlands across the grazing season.

	Cattle	Equines	Sheep	Goats	SEM[1]	Effect[2]
Initial LW (kg)	479^{a}	350^{b}	41^{c}	37^{c}	8.6	***
Initial LW per LU (kg/LU)	479^{a}	350^{b}	283^{bc}	256^{c}	25.1	***
LW change per LU (g/day/LU)						
Spring	859^{a}	842^{a}	377^{b}	308^{b}	103.0	***
Summer	-345^{b}	-317^{b}	20^{a}	-125^{b}	95.7	***
Autumn	-32^{b}	-71^{b}	237^{a}	-19^{b}	65.0	***
Winter	$-1,878^{c}$	-604^{b}	-290^{b}	41^{a}	121.7	***
Overall	-170^{b}	63^{a}	64^{a}	76^{a}	55.9	**

Values in the same line with different superscript letters are significantly different ($P<0.05$).

[1] SEM maximum standard error of mean.

[2] **$P<0.01$; ***$P<0.001$.

The results indicate that the LWC (g/day/LU) differed significantly ($P<0.05$) between animal species in any period of the grazing season. All four species increased their LW in the spring (859, 842, 377 and 308 g/day/LU in cows, mares, ewes and goats, respectively) as a result of a higher improved pasture availability (6.6 cm, Figure 1), though large herbivores did at a significantly ($P<0.05$) higher degree than small ruminants.

A significant ($P<0.001$) interaction between livestock species and grazing season was observed. In the summer and autumn, cows, mares and goats lost weight (average of -189, -194 and -72 g/day/LU), as a consequence of a lower sward height and nutritive value of the vegetation available in the improved pasture area. However, the losses were higher during summer than during autumn, in spite of the higher sward height during the former (4.6 and 3.0 cm, respectively). On the contrary, ewes were able to increase their LW (129 g/day/LU). In the winter, cows, mares and ewes lost LW (-1,878, -604 and -290 g/day/LU), while goats showed positive LWC (64 g/day/LU).

Table 2 shows the offspring daily LW changes of the four herbivore species across the rearing period. Lambs showed significant ($P<0.05$) higher LW gains (1,083 g/day/LU). Calves, foals and kids presented similar LWC during the rearing period (804, 785 and 696 g/day/LU, respectively).

Table 2. Daily live-weight (LW) changes per livestock unit (LU) in calves, foals, lambs and kids during the rearing period.

Offspring	Calves	Foals	Lambs	Kids	SEM[1]	Effect[2]
Initial LW (kg)	149^{a}	43^{b}	11^{c}	8^{c}	3.9	***
Initial LW per LU (kg/LU)	149^{a}	43^{c}	77^{b}	56^{c}	8.4	***
LW change per LU (g/day/LU)	804^{b}	785^{b}	$1,083^{a}$	696^{b}	91.5	***

Values in the same line with different superscript letters are significantly different ($P<0.05$).

[1] SEM maximum standard error of mean.

[2] ***$P<0.001$.

Discussion

The results obtained in the present study show that cattle presented the poorest performance among all animal species, with cows losing weight during the whole grazing season (-170 g/day/LU), while mares, ewes and goats showed positive LWC. It should be noted that the positive LWC observed in equines was mainly due to the high level of weight increase observed in spring, similar to that achieved by cattle. LWC differed according to both animal species and grazing season. In our opinion, the variations on pasture availability and its nutritive value, intake capacity and grazing behaviour between seasons and animal species could explain this interaction. In fact, cattle and equines presented similar LWC in the spring, indicating a higher intake capacity and growth potential of these animal species when the sward height is higher (Figure 1), as observed by Wright and Whyte (1989). Similar LWC between cattle and equines during spring was observed by Osoro *et al.* (2005) when pasture availability was high (>5 cm).

In the summer and autumn controls, negative and similar LWC were observed in cattle, equines and goats as a consequence of a decrease in the sward height (<5 cm). Previous studies (Forbes and Hodgson, 1985; Osoro *et al.*, 2000) also showed a negative effect of low sward height (<6 cm) on intake capacity of cattle and its consequent negative effect on LWC. This study also highlights the ability of ewes to maintain their intake and LW in situations where sward height is below 5 cm, as observed by Orr *et al.* (2000) and Osoro *et al.* (2002). Surprisingly, equines and goats showed a similar LWC pattern, although it's recognized the capacity for these animal species, especially goats, to include browse species such as heather (*Erica* spp) and gorse (*Ulex* spp) in their diets when pasture availability is low (Osoro *et al.*, 2005; 2008b). This ability allowed goats to increase their LW (41 g/day/LU) and equines to lose weight at a lower rate than cattle in the winter controls.

Although the mean sward height of the improved pasture was higher in the summer (4.6 cm) than in autumn (3 cm), the level of weight lost of cows, mares and goats was lower in the autumn, while sheep were able to increase their positive LW from 20 g/day observed in the summer to 237 g/day in the autumn. In our opinion the late weaning date, mid-summer for the small ruminant species and late-summer for cows, and a higher nutritive value of the improved pasture in the autumn as a result of its re-growth, could explain the worst performance of these animal species during the summer.

Conclusions

Data obtained in this study showed that on these heathlands with adjacent areas of improved pasture ewes, does and mares presented similar positive levels of performance if the entire season is considered. By contrast, the results indicated that cows would be the animal species less adapted to these marginal areas and the increase of area of the improved pasture should be considered for the achievement of sustainable animal production systems. Regarding the offspring, lambs showed significantly higher LWC than calves, foals and kids whose. In conclusion, in this study sheep presented the best performance and the development of sustainable systems in these vegetation communities should be based on this animal species.

References

Barthram, G.T., 1986. Experimental techniques: the HFRO swardstick. The Hill Farming Research Organisation Biennial Report 1984-85. HFRO, BushEstate, Penicuik, Midlothian, UK, 29-30.

Celaya, R., M. Oliván, L.M.M. Ferreira, A. Martínez, U. Garcia and K. Osoro, 2007. Comparison of grazing behaviour, dietary overlap and performance in non-lactating domestic ruminants grazing on marginal heathland areas. Livestock Science 106: 271-281.

Forbes, T.D.A. and J. Hodgson, 1985. Comparative studies of the influence of sward conditions on the ingestive behaviour of cows and sheep. Grass and Forage Sci. 40: 69-77.

Hofmann, R.R., 1989. Evolutionary steps of ecophysiological adaptation and diversification of ruminants: comparative view of their digestive system. Oecologia, 78: 443-457.

Illius, A.W. and I.A. Gordon, 1993. Diet selection in mammalian herbivores: constraints and tactics. In: R.N. Hughes (ed.), Diet Selection: An Interdisciplinary Approach to Foraging Behaviour. Blackwell Scientific Publications, Oxford, UK, pp. 157-181.

Orr, R.J., A.J. Parsons, P.D. Penning and T.T. Treacher, 1990. Sward composition, animal performance and the potential production of grass/white clover swards continuously stocked by sheep. Grass and Forage Sci. 45: 325-336.

Osoro, K., A. Martínez and R. Celaya, 2002. The effect of breed and sward height on sheep performance and production per hectare during the spring and autumn in Northern Spain. Grass and Forage Sci. 57: 137-146.

Osoro, K., L.M.M. Ferreira, U. Garcia and R. Celaya, 2008a. Performance of domestic herbivores in marginal heathlands. International Mediterranean Symposium of the European Association for Animal Production, 6-8 of November, Corte, France.

Osoro, K., L.M.M. Ferreira, U. Garcia and R. Celaya, 2008b. Differences in diet selection and grazing behaviour between equines and cattle grazing on upland vegetation communities. Book of Abstracts of the 59th Annual Meeting of the European Association for Animal Production, 24-27 of August, Vilnius, Lithuania, p. 140.

Osoro, K., U. Garcia, B.M. Jáuregui, L.M.M., Ferreira, M. Oliván and R. Celaya, 2005. Conducta de pastoreo y variaciones de peso de los herbívoros domésticos en un brezal-tojal parcialmente mejorado. In: Producciones agroganaderas: gestión eficiente y conservación del medio natural (Volumen 1). K. Osoro Otadui, A. Argamendia Gutiérrez, A. Larraceleta González (eds.), Servicio Regional de Investigación y Desarrollo Agroalimentario (SERIDA), XLV Reunión Científica de la Sociedad Española para el Estudio de los Pastos, Gijón (Astúrias) 28 de Maio a 3 Junho de 2005, pp. 252-258.

Osoro, K., J.J. Ormazabal and R. Celaya, 2000. Suckler cows performance during the spring under different sward heights and grazing systems. In: Beef from Grass and Forage, D. Pullar (editor), BGS Occ. Symp. No. 35, pp. 159-163.

Wright, I.A. and T.K. Whyte, 1989. Effects of sward surface height on the performance of continuously stocked spring-calving beef cows and their calves. Grass and Forage Sci. 44: 259-266.

Performance of domestic herbivores in marginal heathlands

K. Osoro[1], L.M.M. Ferreira[2], U. García[1] and R. Celaya[1]
[1]Servicio Regional de Investigación y Desarrollo Agroalimentario (SERIDA), P.O. Box 13, 33300 Villaviciosa, Asturias, Spain
[2]CECAV, Departamento de Zootecnia, Universidade de Trás-os-Montes e Alto Douro, P.O. Box 1013, 5001-801 Vila Real, Portugal

Abstract

In the north-west of Spain, on a marginal mountain area located at 850-950 m.a.s.l. and dominated by heather species (*Erica umbellata*, *Erica cinerea*, *Calluna vulgaris*), non-lactating goats, ewes, cows and mares were managed for spring-summer grazing during two years (2006-2007). The objective was to study diet selection and live-weight (LW) changes in these animal species. The results show differences in the diet components and in the individual LW changes between these herbivores. Horse was the animal species with higher ability to modify the diet composition according to the changes in vegetation, while cattle and goats only showed minor changes. In both years, cows lost weight in any period of the grazing season, at an average of 0.48 kg/day. However, mares increased LW during the first period (June-July) and lost in the second (August-September), maintaining LW (0.04 kg/day) for the overall grazing season. There were also some differences between the small ruminants. Both sheep and goats increased their LW during the first period, although goats did it at a higher degree. By contrast, in the summer both species lost weight, though the losses tend to be higher in ewes. Thus, in the overall spring-summer grazing, goats were able to increase their LW at 10 g/day while sheep lost 10 g/day ($P<0.01$). Therefore, areas with improved pasture are necessary for productive animals to maintain or increase their LW, to extend the grazing season, and to achieve the sustainability. The demand for this improved pasture is considerably higher in cattle than in the other species.

Keywords: animal performance, herbivores, diet composition, heathland

Introduction

Heathlands cover large areas in the northwest of the Iberian Peninsula, mainly in marginal lands with poor soils. It is known for the poor nutritive quality of these vegetation components. Nevertheless, domestic herbivores have different abilities to use these resources due to their different grazing behaviour and digestive processes. Consequently, the level of adaptation will be reflected in the animals' live weight (LW) changes. The objective of this work was to study the diet selection and LW changes of domestic herbivores (cattle, equines, sheep and goats) during the spring-summer grazing on heathland vegetation conditions.

Material and methods

The experiment was carried out in years 2006 and 2007 on a hill experimental farm (850 to 950 m a.s.l.) located in Asturias (northwest of Spain). Six crossbred mares (350±50.3 kg LW), six Asturiana de los Valles beef cows (479±60.3 kg LW), 12 Gallega and Lacha crossed ewes (41±2.7 kg LW) and 12 Cashmere does (37±2.4 kg LW) were managed each year on a natural shrubland dominated by dwarf heather species (*Erica umbellata*, *Erica cinerea*, *Calluna vulgaris*). The herbaceous plants were scarce, covering the gaps remaining among the shrubs, and were represented by tough and unpalatable grasses such as *Pseudarrhenatherum longifolium* and *Agrostis curtisii*.

Animals were weighed at the beginning (June), in the middle (early August) and at the end of the grazing season (September-October). The composition of the diet selected (herbaceous species and heather) by the different animal species was estimated using *n*-alkane markers (Ferreira *et al.*, 2005, 2007a,b). Samples of the vegetation components and animal faeces were collected in the first half and at the end of the grazing season. Samples of the plant species were analyzed following the procedures of the Association of Official Analytical Chemists (AOAC, 2006) for ash and nitrogen (N). Crude protein (CP) was calculated as N × 6.25. Neutral detergent fibre (NDF), acid detergent fibre (ADF) and acid detergent lignin (ADL) were analysed by the methods of Van Soest *et al.* (1991). Statistical analysis was performed using SPSS version 11.5 (SPSS Inc., 2002). Diet composition estimates were subjected to a General Linear Model ANOVA, including in the model the effects of animal species (cattle, equines, sheep and goats), grazing season (spring and summer), year (2006 and 2007) and their interactions. Individual daily LW changes during the different periods of the grazing season were analysed separately for big (cattle vs. horses) and small (sheep vs. goats) herbivores, including in the model the effects of animal species, grazing season and their interaction.

Results

Nutritive value and diet composition estimates

As expected, the results on the chemical composition (Table 1) showed a very low nutritive value of heather during the grazing season, with high levels of cell wall components (NDF: 562 g/kg DM), especially ADL (347 g/kg DM), and low levels of CP (68 g/kg DM). The results on the herbaceous species indicate how tough and unpalatable the grasses were (*Pseudarrhenatherum longifolium* and *Agrostis curtisii*) that were found within the natural vegetation, with extremely high levels of NDF (800 g/kg DM) and CP levels (70 g/kg DM) similar to those observed in the heather.

Table 2 shows the diet composition estimates for cattle, equines, sheep and goats in the beginning and end of the grazing season. The results indicate that the animal species had a significant (P<0.001) effect on the proportion of herbaceous species in the diet. Such plants were the main diet component for all animal species in both grazing seasons, with overall means ranging from 0.567 (cattle) to 0.691 (equines). A significant (P<0.01) year effect was also observed. However, this was due to a significant (P<0.05) higher proportion of herb species in the sheep diet in 2007 in both grazing periods. Moreover, the same trend for higher herb proportion during spring was observed in goats. In general, the proportion of herbaceous species in the animals' diet was higher during spring, achieving maximum values of 0.633 (cattle), 0.891 (equines), 0.780 (sheep) and 0.649 (goats). Equines seemed to be more influenced by the decline in the availability of grasses at the end of the grazing season. In fact, the proportion of herbaceous plants in their diet decreased from 0.863 to 0.543 in late summer.

Live-weight changes

Cows vs. mares: live weight changes of cows and mares were significantly (P<0.01) different in the first period (spring) and for the overall grazing season (Table 3). In the summer, the differences were

Table 1. Chemical composition (g/kg DM) of the main vegetation components in heathlands across the grazing season.

Vegetation components	Season	Organic matter	Neutral detergent fibre	Acid detergent fibre	Acid detergent lignin	Crude protein
Grasses	Spring	943	790	415	29	76
	Summer	964	811	422	30	66
Heather	Spring	978	568	450	310	70
	Summer	980	555	498	383	66

Table 2. Proportion of herbaceous species in the diet selected by cattle, equines, sheep and goats grazing on heathlands in two different periods during 2006 and 2007.

Year (Y)	Season (Se)	Cattle	Equines	Sheep	Goats	Effects	Sign.
2006	Spring	0.633	0.891	0.532	0.527	Year (Y)	**
	Late summer	0.546	0.533	0.651	0.599	Season (Se)	***
2007	Spring	0.550	0.835	0.780	0.649	Species (Sp)	***
	Late summer	0.606	0.552	0.767	0.596	Y × Se	NS
	Overall	0.567	0.691	0.683	0.593	Y × Sp	**
						Se × Sp	***
						Y × Se × Sp	**

** $P<0.01$; *** $P<0.001$; NS: not significant ($P>0.05$).

Table 3. Daily live-weight (LW) changes of cows and mares on heathlands during the spring-summer grazing.

Year (Y)	2006		2007		SEM[1]	Effects (P)[2]		
Species (Sp)	Cows	Mares	Cows	Mares		Y	Sp	Y × Sp
Initial LW (kg)	452	312	562	331	26.8	*	***	0.069
LW change (g/day)								
Spring	-285	312	-597	513	169	NS	***	NS
Summer	-223	-306	-809	-276	127	*	0.061	*
Overall	-253	-11	-708	100	116	NS	***	*

[1] SEM: maximum standard error of mean.
[2] * $P<0.05$; *** $P<0.001$; NS not significant ($P>0.10$).

smaller (P=0.061) mainly because of the lower differences in the first year (2006). As a consequence, the interaction between the effects animal species and year was significant (Table 3).

Ewes vs. does: for the overall grazing season, does' LW changes (+10 g/day) were significantly ($P<0.01$) more favourable than those observed in ewes (-10 g/day) in both years (Table 4). During the spring season, goats' LW gains (+74 g/day) were significantly ($P<0.01$) higher than those observed in sheep (+38 g/day). In summer, goats lost less weight (-30 g/day) than sheep (-41 g/day) although the differences were not significant (P=0.132). Year had a significant effect in both spring ($P<0.001$) and summer ($P<0.01$), although they presented opposite trends. As a result, when considering the overall grazing season, the year effect was not significant, nor its interaction with the animal species (Table 4).

Discussion

The results highlight the ability of both goats and horses to maintain their live-weight under these poor and marginal conditions when compared with sheep and, especially, with cattle. It is known the poorer adaptation of those animal species with bigger body size and low diet selection ability as cattle on areas where the vegetation resources present low nutritive value (Rook *et al.*, 2004). The results of diet selection are an indication of this low capacity showed by cattle for selecting green

Table 4. Daily live-weight (LW) changes of ewes and does on heathlands during the spring-summer grazing.

Year (Y)	2006		2007		SEM[1]	Effects (*P*)[2]		
Species (Sp)	Ewes	Does	Ewes	Does		Y	Sp	Y × Sp
Initial LW (kg)	39.0	36.69	40.0	33.8	2.15	NS	*	NS
LW change (g/day)								
Spring	18.5	37.4	57.6	110.4	12.1	***	**	NS
Summer	-34.3	-16.0	-47.6	-44.9	6.9	**	NS	NS
Overall	-13.0	5.6	-7.3	14.5	6.2	NS	**	NS

[1] SEM: maximum standard error of mean.

[2] $*P<0.05$; $**P<0.01$; $***P<0.001$; NS not significant ($P>0.10$).

shoots of shrubs due to their muzzle anatomy (Illius and Gordon, 1993). Mares choose a higher proportion of herbaceous (0.89) at the beginning of the grazing season than cattle (0.63). However, at the end of the grazing season the percentage of herbaceous species in the diet did not differ between both species, which could explain the lower differences in live-weight changes between mares and cows in this season.

In opposition to the horses' results, sheep showed the capacity to select herbaceous species in a similar way across the grazing season, despite the reduction in its availability, while goats maintained high percentages of heather in their diets during all the grazing season. Previous studies (Celaya *et al.*, 2007) also showed the ability of goats to include high levels of woody species in their diet, maintaining a relationship between herbaceous and heather around 0.50:0.50 with small changes between years and seasons. This once again emphasizes the capacity of goats to adapt to different vegetation conditions without suffering large consequences in their live-weight changes, in contrast with that observed for other animal species, especially cattle. This late animal species was able to show high positive live weight gains when pasture availability was high, in both quantity and quality, but losing high quantities of live-weight when the pasture availability was low for the same parameters.

Conclusions

Goats and equines were the animal species better adapted to use these poor and marginal areas. Consequently, it can be suggested the integration of heathlands in the system management for shorter periods in the case of cattle and for longer periods when non-lactating ewes, equines and goats are considered. However, the development of sustainable animal production systems will require the establishment of adjacent areas of improved pasture, increasing its extension when considering goats, equines, sheep and cattle, respectively.

References

Association of Official Analytical Chemists, 2006. Official methods of analysis, 18th edition (1st revision). AOAC, Gaithersburg, MD.

Celaya, R., M. Oliván, L.M.M. Ferreira, A. Martínez, U. García and K. Osoro, 2007. Comparison of grazing behaviour, dietary overlap and performance in non-lactating domestic ruminants grazing on marginal heathland areas. Livest. Sci. 106: 271-281.

Ferreira, L.M.M., M. Oliván, U. Garcia, M.A.M. Rodrigues and K. Osoro, 2005. Validation of the alkane technique to estimate diet selection of goats grazing heather-gorse vegetation communities. J. Sci. Food Agric. 85: 1636-1646.

Ferreira, L.M.M., U. Garcia, M.A.M Rodrigues, R. Celaya, A. Dias-da-Silva and K. Osoro, 2007a. The application of the *n*-alkane technique for estimating the composition of diets consumed by equines and cattle feeding on upland vegetation communities. Anim. Feed Sci. Tech. 138: 47-60.

Ferreira, L.M.M., M. Oliván, R. Celaya, U. Garcia, M.A.M. Rodrigues and K. Osoro, 2007b. The use of the alkane technique to estimate diet selection of sheep grazing grass-clover/heather–gorse vegetation communities. J. Sci. Food Agric. 87: 274-285.

Illius, A.W. and I.A. Gordon, 1993. Diet selection in mammalian herbivores: constraints and tactics. In: Diet Selection: An Interdisciplinary Approach to Foraging Behaviour, R.N. Hughes (editor), Blackwell Scientific Publications, Oxford, UK, pp. 157-181.

Rook, A.J., B. Dumont, J. Isselstein, K. Osoro, M.F. WallisDeVries, G. Parente and J. Mills, 2004. Matching type of grazing animal to desired biodiversity outcomes – a review. Biol. Conserv. 119: 137-150.

Van Soest, P.J., J.B. Robertson and B.A. Lewis, 1991. Methods for dietary fiber, neutral detergent fiber, and nonstarch polysaccharides in relation to animal nutrition. J. Dairy Sci. 74: 3583-3597.

Effect of grass nitrogen protection by *Acacia cyanophylla* on nitrogen balance and sheep milk production

O. Maâmouri[1], N. Atti[1], M. Mahouachi[2] and K. Kraeim[3]
[1]Institut National de la Recherche Agronomique de Tunisie, Laboratoire des Productions Animales et Fourragères, Rue Hédi Karray, 2080 Ariana, Tunisie
[2]Ecole Supérieure d'Agriculture de Kef, Département Productions Animales, Kef, Tunisie
[3]Institut Supérieur Agronomique de Chott Meriem, Sousse, Tunisie

Abstract

Tunisian dairy sheep is localized in sub humid area characterized by a high forage production and prairie growing. Nevertheless, young grass is characterized by a high level of soluble nitrogen which generates an important loss of urinary nitrogen leading to environment pollution. The objective of this work is to study the natural protection of protein by acacia tannins and its effect on urinary losses and milk production. An experimental sheep flock conducted on ray-grass grazing was divided into three homogeneous groups according to lactation rank. The first group (Control, C) received 300 g of concentrate, the second group (C-A) 300 g of concentrate and 100 g acacia, the last group only 100 g acacia (A). The nitrogen balance was realized on individual boxes with a daily cut herb. The dairy control was twice a month. Total milk production was higher for C-A group than for other groups, 48, 45 and 42 l for C-A, C and A respectively. Nitrogen intake varied significantly according to the concentrate and/or acacia supply. The average values were 22, 20 and 15 g/day for C-A, C and A group respectively. Urinary nitrogen excretion decreased from 6.5 g/d for C group to 4.5 g/d for C-A one, consequently to acacia intake. Conversely, the nitrogen retention significantly increased with acacia intake from 6 for C group to 10 g/d for C-A one. In conclusion, the acacia supply is a valid way to increase nitrogen retention and to manage the loss of nitrogen and environment.

Keywords: dairy sheep, pasture, Acacia cyanophylla*, milk yield, nitrogen retention*

Introduction

Sicilo-Sarde sheep nutrition in feedlot or pasture showed that grazing increased milk production, milk fat and milk protein quantities; it improved the dietetic quality of milk by augmenting CLA proportion compared to that produced in feedlot (Atti *et al.*, 2006). But it is also important to take into account the chemical characteristics of grazing grass as high levels of soluble nitrogen. This component is object of an intense degradation in the rumen resulting in a decrease on nitrogen retention and rejection (urine) of important amount of nitrogen regarded as a pollutant by its infiltration into the soil as nitrates. As solution, the nitrogen protection against attack of rumen microorganisms and the development by natural tannins was recommended (Barry *et al.*, 1986; Woodward and Reed, 1997). So, the objective of this work was to study the effect of natural protection of protein by acacia tannins on sheep urinary losses and milk production.

Material and methods

The experiment was carried out with 54 Sicilo-Sarde ewes from the dairy experimental farm of the National Institute of Agricultural Research of Tunisia (INRAT). Ewes were used in a 2 months experiment on ray grass grazing. They were divided into three homogeneous groups according to lactation rank. The first group (Control, C) received 300 g of concentrate, the second group (C-A) 300 g of concentrate and 100 g acacia, the last group only 100 g acacia (A). The dairy control was twice a month. The nitrogen balance was realized on metabolic boxes where ewes stayed seven days for adaptation to the new confinement conditions. In the following week, we measured nitrogen balance (total faecal and urine collection). Samples from rations (a daily cut herb of ray-grass, concentrate and acacia), refusals, and faeces were taken daily. Urine was collected in buckets and preserved with 50 ml of 10% sulphuric acid. After weighing and homogenisation, aliquot fractions were pooled and stored at -15 °C each day for a week. Then nitrogen was analysed in rations, faeces, refusals and urine.

For statistical analysis, a one-way analysis of variance for diet (C, C-A and A) effect on digestibility, nitrogen balance and milk production using GLM procedure in SAS (1989) was applied. Then, the test tdiff was used to compare diet mean effects (α=0.05). The following contrasts were used to compare the effects of the different diets:
- C1: effect of acacia;
- C2: effect of concentrate;
- C3: effect of concentrate and acacia.

Results and discussion

Digestibility and nitrogen balance

Nitrogen ingested ranged from 22.4 and 14.8 g/d; it increased according to crude protein (CP) level in the ration (Table 1). The faecal nitrogen had the same tendency as the ingested one, hence it was higher for C and C-A groups than for group A. However, when reported as percent of nitrogen intake, feacal nitrogen had the similar values with all treatments. The urinary nitrogen decreased significantly (P<0.001) with acacia supplement. This means that acacia played the role of protein protection which resulted in urinary nitrogen reduction. This result confirmed other ones (Monteils *et al.*, 2003). The nitrogen retention increased with meal CP content (Table 1). Furthermore, acacia supply lead to a strongly augmentation for the group C-A (P<0.001). This result agree with those of Barry *et al.* (1986) that awarded an increase in the nitrogen retention in sheep receiving condensed tannins supply.

Milk yield and composition

The daily milk yield was significantly different between C-A and A groups (565 ml/d and 482 ml/d); C group had an intermediate value (521 ml/d). Total milk yield was higher for C-A group than for other groups, 48, 45 and 42 l for C-A, C and A, respectively. So the protection of nitrogen from microbial degradation in the rumen by tannins of the acacia, which resulted in an increase of nitrogen retention lead to augmentation of milk production (Figure 1; R^2=0.998). Those results agree with those of Woodward *et al.* (1999) who reported an increase in the milk yield from dairy cows fed

Table 1. Average nitrogen (N) balance according to experimental diet.

Diets	C	C-A	A	ESM[1]	P	C1[2]	C2	C3
N intake (g/day)	19.6^a	22.4^b	14.8^c	0.59	0.0001	***	***	***
Faecal N (g/day)	7.3^a	7.8^a	5.7^b	0.43	0.0001	***	***	ns
Faecal N (%N intake)	37	35.1	38.8	2.64	0.165	ns	***	ns
Urinary N (g/day)	6.5^a	4.5^b	6.4^a	0.94	0.121	ns	ns	***
Urinary N (%N intake)	32.9^a	20.1^b	43.4^c	5.19	0.0006	***	***	***
N retention (g/day)	5.8^a	10^b	2.6^c	1.06	0.0001	***	***	***
N retention(%N intake)	29.8^a	44.8^b	17.8^a	7.14	0.0018	***	***	***

Means within the same row with different superscript letter differ significantly (P<0.05).

[1] ESM: mean standard error.

[2] C1: effect of acacia, C2: effect of concentrate, C3: effect of concentrate and acacia.

NS: not significant. *** significant at (P<0.05).

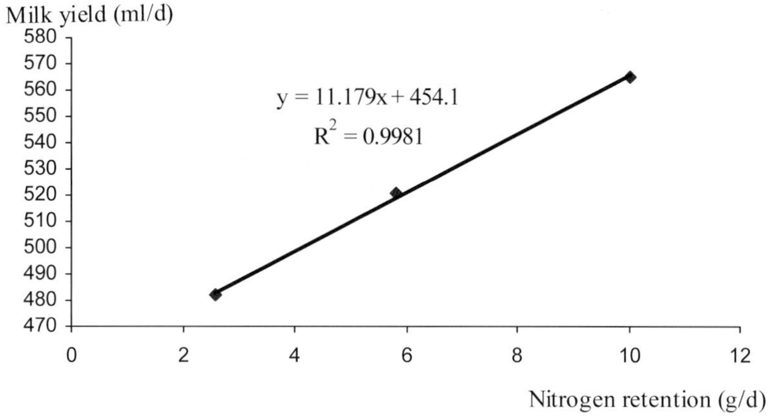

Figure 1. Relationship between milk yield (ml/d) and nitrogen retention (g/d).

with clover corniculatis contained condensed tannins. On the other hand, increasing milk yield for groups who received the concentrate complementation (C, C-A) could be explained by the energy and protein diet, complementation is often favorable to milk, fat and protein production as energy and protein intake is higher (Purroy and Jaime, 1995). However, milk protein content decreased with acacia and concentrate supply which is clear for C-A group (Table 2) and which was also mentioned by Monteils *et al.* (2003) in their work on dairy cows, but the valorisation of nitrogen is slightly decreasing. The milk protein and fat yields were higher for supplemented groups than unsupplemented one but difference was significant only for fat yield (Table 2).

Conclusion

Use of acacia cyanophylla in Sicilo-Sarde sheep nutrition is a valuable way to increase milk yield by their ability to improve nitrogen retention which also permit a reduction in urinary excretion and an environment protection.

Table 2. Average milk yield and composition according to experimental diet.

Diets	C	C-A	A	ESM[1]	*P*	C1[2]	C2	C3
Milk yield (ml/d)	521	565[a]	482[b]	119.7	0.179	ns	***	ns
Total milk yield (l)	45	48[a]	42[b]	8.49	0.179	ns	***	ns
Fat (g/kg or %)	7.8	7.7	7.6	0.64	0.296	***	ns	ns
Protein (g/kg or %)	5.1	4.8	5	0.39	0.234	ns	ns	***
Fat (g/d)	41	44.2[a]	36.8[b]	8.77	0.188	ns	***	ns
Protein (g/d)	26.5	27.4	24	5.57	0.323	ns	***	ns
Urea (mg/kg)	465[a]	447	403[b]	76.2	0.072	ns	***	ns

Means within the same row with different superscript letters differ significantly ($P<0.05$).

[1] ESM: mean standard error.

[2] C1: effect of acacia, C2: effect of concentrate, C3: effect of concentrate and acacia.

Ns: not significant; *** significant at ($P<0.05$).

References

Atti, N., H. Rouissi and M.H. Othmane, 2006. Milk production, milk fatty acid composition and conjucated linoleic acid (CLA) content in dairy ewes raised on feedlot or grazing pasture. Livest. Prod. Sci. 104: 121-127.

Barry, T.N., T. R. Manley and S.J. Duncan, 1986. The role of condensed tannins in the nutritional value of Lotus pedunculatus for sheep 4. Sites of carbohydrate and protein digestion as influenced by dietary reactive tannin concentration. British Journal of Nutrition 55: 123- 137.

Monteils, V., S. Jurjanz., G. Blanchart and F. Laurent, 2003. Effects of crude protein and energy contents of diet on nitrogen distribution between milk, urine and faeces by dairy cows. Renc. Rech. Ruminants, 10: 393.

Purroy, A. and C. Jaime, 1995. The response of lactating and dry ewes to energy intake and protein source in the diet. Small Rum. Res., 17: 17-24.

Woodward, A. and J.D. Reed, 1997. Nitrogen metabolism of sheep and goats consuming Acacia brevispica and Sesbania sesban. Journal of Animal Science 75(4): 1130-1139.

Woodward, S.L., M.J. Auldist., P.J. Laboyrie and E.B.L. Jansen, 2000. Effect of L. corniculatus and condensed tannins on milk yield and milk composition of dairy cows. Proceedings of the New Zealand Society of Animal Production 13 Supplement July 2000 A: 521-525.

Carob (*Ceratonia siliqua*) pulp as an energy source in diet for growing rabbits

A.Gasmi-Boubaker[1], F. Ben Salah[1], H. Hassen[1], R. Bergaoui[1] and A. Khaldi[2]
[1]Institut National Agronomique de Tunisie, Tunis, Tunisie
[2]Institut National des Recherches en Génie Rural, Eaux et Forêts, Tunis, Tunisie

Abstract

An important objective in livestock feeding is to promote the use of local feeds and by-products in order to reduce feeding costs. Carob pulp had been frequently used for feeding ruminants in the Mediterranean areas. This study aimed to evaluate carob pulp as an alternative dietary energy source to barley in a diet for growing rabbits. Three dietary treatments designated as D0 (commercial feed without carob pulp), D8 (commercial feed incorporating 8% carob pulp) and D15 (commercial feed incorporating 15% carob pulp) were fed to 60 rabbits (20 rabbits per group) weaned at 28 d of age to evaluate growth performance (from 29 to 70 days of age). At the end of the growth trial, 10 rabbits were slaughtered in each group, and used for the evaluation of carcass traits. Another 15 rabbits (5 rabbits per group; 2,100±200 g) were used in a digestibility trial to determine digestibility coefficient of diets. No significant effect was detected on the final weight and the daily gain. The feed intake was higher in rabbits fed the commercial diet (D0) than in other rabbits (123.3, 109.5 and 119.1 g/d in rabbits fed the D0, D8 and D15, respectively; $P<0.05$). Similarly, feed conversion tended to be higher (3.33, 3.05 and 3.03 kg/kg in rabbits fed D0, D8 and D15, respectively; $P<0.05$), while dressing out percentage was significantly lower (57.67, 59.08 and 59.89% in rabbits fed D0, D8 and D15, respectively; $P<0.05$) in rabbits fed the commercial diet. There were small significant differences in digestibility of energy and organic matter between the 3 experimental diets. It can be concluded from the finding that incorporation of carob pulp at 15% inclusion level can replace barley grain as an energy source for growing rabbits

Keywords: carob pulp, barley, rabbit, digestibility, growth

Introduction

Climatic conditions and shortage of water resources has increased costs of animal feeds in many countries. The search for alternative feed ingredients becomes the focus of most animal nutritionists. Carob pulp represents an important feed resource for animals in the Mediterranean areas. However, this material is still under-exploited. Noteworthy studies have been conducted in recent years (Boubaker *et al.*, 2007), showing the potential use of this by-product for feeding rabbits. Thus, the aim of this study was to evaluate carob pulp as an alternative energy source to barley in terms of growth performance, carcass traits, and digestibility of nutrients in growing rabbits.

Material and methods

Diets

Three experimental diets were formulated according to the values from NRC (1977) and were pelleted. The carob pulp levels of the three experimental diets were 0, 80 and 150 g/kg original matter basis for D0, D8 and D15, respectively. The chemical composition of the experimental diets is shown in Table 1.

Table 1. Chemical composition (% dry matter) of the experimental diets.

	Carob pulp (% of diet)		
	0	8	15
Moisture	8.00	9.33	11.33
Organic matter (OM)	88.76	87.86	89.47
Ash	11.23	12.13	10.52
Crude protein (CP)	20.46	19.41	18.64
Crude fibre (CF)	15.96	15.98	14.59
Phosphorus	0.63	0.63	0.53
Magnesium	0.33	0.33	0.28
Sodium	0.36	0.36	0.32
Potassium	1.40	1.60	1.50

Growth trial

Sixty New ZealandxCalifornian weanling mixed-sex rabbits (28 day old) were used for the growth performance evaluation. Rabbits were randomly allocated into three groups (20 rabbits per group and 5 rabbits per cage), according to average body weight and fed either of the experimental diets (D0, D8 or D15). During the experiment (from 29 to 70 days of age), the diets and water were offered *ad libitum* to all rabbits. The rabbits were weighed individually at weekly intervals, while the feed consumption was recorded daily and the feed efficiency was estimated as the ratio between consumption and weight gain. At the end of the experiment, 10 rabbits were slaughtered in each group and used for the carcass evaluation according to the methodology of Blasco and Ouhayoun (1996).

Digestibility trial

Another group of 15 New ZealandxCalifornian rabbits (2,100±100 g) was used in a digestibility trial to determine digestibility coefficient of diets following European reference method (Perez *et al.*, 1995). Rabbits were assigned at random to the three experimental diets (5 rabbits per group), and individually housed in digestibility cages (60 × 30 × 60 cm). Diets were offered *ad libitum* and after an adaptation period of 15 days, a 5-day balance period started.

Chemical analysis and statistics

Crude protein, ash and crude fiber content were determined by the Association of Official Analytical Chemists (AOAC) procedure (1990). Dry matter of samples was determined by drying at 105 °C to constant weight. Gross energy of diets and faeces was measured by adiabatic calorimeter. Phosphorus, K and Mg were measured by ashing samples at 550 °C and dissolving the residue in hydrochloric acid. The resultant extracts were analyzed for K and Mg using an atomic spectrophotometer and P using a technicon autoanalyser.

Data on effect of the diets on growth performance, carcass traits, and digestibility coefficient were examined by one-way analysis of variance using the GLM procedure of Statistical Analysis Systems Institute (1998). Sheffe's test was used for mean comparison where appropriate. The statistical significance was considered as $P<0.05$.

Results and discussion

Organic matter (OM), crude fiber (CF) and mineral values of the experimental diets were very close to each other while the crude protein (CP) level decreased as the inclusion level of carob pulp increased (Table 1). However, the average protein content of the 3 diets was within the values recommended for growing rabbits (Lebas, 2004).

Table 2 shows the total tract apparent digestibility of the three experimental diets. Digestibility of energy and organic matter tended to be lower ($P<0.05$) in rabbits fed the diet containing 15% carob pulp than in other rabbits. However, digestibility of dry matter and crude protein were not affected by the treatments. In our previous study (Boubaker *et al.*, 2007) using the same batch of carob pulp but with a higher level (20%), crude protein digestibility was similarly unaffected.

Growth performance and carcass traits

The productive performance data are given in Table 3. No significant effect was detected on final weight and daily gain. However, it was observed that feed intake and feed conversion were higher in rabbits fed the control diet. The slaughter results were affected by the treatments and animals fed on diets containing carob pulp tended to give heavier hot and cold carcass than rabbits receiving the control diet.

Table 2. Total tract apparent digestibility (%) of the experimental diets.

	Carob pulp (% of diet)		
	0	8	15
Energy	61.07±0.81[a]	61.12±0.82[a]	58.83±1.70[b]
Dry matter	61.30±0.74	61.42±0.81	60.60±1.70
Organic matter	62.97±0.72[a]	62.37±0.79[ab]	61.32±1.67[b]
Crude protein	76.26±0.47	74.47±0.54	74.81±1.15

Means with different superscript letters on the same row differ significantly ($P<0.05$).

Table 3. Growth performance and carcass traits of rabbits fed experimental diets.

	Carob pulp (% of diet)		
	0	8	15
Initial weight, g	537	600	619
Final weight, g	2,111	2,141	2,193
Daily weight gain, g	38.10	37.84	38.39
Daily feed intake, g	123[a]	109[b]	119[ab]
Feed efficiency (g gain/g feed)	3.33[a]	3.05[b]	3.03[b]
Slaughter weight, g	2,036	2,061	2,123
Hot carcass yield, %	59.9	60,7	61,5
Cold carcass yield, %	57[a]	59[b]	59[b]
Reference carcass yield, %	48[a]	50[b]	51[b]

Means with different superscript letters on the same row differ significantly ($P<0.05$).

Conclusion

It can be concluded from the finding that incorporation of carob pulp at 15% inclusion level can replace the traditionally used sources of energy such as barley, without adverse effect on growth performance. This study requires to be completed with economic analysis.

Acknowelegements

This study was funded by the Ministry of Scientific Research, Technology and Capacity Building of Tunisia (Laboratoire de Recherche en Economie Agro-Alimentaire, INAT).

References

Association of Official Analytical Chemists, 1990. Official Methods of Analyses. AOAC, Washington, DC, USA.

Blasco A. and J. Ouhayoun, 1996. Harmonization of criteria and terminology in rabbit meat research: revised proposal. World Rabbit Sci., 4, 93-99.

Boubaker A., R. Bergaoui, A. Khaldi, M.R. Mosquera-Losada and A. Ketata, 2007. First attempt to Study Carob Pulp Utilization in Rabbits Feeding. Journal of Animal and Veterinary Advances, 6 (11): 1269-1272.

Lebas F., 2004. Reflections on rabbit nutrition with a special emphasis on feed ingredients utilization. In: Proc. 8th World Rabbit Congress, Mexico, p. 686-736.

Lebas F., T. Gidenne, L. Maertens, G. Xiccato, R. Parigi-Bini, A. Dalle Zotte, M. Cossu, A. Carazzolo, M. Villamide, R. Carabao, M. Fraga, M. Ramos, C. Cervera, E. Blas, J. Fernandez and J. Bengala Freire, 1995. European reference method for in vivo determination of diet digestibility in rabbits. World Rabbit Sci., 3, 41-43.

Statistical Analysis Systems, 1998. User's Guide (release 6.03). SAS Inst.Inc., Cary. NC, USA.

Nutritional status of the endangered Pantelleria donkey breed

B. Chiofalo[1], L. Liotta[1], L. Sanzarello[2] and V. Chiofalo[1,2]
[1]Dept. Morphology, Biochemistry, Physiology and Animal Production, Sect. Animal Production, University of Messina, Polo Universitario Annunziata, 98168 Messina, Italy
[2]Consortium of Meat Research, Polo Universitario Annunziata, 98168 Messina, Italy

Abstract

An evaluation of the nutritional status of 30, on the 58 existing endangered Pantelleria donkeys, fed with meadow hay and oats, independently by the sex, age, and physiological state, was carried out through the study of some haematic parameters, BCS, faecal consistency and undigested fraction. Data were processed in relation to the sex and age (ANCOVA) and physiological state (pregnant vs. non pregnant, ANOVA). Comparing the sex, significant higher values for NEFA (254.09 vs. 108.28 µmol/l; $P<0.001$) were observed in the males probably in relation to their temperament with, consequently, higher lipid mobilisation, as confirmed by the significant ($P=0.006$) higher ALT activity (6.29 vs. 4.90 UI/L). Comparing the age, the most significant ($P<0.01$) differences were observed for urea which showed higher values in the younger animals probably for the worse nutritional *status* of these animals which resort to the amino acid deamination to obtain energy. Comparing the physiological state, the highest values of NEFA (120.21 vs. 99.87 µmol/l), and beta-hydroxybutyrate (0.23 vs. 0.16 mmol/l, $P=0.023$) of the pregnant females could be due to the negative energetic balance and to the unbalancing of the forage/concentrate ratio as also confirmed by the lower BCS value (2.5). Undigested fraction has shown in the faeces the presence of integral seeds of oats, due to the scarce utilisation of the feed, confirmed also by the faecal consistency which was 3.5 (solid faeces), therefore index of an unbalanced ration in favour to the fibre. Results show a not good nutritional *status* of this animals, an indispensable factor for the safeguard of this important animal germplasm.

Keywords: Pantelleria donkey breed, metabolic profile

Introduction

Pantelleria donkey is a Sicilian breed found in the Pantelleria island placed in the Sicilian Canal between the Northern coast of Africa and the Southern part of Sicily (Italy). It represents one of the most ancient donkey breeds (Chicoli, 1870) and, today, it is placed in the 'critical' category for danger of extinction (FAO, 1992).

In order to contribute to the safeguard of this important germplasm (Chiofalo, 2008a) and considering the lack of scientific data about it, the aim of this research was to study some haematic parameters of the Pantelleria breed. In fact, the knowledge of reference values of serum parameters could serve as a guide to improve husbandry systems and welfare of this ancient and innovative species, revalued as 'dairy donkey' (Salimei and Chiofalo, 2006) for the hypoallergenic role of its milk for infants affected by cow's milk protein allergy (CMPA).

Material and methods

The research was carried out in July 2006 on 30 Pantelleria donkeys belonging to the entire nucleus of the 58 existing endangered Pantelleria donkey breed reared in the experimental farm of St. Matteo in Erice (Trapani, Italy) of the Forestal Department Enterprise of Sicily (Italy). The animals, 16 females and 14 males, of age between 2 and 20 years old, registered to the National Native Breeds Register for the equine Races and Populations, were reared in semi-extensive conditions and fed with 3 kg/head/day of meadow hay (dry matter: 93.09%; crude protein: 4.49%; ether extract: 2.38%; crude fiber: 42.57%; ash: 8.71%, on a DM basis) and 4 kg/head/day of oats (dry matter: 92.50%; crude protein: 11.68%; ether extract: 6.10%; crude fiber: 14.70%; ash: 4.13%, on a DM basis), independently by the sex, age, and physiological state. On each subjects, in the morning while fasting, individual blood samples (10 ml) were taken from the jugular vein into a vacutainer® and centrifuged (ASPA, 1999). Sera were analysed using an automatic analyzer ILab300 Plus (Instrumentation Laboratory, IL Company, Lexington, MA, USA) and commercial kits (IL Company, Lexington, MA, USA; RANDOX Laboratories LTD., Crumlin, UK) to determine: glucose, non esterified fatty acids (NEFA), total cholesterol, high density lipoprotein (HDL), triglycerides, beta-hydroxybutyrate (BHBA) for the energetic metabolism; creatinine, blood urea nitrogen (BUN) and creatine kinase (CK) for the protein metabolism; Aspartate aminotransferase (AST) and alanine aminotransferase (ALT) for the hepatic functionality and alkaline phosphatase (AP), calcium (Ca), iron (Fe) for the mineral and bone metabolism. On each subject, the Body Condition Score (BCS), using a scale from 0 to 5 (from the most emaciated to the fattest animal, Martin-Rosset, 1990), and the digestibility (Dell'Orto and Savoini, 2005) were studied; the latter was determined through the study of the Faecal consistency (FC) using a score from 1 (liquid/diarrhoeic faeces) to 5 (solid/hard faeces), and the evaluation of the Undigested Fraction (UF) using a riddle with 1.6 mm of mesh. Data were subjected to ANCOVA in relation to the sex (covariate age) and in relation to the age (covariate sex), and to ANOVA considering the main variable: physiological state (pregnant vs. non pregnant) (SAS, 2001).

Results and discussion

Results are within the reference values reported for the asinine species (Mori *et al.*, 2003). Comparing the sex (Table 1), significant higher values for NEFA and triglycerides were observed in the males probably in relation to their temperament and, therefore, to their higher physical activity with, consequently, higher lipid mobilisation (Ubaldi *et al.*, 1982), as confirmed by the significant higher ALT activity (ASPA, 1999). Comparing the age (Table 2), the most significant differences were observed for BUN which showed higher values in the younger animals (1st Class) probably for the worse nutritional *status* of these animals, as confirmed by the significant lower values of AP and Fe, which resort to the amino acid deamination to obtain energy (Chiofalo *et al.*, 2002); therefore, this parameter could represent a criteria to estimate the nitrogen balance of the diet (Martin-Rosset, 1990). Comparing the physiological state (Table 3), the significant higher values of BHBA of the pregnant females could be due to the negative energetic balance and to the unbalancing of the forage/concentrate ratio (Chiofalo, 2008b).

Table 1. Effect of sex on the haematic parameters (mean±SE, P).

Serum parameters	Female[1]	Male[1]	*P*
NEFA (μmol/l)	108.28±12.63	254.09±13.65	<0.0001
Triglycerides (mmol/l)	0.53±0.04	0.70±0.05	0.015
Total cholesterol (mmol/l)	1.72±0.07	1.70±0.07	0.863
HDL (mmol/l)	1.23±0.05	1.20±0.05	0.735
BHBA (mmol/l)	0.19±0.01	0.21±0.01	0.140
BUN (mmol/l)	3.42±0.13	3.22±0.13	0.305
Creatinine (μmol/l)	101.86±2.94	95.62±3.14	0.159
CK (U/l)	125.64±18.38	147.02±15.16	0.453
ALT (U/l)	4.90±0.34	6.29±0.31	0.006
AST (U/l)	288.04±11.59	254.39±11.59	0.051
AP (U/l)	150.21±8.38	134.31±8.70	0.201
Ca (mmol/l)	2.70±0.07	2.67±0.07	0.749
Fe (μmol/l)	9.31±0.77	11.16±0.83	0.118

[1] Number of the observed animals: 16 females and 14 males.

Table 2. Effect of age on the haematic parameters (mean±SE, P).

Serum parameters	1st class[1]	2nd class[1]	3rd class[1]
NEFA (µmol/l)	166.60±11.29[A]	156.50±13.03[A]	217.75±12.62[B]
Triglycerides (mmol/l)	0.59±0.05	0.66±0.06	0.61±0.06
Total Cholesterol (mmol/l)	1.62±0.09	1.85±0.10	1.75±0.10
HDL (mmol/l)	1.15±0.08	1.33±0.09	1.25±0.08
BHBA (mmol/l)	0.19±0.01	0.21±0.01	0.19±0.01
BUN (mmol/l)	3.86±0.17[aA]	3.00±0.19[bB]	3.20±0.19[b]
Creatinine (µmol/l)	102.90±3.18	97.46±3.71	98.35±3.52
CK (UI/l)	140.20±8.32	134.13±9.61	121.68±8.83
ALT (U/l)	5.60±0.35	5.13±0.44	6.15±0.48
AST (U/l)	274.13±13.67	248.17±15.96	281.03±16.49
AP (U/l)	131.80±8.36[A]	147.75±9.34[a]	179.93±9.65[bB]
Ca (mmol/l)	2.63±0.08	2.69±0.09	2.72±0.09
Fe (µmol/l)	8.08±0.85[a]	11.32±0.99[b]	11.13±1.03[b]

Mean values with different capital or small letters in the row are significantly different for $P<0.01$ and $P<0.05$, respectively.

[1] Number and age of the observed animals: 1st class (11 donkeys of age between 2 and 5 years), 2nd class (10 donkeys of age between 6 and 9 years), 3rd class (9 donkeys of age ≥10 years).

Table 3. Effect of the physiological state on the haematic parameters (mean±SE, P).

Serum parameters	Pregnant[1]	Non-pregnant[1]	*P*
NEFA (µmol/l)	120.21±15.44	99.87±11.41	0.370
Triglycerides (mmol/l)	0.57±0.07	0.49±0.05	0.439
Total cholesterol (mmol/l)	1.58±0.21	1.90±0.14	0.305
HDL (mmol/l)	1.43±0.21	1.37±0.14	0.826
BHBA (mmol/l)	0.23±0.02	0.16±0.02	0.023
BUN (mmol/l)	3.63±0.35	3.44±0.27	0.711
Creatinine (µmol/l)	105.53±7.08	99.57±5.46	0.565
CK (U/l)	125.64±18.38	147.02±15.16	0.453
ALT (U/l)	4.68±0.88	4.85±0.51	0.879
AST (U/l)	306.96±22.92	273.02±22.40	0.430
AP (U/l)	173.01±22.60	160.62±16.75	0.703
Ca (mmol/l)	2.66±0.15	2.65±0.12	0.966
Fe (µmol/l)	10.19±1.26	10.33±1.02	0.932

[1] Number of the observed animals: 7 pregnant and 9 non-pregnant females.

Undigested fractions of the 30 animals have shown in the faeces the presence of integral seeds of oats, probably due to the scarce utilisation of the feed correlated to the low level of physically efficacious fibre (peNDF) for stimulating the chewing (Dell'Orto and Savoini, 2005), with low permanence of the vegetable walls in the large intestine and, therefore, low cellulolytic bacteria activity, as reported in equine species by Martin-Rosset (1990); this is also confirmed by the Faecal Consistency Score which was 3.5, therefore index of an unbalanced diet in favour to the fibre, and by the BCS value of 2.5.

Conclusions

Results represent a starting point to define a database of blood parameters for Pantelleria donkey breed. On the whole, they show a not good nutritional *status* of these animals, and confirm the importance of a constant monitoring of the metabolic-nutritional *status* in order to rationalize the breeding system and to improve the productive and reproductive performances, indispensable factor to conserve this important animal germplasm.

Acknowledgements

Research financed by University of Messina (PRA 2005, Dr. Biagina Chiofalo). Authors wish to thank the experimental farm of St. Matteo in Erice (Trapani, Italy) of the Forestal Department Enterprise of Sicily (Italy) and Mariangela Bellinvia for the cooperation.

References

ASPA, 1999. Guida all'interpretazione dei profili metabolici. Univ. Perugia Publ., 135.

Chicoli, N., 1870. In: Riproduzione, Allevamento e miglioramento degli animali domestici in Sicilia. Stamperia di G. Lorsnaider – Palermo, 131-132.

Chiofalo, L., 2008a. Attualità dell'asino come animale in produzione bio-zootecnica. In: L'asino all'attenzione della comunità scientifica e del territorio, Chiriotti (eds.), Pinerolo (Italy), 11-18.

Chiofalo, B., 2008b. Strategie nutrizionali per l'ottimizzazione dell'effetto probiotico del latte di asina. In: L'asino all'attenzione della comunità scientifica e del territorio, Chiriotti eds, Pinerolo (Italy), 27-31.

Chiofalo, B., L. Liotta, G. Ferrarolo and V. Chiofalo, 2002. Influence of the lactation on some haematic parameters of the Sanfratellana mares living in extensive condition. In: proceedings of the 4° Congress New findings in equine practice, pp. 73-79.

Dell'Orto, V. and G. Savoini, 2005. Alimentazione della vacca da latte. Edagricole - Edizioni Agricole de Il Sole 24 ORE Edagricole S.r.l., Bologna (Italy), 268.

FAO, 1992. Recommendations of the FAO Expert Consultation. In: J. Hodges (ed.), The management of global animal genetic resources, 1-24.

Martin-Rosset, W., 1990. Bases du rationnement. In: L'alimentation des chevaux, INRA, Parigi, 1990.

Mori, E., W.R. Fernandas, R.M.S. Mirandola, G. Kubo, R.R. Ferriera, J.V. Oliveira and F. Gacek, 2003. Reference values on serum biochemical parameters of Brazilian donkey (*Equus asinus*) breed. J. Equine Vet. Sci., 23: 358-364.

Salimei, E. and B. Chiofalo, 2006. Asses: milk yield and composition. In: Miraglia N. and Martin-Rosset W. (eds.), Nutrition and Feeding of the broodmare. Wageningen Academic Publishers. The Netherlands, 117-131.

SAS, 2001. User's Guide: Statistics. Version 8.2 Ed. SAS Inst., Care, NC, USA.

Ubaldi, A., E. Corbella and P. Montanari, 1982. Diagnostica chimico-clinica veterinaria. Ambrosiana ed., Milano (Italy), 199.

Study of some haematic parameters as indicators of the metabolic nutritional status in Nero Siciliano growing pigs

L. Liotta[1], B. Chiofalo[1], E. D'Alessandro[1] and V. Chiofalo[1,2]
[1]Department MOBIFIPA, Animal Production, University of Messina, Polo Universitario Annunziata, 98168 Messina, Italy
[2]Consortium of Meat Research, Polo Universitario Annunziata, 98168 Messina, Italy

Abstract

The effect of the two feeding systems on some biochemical parameters of 30 Nero Siciliano growing pigs was studied. The animals, 16 castrated males and 14 females, were divided into two groups of 15 each, homogeneous for live weight (32.47±1.74 kg), sex (8 males and 7 females) and age (4 months), called 'traditional' and 'experimental'. The pigs of the 'traditional' group were fed with barleycorn (2 kg/head/day) and citrus pulp (1 kg/head/day) whereas, those of the 'experimental' group received 1.3 kg/head/day of a pelletted complete feed (dry matter: 87.5%, on a DM basis: 18.29% crude protein, 3.43% ether extract, 6.06% crude fiber, 7.43% ash) appropriately formulated to be isoenergetic and isonitrogenous with the traditional diet. The trial lasted 120 days, preceded by 15-day adaptation period. Every thirty days, blood samples were collected for the determination of triglycerides, total cholesterol, high density lipoprotein (HDL), creatinine, blood urea nitrogen (BUN), activity of aspartate aminotransferase (AST), alanine aminotransferase (ALT) and alkaline phosphatase (ALP). Data were subjected to ANOVA. The 'traditional' group showed significant lower levels for Triglyceride (0.51 vs. 0.64 mmol/l; P=0.0001), as regards the energetic metabolism, and for BUN (4.19 vs. 5.35 mmol/l; P<0.001), regarding the proteic metabolism, probably in relation to the lower digestibility of the diet (barleycorn and citrus pulp). The scarce utilisation of feed was also testified by the Faecal consistency and Undigested fraction which has shown in the faeces a presence of integral seeds of barleycorns. No significant differences were observed for the other blood parameters.

Keywords: haematic parameters, growing pig, metabolic nutritional status

Introduction

The knowledge of dietary utilisation during growth in Nero Siciliano pig is essential to implement dietary programs in this breed. Chiofalo *et al.* (2001) have observed in Nero Siciliano pig reared in outdoor conditions an inadequacy of food resource with, consequently, undernutrition and metabolic disorders; therefore, the study of metabolic profile improves the information about the requirements of reared animals.

The aim of this study was to verify the relationship among the indicators of nutritional status in growing Nero Siciliano pig, fed diets with different ingredients.

Material and methods

The research was carried out on 30 Nero Siciliano growing pigs, 16 castrated males and 14 females, clinically healthy, divided into two groups of 15 each, homogeneous for live weight (32.47±1.74 kg), sex (8 males and 7 females) and age (4 months), called 'traditional' and 'experimental'.

The pigs of the 'traditional' group were fed barleycorn (2 kg/head/day) and citrus pulp (1 kg/head/day) (dry matter: 16.27%; 6.74% crude protein, 9.86% ether extract, 13.24% crude fiber, 5.65% ash, on a DM basis) whereas, those of the 'experimental' group received 1.3 kg/head/day of a pelletted complete feed (dry matter: 87.5%; 18.29% crude protein, 3.43% ether extract, 6.06% crude fiber,

7.43% ash, on a DM basis) appropriately formulated to be isoenergetic and isonitrogenous with the traditional diet.

The trial lasted 120 days, preceded by 15-day adaptation period. Every thirty days, individual blood samples (10 ml) were taken from the jugular vein into a vacutainer. Blood samples were centrifuged (ALC 4237R) at 3,500 × g for 15 min within two hours of drawing and the sera were frozen at -20 °C until analysis (ASPA, 1999).

The blood parameters were determined on each individual sample of serum by an automatic analyzer ILab300 Plus (Instrumentation Laboratory, IL Company, Lexington, MA, USA) using commercial kits (IL Company, Lexington, MA, USA; RANDOX Laboratories LTD, Crumlin, United Kingdom). The following energetic metabolism parameters were determined: total cholesterol, high density lipoprotein, triglycerides. The protein metabolism parameters were the following: creatinine and blood urea nitrogen (BUN). The hepatic functionality parameters were the following: aspartate aminotransferase (AST), alanine aminotransferase (ALT) and alkaline phosphatase (ALP).

During the trial individual faecal consistency using a score from 1 (liquid/diarrhoeic faeces) to 5 (solid/hard faeces) and undigested fraction (UF) using a riddle with 1.6 mm of mesh, were evaluated (Dell'Orto and Savoni, 2005). Data were subjected to ANOVA (Proc. GLM SAS, 2001).

Results and discussion

The haematic parameters tested were within the physiological ranges reported on the Nero Siciliano pigs (Chiofalo *et al.*, 2001; Liotta *et al.*, 2001a). During the trial, significant lower haematic levels for triglyceride (Figure 1) and BUN (Figure 2) in the animals of the traditional group were observed; the formers could be related to an energetic deficit, as already observed by Liotta *et al.* (2001b) and Chiofalo *et al.* (2003); the BUN trend could be due to a lower protein utilisation. On the whole, the behaviour of both parameters could be correlated to the lower digestibility of the diet administered to the animals of the Traditional group, as testified by the Faecal consistency score (experimental 5 vs. traditional 3) and by the observation of the undigested fractions in the faeces which have shown a presence of integral seeds of barleycorns. No significant differences were observed for the other blood parameters (Table 1).

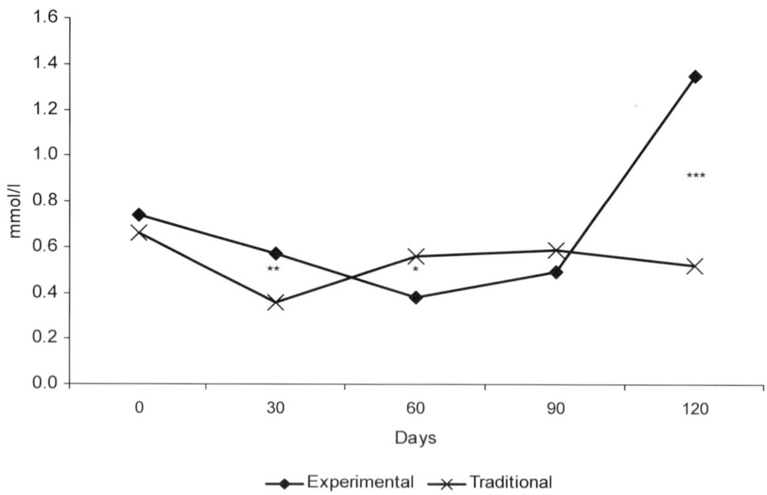

*Figure 1. Trend of triglycerides serum levels in two groups during the experimental period (**P<0.05; **P<0.01; ***P<0.001*).*

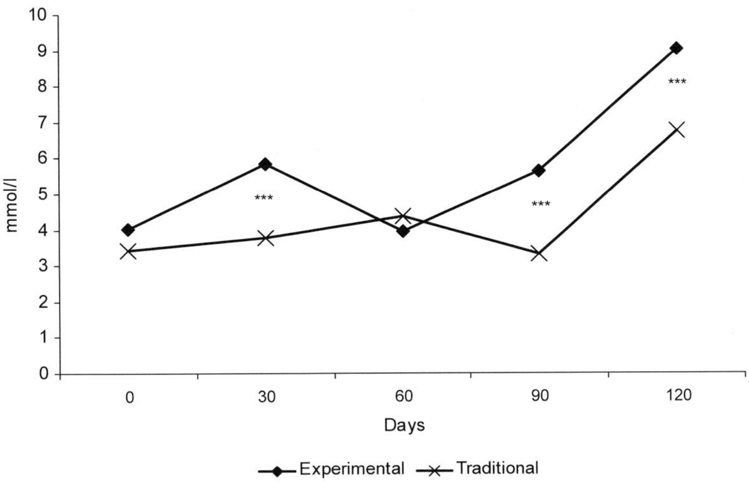

*Figure 2. Trend of blood urea nitrogen levels in two groups during the experimental period (***P<0.001).*

Conclusions

The results showed that the traditional feeding system sometimes does not satisfy the animal requirements and consequently it does not allow optimization of the productive performances (D'Alessandro *et al.*, 2007).

Therefore, the study of some metabolic parameters for the management of the autochthonous breeds could be an innovative instrument to identify a suitable nutritional strategy for animal welfare, productive performances, product quality as well as for the economic benefits of farms.

Table 1. Concentration of serum metabolites in pig sampled (means±SD, P*)*

Parameters	Experimental group	Traditional group	*P*
Hepatic metabolism			
AST ($UI\cdot L^{-1}$)	42.54±9.21	49.54±15.83	0.19
ALT ($UI\cdot L^{-1}$)	53.35±14.76	57.09±15.32	0.29
ALP ($UI\cdot L^{-1}$)	80.71±20.56	80.76±21.01	0.10
Protein metabolism			
Creatinine ($\mu mol\cdot l^{-1}$)	94.91±14.63	97.98±19.65	0.36
Energetic metabolism			
Total cholesterol ($mmol\cdot l^{-1}$)	2.29±0.38	2.30±0.43	0.27
HDL ($mmol\cdot l^{-1}$)	0.70±0.13	0.72±0.18	0.61

Acknowledgements

Research financed by Ministry of University of Research (PRIN - 2003, Prof. V. Chiofalo).

References

ASPA (Scientific Association of Animal Production), 1999. Guida all'interpretazione dei profili metabolici (Guide for interpretation of the metabolic profiles). Università di Perugia Publ., 135 pp.

Chiofalo, B., L. Liotta, A., Zumbo and L. Chiofalo, 2001. Hematic parameters of 'Nero dei Nebrodi' pigs: Effect of season. X International Symposium of Veterinary Laboratory Diagnosticians. Salsomaggiore, Parma (Italy), 4-7 july, 258-259.

Chiofalo B., L. Liotta, A. Zumbo and L. Chiofalo, 2003. Seasonal variations of free fatty acids in plasma of 'Nero dei Nebrodi' pigs in extensive condition. Vet. Res. Com. 27 (Suppl. 1): 253-255.

D'Alessandro, E., L. Liotta, M. Pagliaro and V. Chiofalo, 2007. Influence of the feeding system on in vitam and post mortem performances of Nero Siciliano pigs. Ital. J. Anim. Sci. 6 (Suppl. 1): 683.

Dell'Orto, V. and S. Savoini, 2005. Valutazione digeribilità. In: Alimentazione della vacca da latte, Edagricole (editors), Ozzano dell'Emilia (BO), Italy, 263-266.

Liotta, L., B. Chiofalo, A. Zumbo and V. Chiofalo, 2001a. In vitam parameters and energetic metabolism of Nero dei Nebrodi pigs living in exetensive condition. Preliminary remarks. LV Conv. Naz. S.I.S.Vet., Rimini 20-22 settembre 2001, pp 401-402.

Liotta, L., B. Chiofalo, A. Zumbo and V. Chiofalo, 2001b. Nero dei Nebrodi pig: exploitation of the productions and protection of the territory. Folium. 3 (Suppl.): 517-523.

SAS, 2001. User guide Version 8.2 Statistical Analysis System Inst., Cary, NC, USA.

Net transfer of nutrients to duodenum, and disappearance of n-alkanes in the reticulo-rumen and the hindgut of sheep fed grass/legume combinations

A. Keli[1,2]*, G. Olmos*[2,3]*, A. de Vega*[2] *and J.A. Guada*[2]
[1]*Unité de Recherche sur les Productions Animales, INRA-Centre Régional de Tanger, 78 Bd. Sidi Mohamed Ben Abdellah, 90000 Tanger, Morocco*
[2]*Departamento de Producción Animal y Ciencia de los Alimentos, Universidad de Zaragoza, Miguel Servet 177, 50013 Zaragoza, Spain*
[3]*Present address: Colegio de Postgraduados, Campus SLP, Iturbide # 73 Salinas de Hidalgo, SLP, 78600 México*

Abstract

Four ewes were fed different proportions of lucerne (*Medicago sativa*) and Wimmera ryegrass (*Lolium rigidum*) (100:0, 33:67, 67:33 and 0:100), following a latin square design, to study the influence of the diet on the net transfer of nutrients to the intestines, and on duodenal and faecal recoveries of n-alkanes. Duodenal flows of dry matter, organic matter, neutral detergent fibre, non-ammonia nitrogen (NAN) and microbial nitrogen, and the efficiency of microbial synthesis, were not affected by diet, although the microbial contribution to the duodenal flow of nitrogen, and the ratio NAN/N intake were affected. Duodenal recovery of n-alkanes was not affected by diet, and was complete for those present in higher concentrations in the diets. Isolated rumen bacteria contained significant amounts of n-alkanes, contributing to their duodenal flow in variable proportions depending on the diet consumed. The mechanisms of and the reasons for this incorporation are unknown. The faecal recovery of the n-alkanes was affected by diet only in some cases.

Keywords: n-alkanes, duodenum, nutrients, rumen bacteria, sheep

Introduction

Net transfer of ingested protein to the duodenum is complete only when the plants contain ca. 210 g crude protein (CP)/kg digestible organic matter (DOM) or less (Poppi and McLennan, 1995). Above this threshold large amounts of nitrogen are lost as ammonia across the rumen wall. Legumes usually exceed these CP content, and hence including a grass in the diet might minimize nitrogen losses. The amount of protein reaching the abomasum when animals are fed different grass/legume mixtures (e.g. Janovick *et al.*, 2005) is largely unknown. To this respect, lucerne and Wimmera ryegrass are important resources of integrated sheep farming systems in semiarid areas of the Mediterranean basin, but there is no information about net transfer of nitrogen to the duodenum when fed in combination.

N-alkanes have been widely used to estimate diet selection, intake and digestibility in grazing systems, although it is necessary to adjust the faecal alkane concentration for incomplete recovery. This latter can be due to losses in the reticulo-rumen, in post-ruminal compartments or in both. Papers dealing with ruminal disappearance of n-alkanes are scarce (e.g. Ohajuruka and Palmquist, 1991) but they suggest that losses in the reticulo-rumen are minimal, if any. If this is the case, these hydrocarbons could be confidently used as duodenal flow or rumen transit markers regardless their faecal recovery. Feeding conditions may affect faecal recovery (Mayes *et al.*, 1986), although the effect of diet is unclear. To this respect, the question remains whether the rumen microbes are able to degrade long-chain alkanes, and to what extent this degradation may be affected by the conditions of rumen fermentation.

The aim of this work was to examine the effect of different proportions of lucerne and Wimmera ryegrass in the diet on net transfer of nutrients, particularly protein, to the intestines, and on the disappearance of n-alkanes in the reticulo-rumen and the hindgut of sheep.

Material and methods

Four adult ewes, fitted with ruminal and T-type duodenal cannulae, were allocated to four treatments consisting of different proportions (100:0, 67:33, 33:67 and 0:100) of lucerne (16% CP) and Wimmera ryegrass (4.3% CP) hays chopped to 5 cm, following a latin square design. Diets were offered every 4 hours at 95% of *ad libitum* intake in separate troughs.

As part of a broader experiment, a digestibility trial was carried out followed by a 2-day duodenal sampling for each period of the Latin square. Chemical composition of diets (organic matter (OM), CP, neutral detergent fibre (NDF) and n-alkanes) and faeces (OM and n-alkanes) was determined using the methods described by Keli *et al.* (2008).

Cr-EDTA and $YbCl_3$ solutions were infused as duodenal flow markers, and labelled ammonium sulphate (10% atoms ^{15}N) as microbial marker. After duodenal sampling rumen content was removed to obtain reference samples of solid (SAB) and liquid (LAB) adherent bacteria. Isolation of SAB and LAB was carried out as described by Martin *et al.* (1994) to determine their chemical composition (OM and N), ^{15}N abundance and the concentration of n-alkanes in LAB. The isotope abundance (^{15}N) of bacterial N (LAB and SAB fractions), and NAN in the solid and liquid phases of duodenal samples were determined by mass spectrometry.

The analysis of Cr and Yb in duodenal digesta samples was carried out as described by de Vega and Poppi (1997), and that of n-alkanes in diets, faeces, duodenal digesta samples and LAB as described by Keli *et al.* (2008). Duodenal flow of dry matter (DM) and its fractions (including n-alkanes) was estimated following the procedures described by Faichney (1975).

The microbial contribution to duodenal flow of NAN was calculated as described by Askar *et al.* (2005), and the proportion of n-alkanes of bacterial origin (only LAB fraction) reaching the duodenum was calculated as described by Keli (2006).

Results and discussion

Table 1 shows the duodenal flows of DM, OM, NDF, NAN and microbial N, together with the efficiency of microbial synthesis and its contribution to the NAN flow.

Duodenal flow of nutrients decreased numerically ($P>0.05$) with increasing proportions of ryegrass in the diet, except for LAB-N, which was higher ($P>0.05$) in animals fed mixed diets. Microbial contribution to duodenal flow of NAN was not affected by diet, except when the animals were fed 100% alfalfa. This is in agreement with previous results on the higher contribution of dietary N to duodenal NAN flow in animals fed forages low in NDF and high in N (Pérez *et al.*, 1997). Efficiency of microbial protein synthesis was not affected either by diet type.

Duodenal recoveries of the n-alkanes which presented higher concentrations in the diet were not affected by diet (Table 2), with mean values of 107%, 112% and 107% for C_{27}, C_{29} and C_{31}, respectively. However their faecal recovery was much lower and differed between diets ($P<0.05$) only for C_{27}, with higher values in sheep fed 100% lucerne, and lower in those fed 100% ryegrass.

The contribution of LAB-alkanes to the duodenal flow of n-alkanes is shown in Table 3. Diet effect was highly significant ($P<0.01$) for C_{27}, C_{29}, C_{31} and C_{33}, with microbes from sheep fed diets high in ryegrass (67% or 100%) contributing more alkanes to the duodenum (7.6% on average) than those from lucerne-based diets (4.2%). Even-chain alkanes were not affected by diet type (except c_{30}).

*Table 1. Duodenal flow of dry matter, organic matter, neutral detergent fibre, non-ammonia nitrogen (NAN), solid-adherent bacteria nitrogen (SAB-N) and liquid-adherent bacteria nitrogen (LAB-N), NAN/N intake ratio (NAN/NI), microbial contribution to duodenal flow of nitrogen (%) (SAB-N/NAN and LAB-N/NAN), and efficiency of synthesis of microbial protein (g Nmic/kg OMADR) in sheep fed different proportions of lucerne and Wimmera ryegrass (*Lolium rigidum*) hays.*

	Diet (% lucerne/% ryegrass)				SEM[1]	P^2
	100:0	67:33	33:67	0:100		
Duodenal flow (g/d)						
Dry matter	466	425	391	353	63.9	0.6646
Organic matter	372	318	290	254	41.4	0.3436
Neutral detergent fibre	183	175	141	129	21.2	0.3602
NAN	9.10	6.92	6.83	5.30	0.960	0.1762
SAB-N	4.81	4.32	4.83	3.12	0.410	0.1644
LAB-N	3.75	4.23	4.67	3.10	0.490	0.2865
NAN/NI	0.44^{b}	0.53^{b}	0.88^{ab}	1.28^{a}	0.089	0.0112
SAB-N/NAN (%)	53.5^{b}	74.4^{a}	70.8^{a}	64.4^{ab}	2.08	0.0332
LAB-N/NAN (%)	41.7^{b}	60.8^{a}	68.4^{a}	63.6^{a}	2.40	0.0034
Efficiency of microbial protein synthesis (g Nmic/kg OMADR)						
SAB	13.2	10.2	18.0	11.9	1.76	0.1593
LAB	10.3	11.2	17.5	12.0	2.20	0.1772

[1] SEM: standard error of the mean.
[2] *P*: probability of the differences: mean values within rows with different superscripts are significantly different ($P<0.05$).

*Table 2. Duodenal and faecal recoveries (% of intake) of C_{27}, C_{29} and C_{31} n-alkanes in sheep fed different proportions of lucerne and Wimmera ryegrass (*Lolium rigidum*) hays.*

	Duodenum					Faeces				
	100:0	67:33	33:67	0:100	SEM[1]	100:0	67:33	33:67	0:100	SEM
C_{27}	99	106	112	110	10.8	69^{a}	63^{a}	60^{a}	40^{b}	3.6
C_{29}	107	98	114	128	16.7	79	74	80	60	3.7
C_{31}	101	94	110	124	14.5	82	79	88	70	3.9

Mean values within rows with different superscripts (duodenum or faeces) are significantly different ($P<0.05$).
[1] SEM: standard error of the mean.

This is the first time that incorporation of significant amounts of n-alkanes in ruminal bacteria from sheep is demonstrated, our results supporting the use of these hydrocarbons as duodenal flow internal markers. This is highly innovative, as it may avoid the need of continuous infusion of external markers in digestion studies. On the other hand, the proportion of lucerne and Wimmera ryegrass in the diet did not affect the net transfer of nutrients to the duodenum but the microbial contribution to duodenal nitrogen and the proportion of ingested nitrogen lost as ammonia.

*Table 3. Contribution of LAB (liquid adherent bacteria)-alkanes (%) to the duodenal flow of n-alkanes in sheep fed different proportions of lucerne and Wimmera ryegrass (*Lolium rigidum*) hays.*

	Diet (% lucerne/ryegrass)				SEM[1]	P^2
	100:0	67:33	33:67	0:100		
C_{25}	6.46	5.06	8.95	8.37	0.86	0.1338
C_{27}	4.39[b]	4.27[b]	7.84[a]	7.12[a]	0.21	0.0022
C_{28}	4.23	3.48	7.00	5.23	0.55	0.0999
C_{29}	4.03[b]	4.23[b]	7.69[a]	7.79[a]	0.18	0.0010
C_{30}	3.44[b]	2.61[b]	5.08[a]	4.09[ab]	0.35	0.0477
C_{31}	4.16[b]	4.19[b]	7.53[a]	8.11[a]	0.28	0.0043
C_{32}	1.71	5.53	5.42	5.59	2.17	0.6633
C_{33}	4.13[b]	3.94[b]	7.58[a]	7.38[a]	0.36	0.0094
C_{36}	2.78	1.90	5.30	5.80	2.79	0.9584

Means within rows with different superscript are significantly different ($P<0.05$); C_{23}, C_{24}, C_{26} and C_{35} were not presented in the table due to missing data and the non-detection in the LAB in some cases.

[1] SEM: standard error of the mean.

[2] *P*: probability of the differences.

References

Askar, A.R., J.A. Guada, J. Balcells, A. de Vega and C. Castrillo, 2005. Validation of use of purine bases as a microbial marker by ^{15}N labelling in growing lambs given high-concentrate diets: effects of grain processing, animal age and digesta sampling site. Anim. Sci. 81: 57-65.

De Vega, A. and D.P. Poppi, 1997. Extent of digestion and rumen condition as factor affecting passage of liquid and digesta particles in sheep. J. Agri. Sci. 128: 207-215.

Faichney, G.J., 1975. The use of markers to partition digestion within the gastro-intestinal tract of ruminants. In: Digestion and Metabolism in the Ruminant, I.W. McDonald and A.C.I. Warner (editors), University of New England, Armidale, Australia, 277-291.

Janovick, N.A., J.R. Russell, D.R. Strohbehn and D.G. Morrical, 2005. Grazing and feedlot performance of yearling stocker cattle integrated with spring- and fall-calving beef cows in a year-round grazing system. J. Anim. Sci. 83: 2696-2704.

Keli, A., D. Andueza, A. de vega and J.A. Guada, 2008. Validation of the n-alkane and NIRS techniques to estimate intake, digestibility and diet composition in sheep fed mixed lucerne: ryegrass diets. Livest. Sci. 119: 42-54.

Keli, A., 2006. Use of alkanes in sheep: their behaviour in the digestive tract and possible alternatives. PhD, Zaragoza University, Zaragoza, Spain.

Martin, C., A.G. Williams and B.M. Doreau, 1994. Isolation and characteristics of the protozoal and bacterial fractions from bovine ruminal contents. J. Anim. Sci. 72: 2962-2968.

Mayes, R.W., C.S. Lamb and P.M. Colgrove, 1986. The use of dosed and herbage n-alkanes as markers for the determination of herbage intake. J. Agric. Sci. 107: 161-170.

Ohajuruka, O.A. and D.L. Palmquist, 1991. Evaluation of n-alkanes as digesta markers in Dairy cows. J. Anim. Sci. 69: 1726-1732.

Pérez, J.F., J. Balcells, J.A. Guada and C. Castrillo, 1997. Rumen microbial production estimated either from urinary purine derivative excretion or from direct measurements of ^{15}N and purine bases as microbial markers: effect of protein source and rumen bacterial isolates. Anim. Sci. 65: 225-236.

Poppi, D.P. and S.R. McLennan, 1995. Protein and energy utilization by ruminants at pasture. J. Anim. Sci. 73: 278-290.

Extruded linseed in the diet of dairy ewes: effects on fatty acid composition of milk and cheese

M. Antongiovanni[1], M. Mele[2], A. Buccioni[1], A. Serra[2], S. Minieri[1] and P. Secchiari[2]
[1]Dipartimento di Scienze Zootecniche, Via delle Cascine 5, 50144 Firenze, Italy
[2]Dipartimento di Agronomia e Gestione dell'Agroecosistema, Via del Borghetto 80, 56124 Pisa, Italy

Abstract

Ovine milk is naturally richer in short chain but poorer in mid and long chain saturated fatty acids than bovine milk. It results healthier also because as a human food it is characterized by more vaccenic and the conjugated linoleic acid isomer, rumenic acid. The concentration of the major beneficial components, rumenic (RA) and vaccenic (VA) acids, is very sensitive to the quality of the diet, because both RA and VA are metabolic intermediates of the rumen biohydrogenation of linoleic acid (LA) and VA only of α-linolenic acid (α-LNA). Aim of the trial was to study the effect of extruded linseed in the diet of ewes as a means to enhance the concentration of beneficial fatty acids in milk and in the derived cheese. Two groups of 24 Sarda ewes each have been monitored for milk fatty acid composition for 10 weeks after weaning. The diet of the experimental group was supplemented with 200 g/head/day extruded linseed. Butyric acid (BA) was not influenced by the dietary supplement. Monounsaturated acids (MUFA) resulted depressed (58 vs. 48%) by linseed, whereas ω-3 polyunsaturated acids (PUFA n-3) were increased (1 vs. 2.1%), both starting from the second week of lactation. VA was increased more than threefold from the third week on (1.7 vs. 5.9%) and RA almost threefold from the second week on (0.9 vs. 2.3%). All differences resulted significant (P<0.01). The beneficial acids were slightly lost with cheese making, still the cheese from treated ewes resulted decidedly healthier.

Keywords: ovine milk, fatty acids, rumenic acid, vaccenic acid

Introduction

The quality of the diet of dairy ruminants is highly correlated with the fatty acid composition of milk fat because of the fate of dietary unsaturated fatty acids (UFA) in the biohydrogenation pathway in the rumen: fibrous forages, better if grazed, lead to the beneficial *trans* 11 UFA such as the conjugated linoleic acid (CLA) isomer rumenic acid (RA), but starchy grains lead to the less desirable *trans* 10 isomers (Harfoot and Hazlewood, 1988; Griinari *et al.*, 1998).

Many milk fatty acids are acknowledged as beneficial nutraceutical components: butyric acid (BA) is an important energy source for colonic epithelial cells and exerts a protective role in relation to colonic disease. It is also a positive selecting agent towards the gut microbial population (Pryde *et al.*, 2003); RA, the *cis* 9 *trans* 11 CLA isomer, possesses anticarcinogenic, antiatherogenic, anti-diabetic and immune modulation effects in animal models; the other *trans* 10 *cis* 12 CLA isomer is a lipogenesis inhibitor. Yet, the specific effects of the single isomers need further investigation because some other isomers may be harmful (McGuire and McGuire, 2000; Tricon *et al.*, 2005).

Hence, the positive impact of sheep farming activities on the environment, because of the possible re-population of abandoned rural areas, and the important social and economic effects both on the farmers' life, motivated by the awareness of producing a healthy food, and on human health welfare.

Supplementing the diet with lipid ingredients such as extruded linseed, rich in the octa-decenoic acids linoleic (LA) and α-linolenic (α-LNA) is a feeding strategy to improve the quality of milk fat of dairy cows (Griinari, 2001), not yet deeply studied in the dairy ewe, whose milk is completely transformed into cheese, broadly consumed in the Mediterranean countries. The result with ewes are indeed more remarkable than with cows, simply because the fat content of ovine milk is twice as much, on average.

Material and methods

The experimental animals were 48 Sarda ewes, all of them immediately after the lamb weaning, randomly allotted to two groups of 24 ewes each. The trial lasted 10 weeks, January through March 2006. During the first 6 weeks, due to adverse weather conditions, the animals were fed grass hay *ad lib* indoor, then they were allowed outside on a mixed pasture of *Avena sativa, Lolium italicum* and *Trifolium repens.* In addition, they received a concentrate ration: the treated ewes with 30% extruded linseed on the DM basis, 700 g/head/day, and the control ewes without linseed, 800 g/head/day, in order to feed approximately the same amounts of energy (UFL), digestible proteins (PDI), structural (NDF) and non-structural (NSC) carbohydrates (Table 1),

At the end of each week, milk samples of about 50 ml were collected from the single ewes for the analysis of fatty acids. The extraction technique and the gas-chromatographic method were those indicated by Secchiari *et al.* (2003).

The resulting data were statistically analysed by using the following linear model:

$$y_{inm} = \mu+D_i+W_n+E_m(D_i)+D_i xW_n+e_{inm} \quad (1)$$

with D_i= diet effect (i=1 to 2); W_n= sampling week effect (n= 1 to 10); $E_m(D_i)$= ewe within diet random effect (m= 1 to 48); $D_i xW_n$= interaction effect between diet and sampling week.

Results

As expected, the linseed supplement resulted in marked differences in most fatty acids of milk fat (Table 2). BA was not different, but the harmful myristic (MA) and palmitic acids (PA) were significantly decreased in the milk of treated animals. VA was about four fold higher and RA about three fold. LA was slightly, but significantly, less and α-LNA was more than doubled.

Table 3 summarises all the analysed acids, grouped in classes, together with the milk indexes of the udder Δ^9 desaturase activity (Bauman *et al.,* 2003). The undesirable harmful saturated fatty acids (SFA) resulted significantly decreased, whereas the beneficial monounsaturated (MUFA) and ω-3 polyunsaturated ones (PUFA n-3) were increased. The branched (BCFA) and odd chain (OCFA) acids were depressed, but the *trans* acids (TFA) were enhanced. All the milk indexes were slightly, but significantly depressed.

Of all the desirable acids, the beneficial ones were transferred into cheese without losses (Table 4). On the contrary, they were transferred with gains at the expenses of other components of the lipid fraction.

Table 1. Estimated average nutritional traits of the two diets.

Diet	UFL/kg DM	PDI, g/kg DM	NDF, g/kg DM	NSC, g/kg DM
Control	0.82	109	410	360
Treated	0.86	108	410	335

Table 2. Major fatty acids composition of milk fat (g/100 g total lipids, means of 10 samples).

Fatty acid	Control	Treated	*P*
Butyric (4:0)	4.88	4.09	0.10
Myristic (14:0)	9.85	8.08	<0.01
Palmitic (16:0)	22.87	18.47	<0.01
Stearic (18:0)	8.85	10.45	<0.01
Oleic (18:1 c9)	17.39	17.84	0.26
Vaccenic (18:1 t11)	1.72	5.87	<0.01
Octadecenoic (18:1 t10)	0.50	0.72	0.01
Linoleic (18:2 c9 c12)	2.18	2.00	<0.01
Rumenic (18:2 c9 t11)	0.85	2.26	<0.01
α-Linolenic (18:3 c9 c12 c15)	0.87	1.89	<0.01

Table 3. Classes of fatty acids (g/100 g total lipids) and milk indexes (maens of 10 samples).

Fatty acid class	Control	Treated	*P*
Saturated (SFA)	58.60	48.00	<0.01
Monounsaturated (MUFA)	23.61	29.57	<0.01
ω-6 Polyunsaturated (PUFA n-6)	2.39	2.35	0.32
ω-3 Polyunsaturated (PUFA n-3)	1.04	2.12	<0.01
Branched (BCFA)	1.21	1.02	<0.01
Odd chain (OCFA)	1.63	1.39	<0.01
Trans (TFA)	3.42	8.83	<0.01
ID_{14} (14:1/(14:0+14:1))	1.74	1.31	<0.01
ID_{16} (16:1/(16:0+16:1))	3.53	2.73	<0.01
ID_{18} (18:1/(18:0+18:1)) (cis 9)	66.22	63.10	<0.01

Table 4. Efficiency of transfer of fatty acids from milk into cheese.

Fatty acid	Effect
Butyric (4:0)	Unchanged
Myristic (14:0)	Unchanged
Palmitic (16:0)	Increased by 5%
Oleic (18:1 c9)	Increased by 12%
Vaccenic (18:1 t11)	Increased by 17%
Rumenic (18:2 c9 t11)	Increased by 30%

Discussion

Since MA and PA, particularly the former, are considered susceptible of inducing cardio-vascular disease (Ulbricht and Southgate, 1991), the fact that they appear decreased in the milk of treated ewes is a positive result in terms of food safety. Stearic acid (SA) is not influent in this sense and its increase is not to be considered negatively.

In addition, if we look at RA, which was increased by 170%, and VA, increased by 240%, the comment is that the dietary supplementation with extruded linseed was a successful result: in particular, the most important beneficial component was largely enhanced and the milk and the resulting cheese greatly upgraded as 'functional' foods.

LA resulted lower in treated milk, obviously due to the microbial biohydrogenation activity and the same may be said of α-LNA which, in spite of being higher, is not so high as it should be due to the presence in the diet of whole linseed, whose oil is more than a half represented by LNA.

The figures of Table 3 are altogether comforting as well: less SFA and more MUFA and PUFA n-3. The slight differences of BCFA and OCFA are the consequence of the obvious different microbial strains composition, but without appreciable effects on milk fat quality.

As far as the Δ^9 desaturase indexes are concerned, this enzymatic activity of the mammary gland appeared diminished with linseed, maybe due to the higher amount of SA.

Cheese making resulted in no losses of the beneficial acids, on the contrary with improvements, so reinforcing the classification of sheep cheese as a 'functional' healthy food.

Conclusion

It was confirmed that the diet supplementation with extruded linseed changes significantly the fatty acid profile of milk fat of ewes. Practically, all the recorded modifications are to be considered as improvements of the quality of the produced milk and of the derived cheese, in terms of functional healthy characteristics for a safer and better food.

References

Bauman, D.E. and M.J. Griinari, 2003. Nutritional regulation of milk fat synthesis. Ann. Rev. Nutr. 23: 203-227.

Griinari, M.J., 2001. Development of animal feeding strategies for the production of high CLA milk fat. Proc. 1st International Conference on CLA, Alesund, Norway, 25.

Griinari, M.J., D.A. Dwyer, M.A. McGuire, D.E. Bauman, D.L. Palmquist and K.V.V. Nurmela, 1998. Trans octa-decenoic acids and milk fat depression in lactating dairy cows. J. Dairy Sci. 81: 1251-1261.

Harfoot, C.G. and G.P. Hazlewood, 1988. Lipid metabolism in the rumen. In: P.N. Hobson (ed.), The rumen microbial ecosystem, Elsevier Applied Science Publishers, London, UK, 285-322.

McGuire, M.A. and M.K. McGuire, 2000. CLA: a ruminant fatty acid with beneficial effects on human health. J. Anim. Sci. 77: 1-8.

Pryde, S.E., S.H. Duncan, G.L. Hold, C.S. Stewart, H.J. Flint. 2002. The microbiology of butyrate formation in the human colon. FEMS Microbiology Letters. 217: 133-139.

Secchiari, P., M. Antongiovanni, M. Mele, A. Serra, A. Buccioni, G. Ferruzzi, F. Paoletti, F. Petacchi. 2003. Effect of kind of dietary faton yhe quality of milk fat from Italian Friesian cows. Livestock Prod. Sci. 83: 43-52.

Tricon, S., G.C. Burdge, C.M. Williams, P.C. Calder and P. Yaqoob, 2005. The effects of CLA on human health outcomes. Proc. of the Nutr. Soc. 64: 171-182.

Ulbricht, T.L.V. and D.A.T. Southgate, 1991. Coronary heart disease: seven dietary factors. Lancet. 338: 985-992.

Part 2.

Innovations improving the efficiency of livestock systems in the Mediterranean area

C. Animal products, system traceability

SIEOL, information system for genetic, technical and economic support for French dairy sheep breeders

E. Morin[1], *J.M. Astruc*[2] *and G. Lagriffoul*[3]
[1]*Institut de l'Élevage, 31321 Castanet-Tolosan, France*
[2]*Institut de l'Élevage, INRA-SAGA, 31326 Castanet-Tolosan, France*
[3]*Institut de l'Elevage-CNBL, INRA-SAGA, 31326 Castanet-Tolosan, France*

Abstrract

In France, the 5,200 sheep dairy farms are mainly located in three traditional sheep milk production areas: Roquefort, in the south of the Massif Central (43% of the farms), the French western Pyrenees (41%), and Corsica (8%). Production is characterised by five local breeds, a traditional management system, and milk processing into PDO cheeses. During the last 20 years, the decrease in the number of dairy farms did not prevent milk production from increasing from 120 million litres in 1985 to 260 million litres in 2006. Both the improvement of the management systems and the efficiency of the selection scheme contributed to this evolution. On the farm, the same technician records milk yields and provides genetic, technical and economic advice to the breeders. To help the technicians, since 2005 an information system called SIEOL has been implemented by the technical organisations. SIEOL manages genetic, technical and technical-economic data by collecting all essential information for collective use (e.g. genetic evaluation for the management of selection) but also for use by the farm technician (e.g. to track down the least productive ewes or to improve breeder's income). Today, the SIEOL system is used by 48% of all dairy sheep farms of France and for 61% of dairy ewes.

Keywords: dairy sheep, information system, genetics, technical and economic support

Introduction

In the livestock sector, as in other sectors, management of data is a key stake. For example, in the field of genetic improvement, a large amount of data has to be managed collectively every year including inventory, pedigree information, lambings and matings, milk recording data, qualification of reproducers, estimated breeding values (EBVs) and various assessments. The new French Breeding Law provides a frame for national genetic information systems (JORF, 2006, 2007).

Taking these information systems into account in the national legislation is a significant change which underlines their importance. In the dairy sheep sector, the project of re-building the Dairy Sheep Information System for Breeders (SIEOL in French) was initiated before the recent reform of the French genetic organization. This project was set up in response to a demand from the breeders themselves to build a shared and collective information system for all the technical organizations, in order to provide technical support for all dairy sheep breeders. The objective was to design SIEOL around the job of the field technician who provides a *global service* for the breeders by recording performance, providing genetic, technical and economic advice on matings and lambings, feeding etc. To help dairy sheep technicians provide this *global service*, all the necessary data are recorded in SIEOL and all the necessary interfaces with other systems are available.

The purpose of this paper is to give an overview of dairy sheep production in France, and to describe how the technical management of dairy sheep is organized in order to understand the stakes and the importance of SIEOL.

Dairy sheep production in France

Unlike other French livestock production which is scattered throughout France (Table 1, Figure 1), dairy sheep farms are mainly located in three traditional areas: the Roquefort area in the south of Massif Central (43% of the farms), the French western Pyrenees (43%) and Corsica (9%). In all three areas, dairy sheep are raised on difficult territories: more than 75% of the farms are located in mountains and 10% in the high mountains (Morin *et al.*, 2004).

In the Roquefort area, the flocks are relatively large with a little more than 350 dairy ewes on average. In the western Pyrenees and in Corsica, the flocks are smaller, with fewer than 250 ewes. In all three areas, the local breeds are well adapted to their environment and their production system: the Lacaune breed in the Roquefort area and Corsica breed in Corsica. In the French western Pyrenees, there are three breeds: the Basco-Béarnaise in the Bearn area, the Manech Black Face in the Basque mountains and the Manech Red Face in the Basque hillsides.

Milk sheep production is a seasonal activity. Lambing takes place in the autumn or early winter. The suckling period lasts four weeks. Milking starts after the weaning of the lambs and ends in late spring or summer.

Table 1. Dairy sheep production in France in 2006.

	Number of farms	Number of ewes	Milk production (million liters)	On-farm milk processing
Roquefort area	2,244	850,000	179.6	-
French western Pyrenees	2,150	480,000	57.1	14%
Corsica	420	95,000	9.4	21%
Other areas	350-400	70,000-80,000	15-20	20-30%

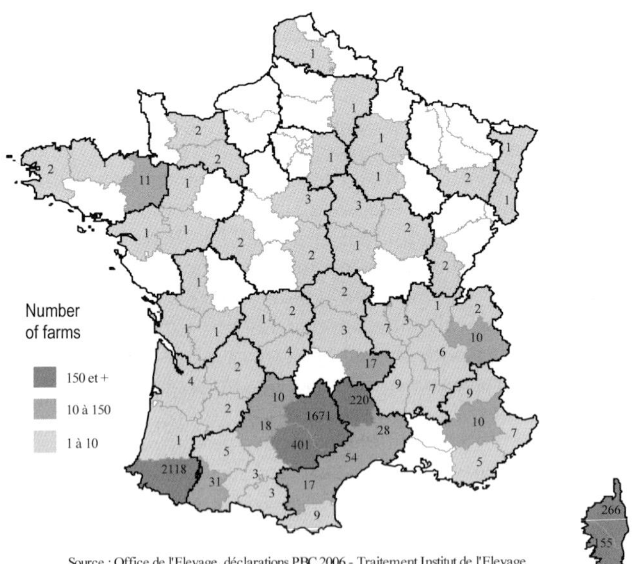

Source : Office de l'Elevage, déclarations PBC 2006 - Traitement Institut de l'Elevage
Fonds cartographiques Articque© Tous droits réservés

Figure 1. Location of dairy sheep farms in France.

Most farmers are involved in PDO production: Roquefort cheese in the Roquefort area, Ossau-Iraty cheese in the French western Pyrenees and Brocciu in Corsica. Processing the milk into PDO cheeses requires that the farmers comply with PDO rules concerning production conditions.

In 2006, sheep milk production was just over 260 million liters. After increasing very rapidly from the 1970s until the mid-90s, milk production is now relatively stable. Nevertheless, over the past 10 years, production has increased by 47% in the French western Pyrenees area, and by a little less than 5% in the Roquefort area (Figure 2).

Technical support for breeders

The evolution of milk production can be explained by the importance of the technical advice given to breeders: the SIEOL system manages data from 55% of French dairy sheep farms and 70% of the dairy ewes (Table 2) (Astruc, 2008).

In the Roquefort area, recording milk yields is very common and covers 80% of dairy sheep flocks and 85% of dairy ewes. In the early 1970s, cheese making factories needed more milk and so technical organizations started recording milk yields and artificial insemination. The primary objective was

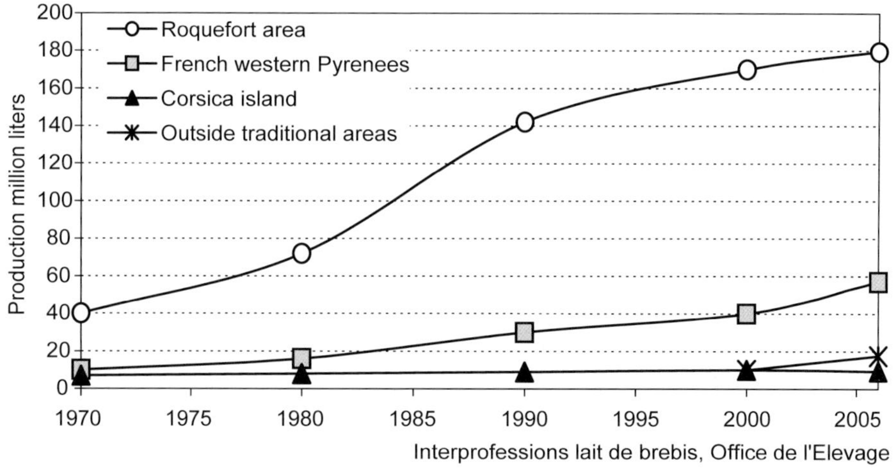

Figure 2. Evolution of dairy sheep production in France.

Table 2. Impact of SIEOL in France in 2007.

Areas	Official milk records		Unofficial milk records		Inventory and lambing	
	Number of farms	Number of ewes	Number of farms	Number of ewes	Number of farms	Number of ewes
Roquefort area	389	172,700	1,391	533,400	148	56,000
French western Pyrenees	344	108,400	125	35,600	23	6,400
Corsica	73	21,100	7	1,700	162	43,100
France	806	302,200	1,523	570,700	333	105,100

(Institut de l'Élevage, CNBL)

to increase the milk yield per ewe. Subsequently, tools were developed to assess the technical and economic results of the flocks. Recently, the dairy industry decided to pay for milk quality, and technical organizations developed tools to work on these new criteria. Early on, using a computer helped technicians to consolidate and analyze technical and economic results. Since the 1960s, three successive national genetic information systems for dairy sheep have been implemented in France mainly for genetic evaluation. The last of the three systems was used from 1984 to 2005, with major adaptations made during its lifetime such as the inclusion of a simplified design for recording milk quality, management of data from breed and AI organizations, and on-farm PC computerization of milk records.

Besides the genetic information system, in the early 1990s a system was designed to provide technical and economic support data (margin on feeding costs, gross margin, analysis of chemical and hygienic quality of milk) for extension services to the breeder. In 2000, the field organizations involved in the breeding of dairy sheep in the National Committee for Dairy Sheep (CNBL) decided to design the SIEOL information system. The project first involved the definition of a data model and a data dictionary, then the specifications, and finally the development of SIEOL.

A project group including technicians from technical organizations, engineers from the French Livestock Institute and INRA wrote the specifications. ARSOE located in the town of Soual and CTIG located in Jouy-en-Josas were in charge of software development. Financial support was provided by the French Ministry of Agriculture.

SIEOL was first implemented in September 2005 and is now used by 140 technicians. The system helps farmers improve the performance of their flock by recording artificial insemination and reproduction results, milk records, providing nutritional advice, and analyzing technical and economic results. SIEOL helps technicians provide relevant advice to dairy sheep farmers.

Organisation of SIEOL

SIEOL is based on a pyramidal organization with three levels:

1. At the local level, the technician responsible for milk records inputs on-farm data on a laptop (identification and registration, matings and lambings, milk records, the data required for technical and economic assessments), using SIEOL software (written in Visual Basic Language) to check input data and propose on-farm correction (if needed) and valorisation for the breeder.
2. The field technician regularly sends all the data concerning the flocks he/she is in charge of via intranet to a regional site located in the town of Soual (Midi-Pyrénées).
3. A national site located at INRA Jouy-en-Josas has all the available data needed for genetic evaluations (data on all the flocks whose milk yields have been recorded since 1950). The exchanges between the regional and national levels are based on replications from one database server to the other.

The same data model is used at each site. But while the national and regional databases contain all the data, the database of each technician only includes the data corresponding to the flocks he/she manages.

SIEOL is supplied by all the organizations involved in both genetic, technical and economic support: milk recording organizations, technical advice organizations, breeders organizations, AI organizations, milk analysis laboratories, molecular analysis laboratories, milk plants and inter-professional organizations, the French Livestock Institute, and INRA for EBVs and research purposes. Each organization can, through specific access rights, update the database and visualize data through a client x server solution (either via an input window or by submitting standardized messages).

Whatever the organization concerned, updating of data is only authorized via national software (collection of 'functions' written in C language, portable and usable on the different databases and systems: SQL Anywhere and Windows, DB2 and AS400, DB2 and MVS or UNIX) which checks the data before updating them and consequently increases the security and quality of the data.

Milk records and genetic improvement

Estimation of genetic value

Estimation of breeding values is one of the main purposes of the SIEOL information system. Regarding the French Breeding Law (JORF, 2006, 2007), genetic evaluations are designed and operated by INRA and results are passed on to the field organizations by the French Livestock Institute. At each run (3 times a year), SIEOL provides relevant data for the evaluations, whose results (EBVs, fixed effects) are fully stored. The computed EBVs to be published (regarding the reliability of the EBVs) are sent back to SIEOL. To achieve this goal, a specific evaluation and research information system, including scripts, programs and database, has been established to organize the interface between SIEOL and the genetic evaluations (Table 3). Scripts and software extract the relevant data from SIEOL, organize and calculate the data in operational tables (pedigrees, performances) to be used for evaluations (extension of lactation, editing, other calculations, etc), codify the effects and the animals, launch the evaluation, get back the raw data from the evaluations and organize them properly.

This set of tables and software can easily be used to implement experimental evaluations (to test new models for example), analyse data, estimate genetic parameters and is thus a useful tool for genetic purposes.

Genetic advice for breeders

Advice to breeders for the within-flock selection choices is the other main objective of the SIEOL information system. This advice is made possible by the SIEOL software at the level of the technician, through different valorisations and decision assistance programs.

These valorisations concern culling, replacement and the mating plan. To this end, the relevant data are edited on the screen (and can be printed out), thus enabling both the breeder and the technician to make the appropriate choices.

After milk recording, SIEOL may valorise data from the most recent test-day and from previous test-days (milk yield, somatic cell count, and sanitary anomalies), data from lambings and EBVs. The field technician can sort and rank the data in different ways, according to milk yield, somatic cell count, etc. This tool makes it easier to choose the ewes to cull.

Table 3. Dairy sheep population in SIEOL, genetic evaluation and PrP information system.

	SIEOL	Data included in the evaluations (2008)	Published EBVs in 2008 (active animals and reliability threshold)	PrP molecular data
Sheep (ewes and rams)	4,149,041	4,016,316	8,715 rams; 317,454 ewes	126,748
Lactations	7,511,547	6,414,737		

EBV: estimated breeding value.

As soon as data on lambings are collected, SIEOL ranks all the lambings, ordered by sires and by the genetic level of the dams, making it easier to choose the ewe lambs to be kept for replacement.

For the mating plan (choice of ewes to be inseminated, planning of the control mating), a set of data concerning all the previous reproduction results of the ewes, EBVs and PrP genotyping for resistance to scrapie, are used for decision support.

These have been just three examples of how the SIEOL system can be used as a decision support tool. However, SIEOL never imposes a decision: it suggests choices and encourages discussions between the technician and the breeder.

Technical and economic advice

Based on available information and results that the farmer wants to analyse, SIEOL suggests the technician should edit three kinds of assessments: technical, economic, and milk quality. These assessments enable the farmer and the technician to identify the strengths and weaknesses of the herd and to draw up a plan for improvement. Over several years, these assessments provide a useful record of technical and economic results.

The first edition proposed by SIEOL is a 'technical assessment'. It details reproduction and production results: fertility rate, level of prolificacy, numerical productivity, milk production per ewe.

In the 'technical and economic balance' the two main criteria are the margin on feeding cost (MCA) and stock level. The margin on feeding cost is equal to the sales of milk minus the price of purchased feed (concentrates and fodder) and the value of homegrown feed. Finally, the 'milk quality assessment' analyses results such as fat content, protein content, and somatic cell counts.

To help the technician evaluate the results of the flock and the progress of margins, the results of the flock can be compared with results of the four previous years as well as with a set of technical references.

For each kind of assessment, the technician has to enter the information needed to calculate the different criteria to be edited. However, as soon as possible, the software proposes to use the data entered for milk recording. For example, 'animal movements' have to be recorded monthly and causes of movements (entry/exit) for adults and yearling ewes; these data can also be calculated by valorizing the dates and causes of movements for each female of the flock.

SIEOL can also record data sent by dairy plants. This avoids mistakes in collecting information, eliminates tedious data entry and allows the technician to spend more time analysing the results with the farmer.

At the end of the year, the results of the individual farms are centralized in a national database. The analysis of these results enables any changes in the dairy sheep farms to be monitored and many of the questions asked by managers of the production chain to be answered (Morin *et al.*, 2005)

Conclusion

The information system is a key element in the development and technical management of a livestock production like dairy sheep production in France. It is significant that the National Committee for Dairy Sheep (CNBL, the technical organization that federates the technical organizations who advise French dairy sheep breeders), was created in the 1970s to implement a unified genetic information system for all daity sheep production areas in France.

Moreover, the specific organization of dairy sheep production in France and cost constraints led the technical organizations to design an original system to provide advice to breeders. This 'global system' is based on multi-function field technicians who are responsible for milk records, mating plans, feeding plans, as well as technical and economic assessments.

It was natural that the SIEOL information system should integrate these characteristics of dairy sheep production and combine on a single computing platform and a single information system the different technical management systems, and primarily genetic, technical and economic assessments. The organization of SIEOL makes data collection easier, as well as improves the homogeneity and quality of the data. It also allows constant updating of data, both collectively as well as individually, in order to provide more relevant advice to the breeder.

The SIEOL information system is a key tool in the current changes in the French Breeding Law. Indeed, based on this Law, the new genetic organization gives the profession an increasing role to play in terms of genetic orientation at a time when State subsidies are decreasing. As the information system is expensive to build and maintain, it was important that SIEOL should combine all the different services to the breeders, in order to actively involve them in the management of SIEOL and in its financing.

Acknowledgements

The authors would like to thank Luc Estienne, president of the steering committee of SIEOL, as well as all the members of the working group of SIEOL: Xavier Aguerre, Françoise Bouillon, Marie-Claude Carivenc, Eric Carrara, Patrice Gasc, Michèle Lafont, Jean-Baptiste Letxauréguy, Bruno Liquière, Jean-Claude Mathieu, Gilles Noubel and Odile Sallato.

References

Astruc, J.M., G. Lagriffoul, E. Morin, F. Barillet, B. Bonaiti and E. Rehben, 2008. SIEOL: implementing a global information system for genetic and techno-economic support in dairy sheep in France. Proc. 36th Session of ICAR, Niagara Falls, USA, 18-20 June 2008.

JORF, 2006. Décret N° 2006-1662 du 21 décembre 2006 relatif à l'identification et à l'amélioration génétique des animaux Journal Officiel de la République Française du 23 décembre 2006.

JORF, 2007. Arrêté du 12 septembre 2007 relatif aux systèmes nationaux d'information génétique des espèces bovine, ovine, caprine et porcine, Journal Officiel de la République Française du 21 septembre 2007.

Morin, E., C. Cassignol and J. Fraysse, 2004. La production de lait de brebis en France, diversité des systèmes d'exploitation, Institut de l'Élevage collection résultats, octobre 2004, 148 pp.

Morin, E., M. Oçafrain, J.M. Arranz and S. Dupin, 2005. Evolution des résultats enregistrés dans le cadre de l'appui technique auprès des éleveurs ovins lait des Pyrénées-Atlantiques. Options méditerranéennes, Numéro 70, Analyse technico-économique des systèmes de production ovine et caprine: méthodologie et valorisation pour le développement et la prospective, 147-156.

Outlook for producing Timahdite lamb with improved dietetic quality reared on rangelands in the Central Middle Atlas (Morocco)

A. Araba[1], P. Morand-Fehr[2], A. El Aïch[1], M. Bouarour[1], P. Schmidely[2] and V. Berthelot[2]
[1]Institut Agronomique et Vétérinaire Hassan II, Rabat, Morocco
[2]AgroParisTech, 16 rue Claude Bernard, 75231 Paris Cedex 05, France

Abstract

During recent years, the increasing prevalence of cardiovascular diseases in Morocco has very often been associated with consumption of lamb. PRAD Project 04-05 thus aimed to identify methods of rearing to improve the dietetic quality of the meat of Timahdite breed lambs raised in the area of Azrou-Khenifra (Central Middle Atlas of Morocco). Within the framework of this project, two experiments were conducted in two consecutive years to compare three different feeding systems: sheep raised indoors fed with concentrate and hay (CC, currently the most frequently used system), a feeding system based exclusively on pasture (PP, which is subject to climatic risks from a year to another), a feeding system based on pasture supplemented with concentrate (PC, to reduce the effect of the risks of climatic conditions and of limited feed availability on animal growth). Feed intake for animals in the CC and PC groups was adjusted to achieve a daily gain similar to that of animals in the PP group to avoid a possible effect of growth rate on meat quality. The growth rate of the PP group was satisfactory because of the favorable climatic conditions during the course of the trial. Lambs in the PC group tended to have higher growth rate than those in the CC and PP groups. The carcasses and the meat from lambs raised on pasture (PP group) tended to be leaner than those of CC and PC groups. Several lipid parameters related to dietetic quality of muscle and fat were improved in lambs in the PP and PC groups compared to CC, in particular the relationship between the fatty acids of the Omega-6 and Omega-3 families. The most important point is that the addition of concentrate to the diet of lambs raised on pasture did not impair the dietetic quality of the meat in the PP group. These results are currently being analyzed with the aim of setting up the production of Timahdite lambs with a label based on a breed, a technique of feeding on pasture with or without concentrate, and improved dietetic quality of the product.

Keywords: Timahdite breed, pasture, concentrate, growth, meat quality

Introduction

In the Middle Atlas, the cradle of the Timahdite breed, sheep breeding is the principal agricultural activity of the rural population. The Timahdite breed, which is known for its rusticity and its adaptation to the conditions of the area, makes optimal use of the pastoral resources. However, changes are underway in the production system because of the disappearance of great transhumance, the sedentarization of the stockbreeders, the appropriation of the land, the extension of the crops at the expense of pasture, and drought-enduring plants. These developments have led to more frequent supplementation of animals during a long period of the year. In parallel, more fatteners located in the peripheries of large urban centres have started to purchase sheep in the Middle Atlas in order to finish them indoors before marketing them. Knowing that the quality of the lamb meat depends on a great number of intrinsic and extrinsic factors, of which the diet constitutes one of the bases of the connection of the meat to its origin, it was considered important to undertake a study to evaluate performance, carcass characteristics and meat quality of lambs raised on grass compared to those fed partially or completely on concentrate and hay. The results of this work could be used to dissociate the meat produced on pasture from that produced by sheep fed concentrate and hay.

Material and methods

This experiment was conducted in Khenifra (Middle Atlas) on 39 entire male lambs of Moroccan Timahdite breed, about 6 months of age and about 23 kg of liveweight, originating from the same farm to assess effect of pasture vs. feeding concentrate and hay on growth performance and carcass and meat quality of lambs. The experiment was conducted in March, April and May 2007 when grass was easily available. Animals were randomly allocated to one of three diets: pasture (PP), oat hay and concentrate (CC), and pasture and hay and concentrate (PC). CC lambs were collectively fed indoors, and feed intakes for CC and PC animals were adjusted to obtain a daily gain similar to that of lambs in the PP group to avoid a possible effect of growth rate on meat quality. Animals from the CC group were fed a diet containing about 50% commercial concentrate and 50% oat hay. Animals in the PC group received half the CC diet at night, and spent the whole day on pasture. PP animals grazed all day. Water was available for PP and PC animals when returning from pasture and at all times for the CC group.

Animals were weighed at 30-day intervals. Feed intake of pasture-fed animals was estimated using the hand clipping technique in the middle of the trial. At termination, six lambs from each group were slaughtered. Carcass information was collected, and included warm carcass weight, omental and perirenal adipose tissue weights, fifth quarter (offal) constituents, carcass conformation.

The lipids of the semi-membranous muscle (SM), intermuscular (IM), subcutaneous (SC) and perirenal (PR) fats were extracted from freeze-dried samples according to the method of Bligh and Dyer (1959). The methylation of the lipid extracts was carried out using the method of Christie *et al.* (2001) modified by Bas *et al.* (2007). The methyl esters taken by n-hexane were analyzed by chromatography in liquid phase, in a CPL Varia CP-3900 with a capillary tube (60 ml X 0.25 mm i.d.x. 0.25 μm film thickness) of DW-war on silica (Bas *et al.*, 2007).

All data were analyzed using the SPSS GLM procedure for analysis of variance, analysis of covariance and regression routines, following the model: Yij=m+Mi+(axIWt)+Eij, where Yij is the variable analyzed, m is the overall mean, Mi is the effect of diet (i = 1, 2, 3), IWt is the covariate carcass weight (kg) used when analyzing carcass data, and a is the coefficient of regression. The error term was Eij. The Student Newman Keul's procedure was used to separate least squares means when significant effects were detected.

Results and discussion

Growth performance

Table 1 summarizes the effect of diet on overall growth and performance of lambs. There were no differences (P= 0.96 and P= 0.58) between lambs on the three diets with respect to initial live weight and final live weight, since these weights were set constant by design. Average daily gain (ADG) tended to be higher (P= 0.20) in lambs in the PC group (189, 203 and 186 g for CC, PC, and PP groups respectively). Daily dry matter intake averaged 1.30, 1.39 and 1.29 kg in the PP, PC and CC groups respectively. As a consequence, PP animals seemed to require more feed/gain (7.01 kg DM/kg of gain) than those in the PC group (6.83) and CC group (6.61), because of the higher energy content of CC diet, and the reduced movement of animals fed this diet.

Carcass characteristics

The animals were slaughtered with the aim of obtaining a constant live weight. But despite similar live weight, dressing percentage decreased from concentrate-fed to pasture-fed animals (P= 0.004). The trend observed in dressing percentage difference between groups may be partly due to differences

Table 1. Growth performance of lambs as affected by diet.

Variable	Diet[1]			Pr>F
	CC	PC	PP	
Number of animals	13	13	13	-
Initial live weight	23.4	23.0	23.2	0.96
Final live weight	40.0	41.3	39.7	0.58
Average daily gain, g	189	203	186	0.20
Daily DM intake, kg	1.29	1.39	1.30	-
Feed conversion, kg DM/kg of gain	6.61	6.83	7.01	-

[1] Oat hay and concentrate (CC), pasture (PP) and pasture and hay and concentrate (PC).
DM: dry matter.

in digesta weights linked to the different levels of fiber ingested, and in internal fat weights. Traits reflecting the importance of muscle development and conformation were slightly impacted by pasture (Table 2). Sheep in the CC, PC and PP groups had a round thickness of 21.92, 22.03, and 21.30 cm (P=0.23), a carcass compactness of 0.36, 0.36, and 0.35 kg/cm (P=0.58) and a round compactness of 0.32, 0.31, and 0.30 (P=0.006), respectively. Omental fat and kidney fat tended to be lower in lambs raised on pasture only (Table 2). Fat deposition followed the trend that faster gaining animals deposited more fat. Fat deposition appeared to be linked to growth rate. This result supports the findings of Gaili (1993) who showed a positive and significant relationship between fat and growth rate. This likely reflects the lower priority for partitioning feed energy to omental and kidney fats vs. other tissues until greater rates of gain are achieved.

Fatty acid composition

Table 3 presents the fatty-acid composition of the muscle and the three fats while highlighting individual fatty acids and sums of the same fatty-acid family which are typical of ruminant tissues or which can affect the dietetic value of the meat or the fat consumed by humans.

Table 2. Carcass characteristics of lambs as affected by diet.

Variable	Diet[1]			Pr>F
	CC	PC	PP	
Number of animals	13	13	13	-
Carcass weight	19.4	19.6	19.3	0.69
Dressing %	48.0	47.5	47.4	0.004
Omental fat, kg	413	466	407	0.15
Perirenal fat, kg	398	431	382	0.128
Carcass compactness, kg/cm	0.36	0.36	0.35	0.58
Round thickness, cm	21.92	22.03	21.30	0.23
Round compactness	0.32	0.31	0.30	0.006

[1] Oat hay and concentrate (CC), pasture (PP) and pasture and hay and concentrate (PC).

Table 3. Fatty acid composition of adipose tissue and muscle of lambs in 2007.

Diet	Muscle			Adipose tissue								
				Intermuscular			Subcutanuous			Perirenal		
	PP	PC	CC	PP	PC	CC	PP	PC	CC	PP	PC	CC
FA[1]:												
16:0	25.8	24.9	23.1	21.4	20.7	22.4	24.7^{a}	24.5^{a}	22.9^{b}	22.6^{a}	22.3^{a}	18.8^{b}
18:0	12.2	12.7	12.1	23.6	26.7	20.1	16.9	20.6	18.8	28.8	29.4	30.2
18:1n9	27.1^{a}	28.9^{a}	34.1^{b}	27.1^{a}	27.1^{a}	33.1^{b}	23.9^{a}	23.0^{a}	31.6^{b}	17.1^{a}	18.5^{a}	28.6^{b}
18:2c9t10				0.78	0.89	1.05	0.51^{a}	0.50^{a}	1.16^{b}	0.47^{a}	0.51^{a}	1.15^{b}
18:2c9t11	0.71	0.63	0.67	0.83	0.75	0.82	1.0	0.87	0.89	0.62	0.61	0.51
Sum of:												
Saturated FA	43.5^{a}	41.9^{ab}	39.0^{b}	52.0^{a}	52.8^{a}	48.0^{b}	50.0^{a}	51.1^{a}	47.3^{b}	58.7^{a}	57.2^{a}	53.5^{b}
Odd FA	1.61^{a}	1.50^{a}	1.25^{b}	2.09^{a}	2.07^{a}	1.81^{b}	2.80^{a}	2.82^{a}	2.21^{b}	2.89^{a}	2.87^{a}	1.91^{b}
Branched FA	0.26	0.22	0.20	1.91	1.88	1.85	2.22^{a}	2.23^{a}	2.00^{a}	2.23^{a}	2.21^{a}	1.81^{b}
Cis mono-unsaturated FA	31.3^{a}	33.1^{a}	38.7^{b}	34.1^{a}	34.0^{a}	40.3^{b}	31.0^{a}	29.7^{a}	39.2^{b}	22.9^{a}	24.4^{b}	35.5^{c}
CLA FA	0.73	0.65	0.68	0.86	0.78	0.87	1.05	0.91	0.94	0.64	0.63	0.54
n-6 FA	7.81	8.31	9.57	3.06	3.03	3.18	2.90^{a}	3.00^{a}	3.30^{b}	3.06	3.03	3.18
n-3 FA	2.80^{a}	2.74^{a}	1.27^{b}	0.60	0.49	0.30	1.20^{a}	1.30^{a}	0.42^{b}	1.13^{a}	1.17^{a}	0.18^{b}
n-6/n-3	2.79^{a}	3.03^{a}	7.54^{b}	5.10^{a}	6.18^{a}	10.6^{a}	2.42^{a}	2.31^{a}	7.86^{b}	2.71^{a}	2.59^{a}	17.6^{b}

[1] FA: fatty acid.

Values with different superscript letters differ significantly ($P<0.05$).

The fatty-acid composition of the muscle and the three fats of the lambs in this experiment are in agreement with the compositions already published in the literature. The internal fat (PR) was higher in saturated fatty acids (18:0) and lower in mono or poly-unsaturated fatty acids (18:1 n-9, CLA, n-3) than in subcutaneous fat (SC). The intermuscular fat (IM) was close to PR or intermediary between PR and SC.

The composition obtained in lambs in this experiment carried out in a Mediterranean zone was rather different from that observed in other genotypes in temperate zones (Bas *et al.,* 2007) even if it is always arbitrary to compare compositions of fatty-acids of animals raised under different conditions and with samples analyzed by methods that are not completely identical. In particular, the composition of subcutaneous fat (SC) of the Timahdite lambs was lower in mono- and polyunsaturated fatty acids and higher in saturated fatty acids than that of lambs raised in temperate zones, reflecting the adaptation to the heat of lambs in Mediterranean zones. The fatty-acid composition of subcutaneous fat adapts to the outside temperature so that their fluidity is comparable whatever the latitude and the metabolic processes can proceed under identical conditions.

On the other hand, fatty-acid composition of perirenal fat and of muscle from kids raised in Argan forest (Bas *et al.,* 2005) was rather close to that of the Timahdite lambs even if in Argan forest the nature of the diet did strongly affect the fatty-acid composition of tissues. Only minor fatty-acid percentages, in particular the odd fatty acids were very different in lambs and in kids from Morocco. The higher percentage in the kids can be regarded as a characteristic of the species.

The effect of diet on fatty-acid composition of tissues from the Timahdite lambs was important especially on internal (PR) and subcutaneous (SC) fat whereas the muscle (SM) seemed to be less sensitive to the effect of diet, and the intermuscular fat (IM) was in an intermediate position. The fatty acids that were most sensitive to this effect were saturated fatty acids (in particular palmitic acid (16:0) more than stearic acid (18:0), odd and branched fatty acids).

Some unsaturated fatty acids like CLA (rumenic acid 18:2c9t11) and fatty acids from the n-6 family were rather less sensitive to this effect, whereas the *cis* mono-unsaturated fatty acids (oleic acid 18:1 n-9) and fatty-acids of the n-3 family were more sensitive.

In general, there was no significant difference between lambs in the PP and PC groups (outdoor groups) except the sum of *cis* mono-unsaturated fatty acids of the perirenal fat, which was significantly higher in the PC group than in the PP group.

However, the difference was significant between PP and PC groups on one hand, and the CC group on the other. That means that the characteristics of the fatty-acid composition in lamb tissues due to the exclusive use of pasture (PP diet) persist even if concentrated feed is supplied to the lambs on pasture (PC group).

Compared to the rich concentrate diet (CC), the outdoor groups (PP and PC) were characterized by a significantly higher total saturated fatty-acid percentage mainly because of the variations in palmitic acid (16:0); the same tendency was observed for odd and branched minor fatty acids, which could be explained by a higher fiber ingestion. On the other hand, pasture based diets (PP and PC groups) produced a significant reduction in cis mono-unsaturated fatty acids particularly oleic acid (18:1 n-9) in the four tissues analyzed. It should also be noted that trans mono-unsaturated fatty acid (18:2c9t10), when measured, always tended to be lower with the PP and PC diets. In addition, the nature of the diet did not significantly influence the conjugated linoleic acids (CLA), particularly the 18:2 c9t11.

The sum of the fatty acids of the n-6 family tended to be higher in the lambs on the CC diet, significantly only in subcutaneous fat (SC). On the other hand, the fatty-acid percentages of the n-3 family were significantly higher in the PP and PC groups than in the CC group so that n-6/n-3 ratios were equal or lower than six and generally close to 2.5-3 in the PP and PC groups but were always higher than seven in the CC group.

Some of the fatty acids shown in Table 3 are likely to play a positive or negative dietetic role in human nutrition (Ledoux, 2006; Ailhaud, 2006; Martin, 2007; Legrand 2008).

Nevertheless, the results found on these Timahdite lambs show that pasture-based diets (PP and PC) clearly raised the percentage of the n-3 family acids two to seven fold depending on the tissue. Consequently with these diets, the n-6/n-3 ratio was equal to or lower than 5-6 whereas it can reach up to 17 (in perirenal fat) when the lambs consume 50% concentrate diet (CC group). Identical results have been obtained in kids (Bas *et al.*, 2005).

On the other hand, the nature of the diet did not significantly influence the percentage of CLA observed in kids (Bas *et al.*, 2005). But the increase in the percentage of CLA in kids raised in Argan forest was especially due to the restricted available pasture which limited the growth rate. Therefore, there could be two nutritional factors to distinguish the influence of the percentages of tissue fatty acids with high dietetic value: intake level and the nature of the feed (pasture flora or trough feeding with concentrate).

Lastly, it should be noted that the CC diet resulted in lower saturated fatty acid contents and palmitic acid (often to a significant degree) than the PP and PC diets; which also differ from the result obtained in kids raised in Argan forest.

Conclusions

Growth performance and carcass characteristics tended to be improved with increasing concentrate and hay in the diet. These results were obtained during a period of grass availability when environmental conditions were favorable. However, this feeding situation is limited in time in Mediterranean countries, which makes the supplementary feeding of lambs useful. Such supplementation allows some non-significant improvement in growth performance without deteriorating the quality of the carcasses and some dietetic parameters of meat and fat. Indeed, from the dietetic point of, Timahdite lambs raised on pasture presented clearly higher n-3 fatty acid percentages and an n-3/n-6 ratio of 5-6 or less compared to lambs raised indoors, even when growth rates were statistically similar for all the diets. This effect persisted even when the lambs' diets were supplemented with concentrate.

On the other hand, pasture-based diets did not improve meat and fat dietetic value in terms of saturated fatty-acids and CLA in these lambs, contrary to what has been observed in kids, probably because of the level of intake which was identical for the three groups. Indeed, the ingestion of lambs in the CC group was adjusted to achieve a similar level of growth to that of lambs in the PP and PC groups. Thus it is probable that the reduction in intake in the CC group set by design would be accompanied by an increase in CLA and a reduction in the saturated fatty acids in lamb tissues. In any case, this assumption should be tested in a future experiment. Since in general, pasture feeding allows a lower intake and growth rate than intensive feeding with hay/concentrate, this would lead to an improvement in the dietetic value of meat and fat not only because of the improved n-6/n-3 ratio but also if the later assumption is verified, in terms of CLA and saturated fatty-acids as well as less fat deposition in fat tissues.

References

Ailhaud, G., 2006. Acides gras polyinsaturés et masse adipeuse: une relation qui pose problème. Cholé-Doc, N° 98, 1-3.

Bas, P., E. Dahbi, A. El Aïch, P. Morand-Fehr and A. Araba, 2005. Effect of feeding on fatty acid composition of muscles and adipose tissues in young goats raised in the Argan tree forest of Morocco. Meat Science, 71, 317-326.

Bas, P., V. Berthelot, E. Pottier and J. Normand, 2007. Effect of level of linseed on fatty acid composition of muscles and adipose tissues of lambs with emphasis on trans fatty acids. Meat Science, 77, 678-688.

Bligh, E.G. and W.J. Dyer, 1959. A rapid method to total lipid extraction and purification. Canadian Journal of Biochemistry and Physiology, 37, 911-917.

Christie, W.W., J.L. Sébécio and P. Juanéda, 2001. A practical guide to the analysis of conjugated linoleic acid (CLA). Inform., 12, 147-152.

Gaili, E.S.E., 1993. Breed and sex differences in body composition of sheep in relation to maturity and growth rate. J. Agri. Sci., 118, 121-126.

Ledoux, M., 2006. Les acides linoléiques conjugés: présence dans les aliments et propriétés physiologiques. Cholé-Doc, N° 94, 1-4.

Legrand, Ph., 2008. Intérêt nutritionnel des principaux acides gras des lipides du lait. Cholé-Doc, N° 105, 1-4.

Martin, J.C., 2007. L'acide ruménique: un trans particulier. Cholé-Doc. N° 103, 1 et 3-4.

Effect of different production systems on lamb sensory quality

B. Panea, G. Ripoll, S. Carrasco and M. Joy
Centro de Investigación y Tecnología Agroalimentaria de Aragón, Avenida de Montañana 930, 50059 Zaragoza, Spain

Abstract

The influence of four different production systems (thereafter, treatment) on lamb meat sensory characteristic measured by a trained panel test was analysed. The treatments were: INT, lambs were kept indoor with free access to concentrate while dams were only indoor from 15:00 to 08:00 h. EST, lambs and ewes were kept indoor with free access to concentrate. GR, lambs and ewes grazed alfalfa without concentrate, GR+S, animals grazed alfalfa and lambs had cereal supplement. Lambs from DRL-GRE and DRL were weaned at 45 days old. Six entire male lambs from each treatment were slaughtered at 20-22 kg live weight and theirs carcasses were cooled at 4 °C. At 24 hours *post mortem*, *Longissimus lumborum* were excised, vaccum packed and aged for a total period of three days. Samples were grilled to an internal temperature of 70 °C. A trained nine-member panel valued lamb odour intensity, tenderness, juiciness, lamb flavour intensity, liver flavour intensity, fat flavour intensity, metallic flavour intensity and overall liking using a 10-point scale. An ANOVA procedure, means and standard deviation for each attribute in each treatment and Pearson correlation analysis were carried using SPSS v.13.0 software. No differences between treatments were found for any variable. All variables were closely related one each others. Overall liking was positively correlated with tenderness and juiciness and negatively correlated with liver, fat or metallic flavours. From a sensorial point of view, alfalfa grazing is a workable alternative to traditional concentrate feeding for rearing lambs.

Keywords: feeding system, panel test, lamb

Introduction

In the European Mediterranean countries lambs are traditionally reared with their dams in indoor conditions until weaning (about 45 days of age) and thereafter fattened with concentrate until slaughter, around 20-24 kg live weight. Consumers are used to this kind of meat and they consider it as a high quality product (Sañudo *et al.*, 1998). Nowadays, this production system has been altered because there are many factors that constraint it. First, preferences of consumers are changing requiring meat with low fat content, as it is associated to a healthier meat in comparation to meat rich in fat. Second, EU Common Agricultural Policy promotes the landing of production systems and, third, the recent and constant increases of cost of cereals have aggravated the situation causing many farms to become unprofitable. Under these circumstances, an alternative to indoor production systems would be grazing systems. Nevertheless, feeding system can affect meat characteristics. Therefore, the present trial pretend asses the effect of feeding system on sensorial meat quality.

Material and methods

Twenty-four single male lambs were used. After birth, lambs were provided a complement of selenium and remained drylot with their dams for some days to ensure maternal bonding and randomly allocated to four treatments (n=12):

1. Grazing (GR): lambs and ewes were continuously stocked on a permanent pasture. No concentrate was available to dams or lambs. Lambs suckled their mothers and grazed until slaughter.
2. Grazing with supplement for lambs (GR+S): the same management as GR, but these lambs received concentrate *ad libitum* in lamb creep feeders until slaughter.

3. Indoor lambs with grazing ewes (DRL-GRE): lambs remained indoors and fed creep while ewes grazed during eight hours a day (08:00 to 16:00 h). Thereafter, they received a supplement of 0.5 kg fresh matter of barley meal/day (119 g CP, 248 g NDF, on dry matter basis). This system aimed to reply the commonly used management in that region.
4. Drylot (DRL): lambs and ewes were kept always in confinement. Ewes had free access to dry unifeed (110 g CP, 681 g NDF, on dry matter basis) and lambs had concentrate *ad libitum*.

In treatments DRL-GRE and DRL lambs were weaned at 45 days old, while in GR and GR+S lambs were unweaned. In all treatments, lambs were weighed on weekly intervals and slaughtered, accordingly to the guidelines of the Council Directive 86/609/EEC (European Communities, 1986) when they reached 22-24 kg of live weight. The day after slaughter carcasses were split along the spine and the m. *Longissimus dorsi* from 4th to 6th *lumbar vertebrae* were removed of the left half carcasses, vacuum-packed, aged for a total period of three days and frozen (-20 °C) until their analysis.

The day of the panel session, samples were thawed under tap water until they reached an internal temperature of 17-19 °C, over wrapped by aluminium foil and cooked (grilled) in a preheated double plate grill (SAMMIC P8D-2; 200 °C) to an internal temperature of 70 °C, which was monitored by an internal termocouple (JENWAY 2000). Every loin was then trimmed of any external connective tissue, cut into samples that were 2 × 2 cm, wrapped in codified aluminium foil and stored in a warm cabinet at 50 °C until tasted. A comparative multi-sample test, with four samples each, was used to detect differences in sensory attributes among treatments. Samples were served to a trained (ISO-8586-1) nine-member panel (with panel members in individual booths) under red lighting to mask any differences in meat colour. Panellist was asked to evaluate on a 10-point scale the following attributes: lamb odour intensity, tenderness, juiciness, lamb flavour intensity, liver flavour intensity, fat flavour intensity, metallic flavour intensity and overall liking, with a score of 1 the lowest and 10 the highest possible for each attribute. A total of 54 juices per treatment were obtained.

Statistical analyses were performed using SPSS v.13.0 software. An ANOVA procedure was carried with treatment as main effect. Means and standard deviation for each attribute in each production system were calculated. Finally, Pearson correlation analysis was carried out in order to study variable relationships.

Results and discussion

Table 1 shows significance of the main effect on sensory variables (*P* value), as well as means and standard deviation for each attribute in each treatment. The feeding system did not affect any sensory characteristics of meat lamb. In literature, diet effect on sensory characteristic was not clear and results were frequently contradictory. Regarding tenderness, Paul *et al.* (1964) found slight differences between animals finished on concentrate or grass, presenting the firsts higher values on tenderness. Nevertheless, neither Summers *et al.* (1978) nor Kemp (1981) detected significant differences when comparing animals fed-grass, concentrate or both. Concerning level of energy of the diet, Solomon *et al.* (1986, 1988, 1996) found that meat from lambs reared with low-level of energy diet was more tender that meat from animals reared in a high-level diet, whereas Crouse *et al.* (1978) demonstrated just the opposite. Protein content of the diet was important only if differences between diets were significant (10% vs. 16%), and in this case, higher the protein level, higher the tenderness (Kemp *et al.*, 1976). Nevertheless, Fhamy *et al.* (1992) did not found differences in tenderness when difference between diets was only of 3 points (13 vs. 16%). Finally, neither type of forage (Field al., 1978, Vipond *et al.*, 1995, Hopkins *et al.*, 1999) nor weaned (Summers *et al.*, 1978, Sañudo *et al.*, 1998) affected tenderness.

Similarly, juiciness is not affected by feeding system: neither energy level of the diet (Crouse *et al.*, 1978; Solomon *et al.*, 1986; Bosman *et al.*, 1994), nor protein content (Kemp *et al.*, 1976, 1981) nor

Table 1. Effect of treatment on lamb sensory variables (scale 1-10).

	P	GR[1]		GR+S[2]		DRL-GRE[3]		DRL[4]	
		Mean	St.dev	Mean	St.dev	Mean	St.dev	Mean	St.dev
Lamb odour intensity	0.839	5.1	1.4	4.9	1.5	5.0	1.6	4.8	1.4
Tenderness	0.594	7.2	1.3	7.1	1.0	7.4	1.1	7.2	1.0
Juiciness	0.838	6.4	1.3	6.4	1.2	6.3	1.3	6.2	1.1
Lamb flavour intensity	0.983	6.0	1.4	6.0	1.2	5.9	1.5	5.9	1.2
Liver flavour intensity	0.365	2.9	1.4	2.6	1.3	3.1	1.5	3.0	1.6
Fat flavour intensity	0.980	4.2	1.7	4.3	1.9	4.3	1.8	4.3	1.8
Metallic flavour intensity	0.680	3.3	1.6	3.3	1.7	3.7	1.7	3.4	1.6
Overall liking	0.313	6.4	1.3	6.3	1.3	6.0	1.4	6.1	1.1

[1] Grazing.
[2] Grazing with supplement for lambs.
[3] Indoor lambs with grazing ewes.
[4] Drylot.

type of forage (Field *et al.,* 1978; Vipond *et al.,* 1995; Hopkins *et al.,* 1999) nor weaning (Summers *et al.,* 1978; Vipond *et al.,* 1995; Sañudo *et al.,* 1998) affected juiciness of lamb meat.

As regards odour and flavour, Paul *et al.* (1964) did not find differences between animals finished with pastures or with barley and alfalfa hay concentrate. Similarly, Solomon *et al.* (1996) did not find differences between diets (alfalfa vs. concentrate). Nevertheless, evidence exists that a diet based on pastures promotes odours slightly stronger that maize concentrate (Rousset-Karim *et al.,* 1997) and that the substitution of forage by concentrate reduces the meat flavour intensity (Field *et al.,* 1978). Forage type can affect odour and flavour. Cramer *et al.* (1967) demonstrated that odour and flavour was higher when animals were reared with *Trifolium repens* than when animals were reared with *Lolium perennes*. Nevertheless, Vipond *et al.* (1995) in a study comparing the same species did not find differences on flavour. Park *et al.* (1972) found that meat from animals reared with *Medicago sativa* was more flavoured and less acceptable than meat from animals reared on *Phalaris tuberosa*, but Hopkins *et al.* (1999), comparing meat from animals reared with *Medicago sativa* or *Atriplex nummularia* found that meat from the last diet has a stronger aroma, whereas no differences in acceptability were found. Regarding flavour acceptability, results were also contradictory. Summers *et al.* (1978) or Kemp *et al.* (1981) pointed out that flavour acceptability was superior in animals supplemented concentrate than in animals reared only on pasture, but others authors concluded that neither energy level of the diet (Crouse *et al.,* 1978; Solomon *et al.,* 1986, 1988; Bosman *et al.,* 1994) nor protein content (Kemp *et al.,* 1976, 1981) affected flavour acceptability.

Finally, and considering overall liking, it was demonstrated that neither energy level of the diet (Crouse *et al.*, 1978), nor protein content (Kemp *et al.,* 1976) nor species of pasture (Vipond *et al.,* 1995; Hopkins *et al.,* 1999) nor weaning (Sañudo *et al.,* 1998) affected overall liking.

Table 2 shows Pearson's correlation coefficients. In general, all variables were closely related to each other. Lamb odour was positively related with lamb flavour, which was in agreement with results by Alfonso (2000). Tenderness was positively correlated with juiciness and overall liking as well as juiciness and overall liking were positively correlated among them. Nevertheless, overall liking was negatively correlated with liver flavour, fat flavour and metallic flavour, which were positively

Table 2. Pearson's correlation coefficients between sensory variables from lamb meat. Only significant coefficients were shown[1].

	Tenderness	Juiciness	Lamb flavour	Liver flavour	Fat flavour	Metallic flavour	Overall liking
Lamb odour	-0.18**		0.39**	0.14*	-0.16*		
Tenderness		0.39**					0.37**
Juiciness			0.36**	0.19**	0.14*	0.14*	0.16*
Lamb flavour				0.33**			
Liver flavour					0.25**	0.49**	-0.42**
Fat flavour							-0.30**
Metallic flavour							-0.41**

[1] ** Correlation is significant at 0.01 (bilateral); * correlation is significant at 0.05 (bilateral).

correlated one each other. Hence, it seemed that main problems in lamb meat sensory quality were odour/flavour related.

References

Alfonso, M., 2000. Caracterización sensorial y aceptabilidad de la carne de doce tipos ovinos representativos de distintos sistemas de producción europeos. Doctoral Thesis. University of Zaragoza (Spain). 338 pp.

Bosman, M.J.C., E.C. Webb, N.H. Casey, A.M Van Aardt and N. Silvis, 1994. Die invloed van hoë-en medium-energiediëte in verpild vorm op die eet-envetkwaliteit asook gaarmaakverliese van lamsvleis van verskillende rasse. Tydskrik vir Dieetkundunde en Huishoudkunde 22: 19-25.

Cramer, D.A., R.A. Barton, F.B. Shorland and Z. Czochanska, 1967. A comparison of the effect of white clover (*Trifolium repens)* and of perennial ryegrass (*Lolium perenne)* on fat deposition and flavour of lamb. J. Agr. Sci. 69: 367-373.

Crouse, J.D., R.A. Field, J.L. Chant, C.R. Ferrel, G.M. Smith and V.L. Harrison, 1978. Effect of dietary energy intake and palatability of different weight carcasses from ewe and ram lambs. J. Anim. Sci. 47: 1207-1218.

Fhamy, D.C., D.L. Harrison and L.L. Anderson, 1972. Lamb and beef roast cooked from the frozen state by dry and moist heat. J. Food Sci. 37: 226-229.

Field, R.A., J.C. Williams, C.L. Ferrell, J.D. Couse and J.E. Kunsman, 1978. Dietary alteration of palatability and fatty acids in meat from light and heavy weight ram lambs. J. Anim. Sci. 47: 858-864.

Hopkins, D.L. and A. Nicholson, 1999. Meat quality of wether lambs grazed on either saltbush *(Atriplex nummularia)* plus supplements or lucerne (*Medicago sativa*). Meat Sci. 51: 91-95.

Kemp, J.D., A.E. Johnson, D.F. Stewart, D.G. Ely and J.D. Fox, 1976. Effect of dietary protein, slaughter weight and sex on carcass composition, organoleptic properties and cooking losses of lamb. J. Anim. Sci. 42: 575-583.

Kemp, J.D., M. Mahyuddin, D.G. Ely, J.D. Fox and W.G. Moody, 1981. Effect of feedins systems, slaughter weight and sex on organolpetic properties and fatty acid composition of lamb. J. Anim. Sci. 51:321-330.

Park, R.J., J.L. Corbett and E.P. Funival, 1972. Flavour differences in meat from lambs grazed on lucerne (*Medicago sativa)* or phalaris (*Phalaris tuberosa)* pastures. J. Agric. Sci. 78: 47-52.

Paul, P.C., J. Torten and G.M. Spurlock, 1964. Eating quality of lamb. Food Tech. 18: 121 (1779)-130 (1788).

Rousset-Karim, S., O.A. Young and J.L. Berdagué, 1997. Diet and growth effects in panel assessment of shepmeat odour and flavour. Meat Sci. 45: 169-181.

Sañudo, C., I. Sierra, J.L. Olleta, L. Martin, M.M. Campo, M.P. Santolaria, J.D. Wood and G.R. Nute, 1998. Influence of weaning on carcass quality, fatty acid composition and meat quality in intensive lamb production systems. Anim. Sci. 66: 175-187.

Solomon, M.B., G.P. Lynch and B.W. Berry, 1986. Influence of animal diet and carcass electrical stimulation on the quality of meat from youthful ram lambs. J. Anim. Sci. 62: 139-146.

Solomon, M.B. and G.P. Lynch, 1988. Biochemical, histochemical and palatability characteristics of young ram lambs as affected by diet and electrical stimulation. J. Anim. Sci. 66: 1955-1962.

Solomon, M.B., B.W. Berry, F.L. Fluharty and K.E. Mc Clure, 1996. Effects of energy source and ionophore supplementation on lipid composition and sensory properties of lamb. J. Anim. Sci (88th Annual Meeting Abstracts): 162.

Summers, R.L., J.D. Kemp, D.G. Ely and J.D. Fox, 1978. Effect of weaning, feeding systems and sex of lamb on lamb carcass characteristics and palatability. J. Anim. Sci. 47: 622-629.

Vipond, J.E., S. Marie and E.A. Hunter, 1995. Effect of clover and milk in the diet of grazed lambs on meat quality. Anim. Sci. 60: 231-238.

Carcass and meat quality of Majorcan black pig slaughtered at different live weight

J. Gonzalez[1], J. Jaume[2], M. Gispert[1], P. Rodríguez[1], J. Tibau[3] and M.A. Oliver[1]
[1]IRTA Food Technology, Spain
[2]IBABSA Balearic Institute for Animal Biology, Spain
[3]IRTA Animal Production, Spain

Abstract

The Majorcan black pig is an autochthonous breed from Mallorca, in the Balearic Islands, which is considered in danger of extinction from 1997. Its carcass and meat quality characterization will provide an objective criterion for selecting animals, in order to target specifications according to industry demands. Fifty-seven animals were studied in two groups according to their weight: heavy (H; live weight 152.4 kg; age 14 months; 39 females) and light (L; live weight 100.1 kg; age 9 months; 18 females). Both groups were reared in a typical extensive system in Mallorca. Main carcass variables assessed were killing out percentage (H 80.1 vs. L 72.4%), MLOIN (H 66.5 vs. L 42.3 mm), and flare fat weight (H 6.2 vs. L 2.9 kg). Regarding meat quality, the studied parameters were ultimate pH value (H 5.78 vs. L 5.39), lightness (L*; H 43.87 vs. L 51.3) and intramuscular fat percentage in loin (H 9.1 vs. 7.0%). Significant differences were found in all variables studied.

Keywords: carcass, meat quality, heavy pig, Majorcan Black pig

Introduction

Majorcan black pig (called *Porc Negre Mallorquí* in Mallorca) is nowadays the only autochthonous breed conserved in Mallorca, in the Balearic Islands. This breed is reared in a extensive regime based production system. It has been recognized as an endangered breed by the Spanish Ministry of Agriculture, Fish and Food since 1997. Preservation of Majorcan black pig is mainly sustained by the utilization of pure breed animals in the elaboration of the *Sobrassada de Mallorca de Porc Negre Mallorqui*, a kind of cured sausage with paprika which enjoys a protected geographical indication, and by the commercialization of the *Porcella*, the pig is slaughtered at 7-12 kg live weight approximately, between 1-2 months, and consumed as roasted meat (Jaume, 2000). In 1997 appeared the breed herd book, which is administered by the Majorcan black pig producer's association, and currently controls 1,300 sows and 120 boars.

For many years boars have been selected during popular fairs and events, with the morphological traits as unique decisive factor. As a consequence, a high variability within this breed exists on carcass and meat quality. The first step to identify which attributes are able to be improved by means of the selection programme, in order to introduce any innovation in the production system, is the assessment of carcass and meat quality traits.

The most common practice with Majorcan black pig is slaughtering animals at heavy live weight, around 150 kg. Since it has been observed that the productive efficiency is reduced, and that fat deposition dramatically increases at late age in this breed (unpublished data), carcass and meat quality traits of animals slaughtered at low weight are of interest to assess the differences between lighter and heavier animals.

Therefore, the aim of the present study is to achieve economically profitable systems maintaining acceptable carcass and meat quality, as well as to characterize Majorcan Black Pig slaughtered at two different live weights and provide scientific tools for the farmers in order to look for opportunities to improve the production system meeting the industry demands.

Material and methods

Animals, husbandry system and feed

Fifty-seven purebred Majorcan black pig females were studied and distributed in two groups according to carcass weight: 39 heavy weighted pigs (heavy; body weight at slaughter 152.4±19.07 kg, carcass weight 114.6±15.78 kg) that were reared in several different farms and slaughtered in 3 different days at an average age of 14 months; and 18 light weighted pigs (light; body weight at slaughter 100.1±10.90 kg, carcass weight 72.6±9.67 kg) that were reared in the same farm and slaughtered at an age of 9 months. All of them were reared in extensive conditions, with a feeding regime based on pasture, cereals (barley and rye), legume seeds, figs, almonds, acorns and several Mediterranean shrubs.

Carcass quality

Carcass quality parameters were assessed on the left half of the carcass. Carcass length was measured from the first rib to the beginning of the *simfisis pubica;* minimum back fat thickness over the *Gluteus medius* muscle (MLOIN), back fat thickness at the last rib level (back fat) and at the first rib level (Esp1C) were measured at the mid-line.

Meat quality

Meat quality measurements were taken at the first lumbar vertebra level in a *Longissimus lumborum* (LL) muscle sample. Muscle pH was measured using a Knick pH meter (Knick, Berlin, Germany) at 24 h *post mortem* (ultimate pH). The instrumental color measurement was carried out using a Minolta CR-200 chromameter, obtaining the variables CIE L* (lightness) a* (red to green) b* (yellow to blue) (C.I.E., 1986). Intramuscular fat content was determined in the same muscle using the Near Infrared Transmittance technology (NIT) following the standard procedure described by Gispert *et al.* (1997).

Statistical analysis

Statistical analysis was performed using the computer software Statistical Analysis System (SAS, 2001). Significant differences among meat quality parameters by weight were analyzed using the GLM procedure. Differences among weight were tested using the Tukey test. Significance was fixed at $P<0.05$.

Results

Carcass quality

Animals used in this study, especially heavy pigs, presented a high standard deviation in carcass weight due to heterogeneity in the origin of farms, batch effect, feeding regime and energy requirements depending on the season they were slaughtered.

Descriptive analysis of data is shown in Table 1 according to the two groups of animals. Carcass quality variables were statistically analyzed comparing heavy and light pigs being all of them significantly different. Carcass weight mean was 114.6±15.78 kg and 72.6±9.68 kg in heavy and light animals respectively. Killing rate percentage was 80.1% for heavy pigs and 72.4% for light pigs. Higher killing rate percentages are expected in pigs slaughtered at heavy weight, although the observed values in light pigs are quite lower than those from leaner breeds at same slaughter weight. Flare fat was also higher in heavy pigs (6.2 kg) than light ones (2.9 kg), both higher than values

Table 1. Descriptive analysis of carcass quality traits of heavy and light pigs.

	Heavy			Light		
	n	Mean	S.D.	n	Mean	S.D.
Live weight (kg)	9	152.4	19.07	18	100.1	10.90
Carcass weight (kg)	39	114.6	15.78	18	72.6	9.68
Killing rate (%)	9	80.1	0.79	18	72.4	1.94
Flare fat (kg)	30	6.2	1.62	11	2.9	0.61
Carcass length (cm)	25	81.8	7.41	18	75.9	2.58
MLOIN (mm)[1]	26	66.5	10.56	18	42.3	7.89
Back fat (mm)[2]	26	74.0	13.22	18	42.4	5.41
Esp1C (mm)[3]	26	90.1	14.96	18	61.9	6.53

[1] MLOIN: minimum back fat at *Gluteus medius* level.

[2] Back fat: fat thickness at last rib level.

[3] Esp1C: fat thickness at first rib level.

observed in leaner breeds, which are often lower than 1 kg. Carcass length in heavy pigs was 81.8 kg and 75.9 kg in light pigs. Regarding measurement of back fat depositions, mean values obtained in the three points of measure were notably high in both heavy and light pigs: 66.5 vs. 42.3 (heavy vs. light); 74.0 vs. 42.4 and 90.1 vs. 61.9 at MLOIN, back fat and Esp1C levels respectively.

Meat quality

Main results on meat quality traits assessed in the present study are described in Table 2, classified by animal weight. Ultimate pH presents mean values, in both heavy and light animals, accepted by the meat industry. In spite of everything, some pH at 24 hours *post mortem* in heavy pigs were higher than 6.0, indicating some incidence of DFD meat (13% of observations). Ultimate pH values in heavy pigs were higher (5.78) than those of light animals (5.39), showing a P value lower than 0.001. Regarding objective color measures, lightness was significantly higher (P<0.001) in light pigs than heavy pigs (51.53 vs. 43.87 respectively), and a* and b* components did not show any significant difference. Objective measurement of intramuscular fat content relative to muscular tissue presented values of 9.1% vs. 7.0% in heavy and light animals respectively (P= 0.004).

Table 2. Descriptive analysis of meat quality traits of heavy and light pigs.

	Heavy			Light		
	n	Mean	S.D.	n	Mean	S.D.
Ultimate pH	27	5.78	0.253	18	5.39	0.068
L*	27	43.87	2.082	18	51.53	3.390
a*	27	10.15	1.634	18	10.75	1.023
b*	27	1.81	2.066	18	2.37	1.395
Intramuscular fat (%)	27	9.1	2.70	18	7.0	1.42

Implications

From results obtained in the present study, the most particular attributes of this breed were high adipogenic potential, especially regarding back fat tissue and intramuscular fat content. In order to produce high quality meat and meat products and increasing in parallel the productive efficiency of the system, an adequate selection program could permit the reduction of back fat deposition maintaining high levels of intramuscular fat content. Another possibility to achieve this goal could be to reduce slaughter weight. In present study results on carcass quality show a clear difference between two groups of weight regarding back fat thickness, being higher in heavy than light pigs. Intramuscular content was also higher in heavy pigs, but anyhow meat from lighter animals could be used to produce high quality products, as although intramuscular fat content is higher than values shown by other breeds at similar slaughter weight. Colour and pH values, either in heavy and light pigs, would be also adequate for producing fresh meat.

Meat from Majorcan black pig animals is used mainly to produce *Sobrassada de Mallorca de Porc Negre*. From the industry point of view, and observing the results on carcass quality in present study, animals slaughtered at heavy weights would be the most adequate to produce this product due the high amount of back fat needed to elaborate it. So, even though some animals could be selected to deposit less back fat, heavy pigs with high back fat levels would be still necessary to meet the industry demands for the production of the typical *Sobrassada*.

References

Gispert, M., A. Valero, M.A. Oliver and A. Diestre, 1997. Problemas asociados a la falta de grasa en canales porcinas. Eurocarne, 61:27-32.

Jaume J. and L. Alfonso, 2000. The Majorcan Black pig. Animal Genetic Resources Information 27:53-58.

CIE, 1986. Colorimetrie, 2nd ed. Publication C.I.E. nº15, 2. Viena: Central Bureau of the Comission Internationale de l'Eclairage.

SAS, 2001. SAS for Windows Release v.8.2. SAS Institute Inc., Cary, North Carolina, USA.

Influence of feeding system on carcass and meat quality: fat colour as a tool of classification

G. Ripoll, S. Carrasco, B. Panea, P. Albertí, I. Blasco and M. Joy
Centro de Investigación y Tecnología Agroalimentaria de Aragón, Carretera de Montañana, 930, 50059 Zaragoza

Abstract

Forty-five Rasa Aragonesa lambs were allocated in four treatments: GR, lambs and ewes grazing on alfalfa pasture, no concentrate was available; GR+S, as the previous one, but additionally lambs had free access to concentrate; DRL-GRE, lambs remained indoors with free access to concentrate and ewes grazed during 8 h a day, thereafter remaining with lambs; and DRL, lambs and ewes were kept always in confinement, ewes had free access to dry unifeed and lambs to concentrate. In DRL-GRE and DRL lambs were weaned at 45 days old. Lambs were slaughtered at a live weight of 22-24 kg. The live weight, carcass weight, dressing percentage, subjective carcass classification, and meat quality (pH, instrumental toughness, meat colour) were recorded. Moreover the integral of the translated spectrum of renal and caudal fat (SUM) was studied looking for a non-destructive technique to discriminate carcasses from alfalfa grazing-based systems. Extensive lambs (GR and GR+S) had greater dressing percentage than indoors lambs (DRL-GRE, DRL). Feeding system had influence on subjective and instrumental meat colour, being darker in grazing lambs. Renal fat from grazing treatments (GR and GR+S) was lighter and more yellow. Caudal fat colour and SUM allowed too classify 91.1% of carcasses into grazing or non grazing groups. Hence, this is a feasible method to be used in abattoirs to trace alfalfa-fed lambs. It is concluded that feeding system had slightly influence on meat colour.

Keywords: grazing, colour, meat quality, alfalfa, pH

Introduction

Traditional lamb meat production in Spain is based on light lambs (22-24 kg of body weight, less than 90 days old). The conventional feeding system is suckling lambs supplemented with concentrate until weaning and thereafter only concentrate and straw. Lamb meat coming from this production system is characterized by a pale pink meat colour and white fat, without important variations among farms, with the breed and slaughter weight as exclusive factors of differentiation among other labelled lamb meat.

The constantly increasing price of cereals, demand for healthy and safe meat products and the EU Common Agricultural Policy that favours extensive grazing systems for ruminants are stimulating the interest in a pasture-based production system. Previous studies have shown good prospects for animal performance in spring lambing ewes raising lambs on pastures (Joy *et al.*, 2004). Yellow fat is generally not appreciated by consumers, particularly Spanish consumers. Therefore, any feeding systems that affect fat or muscle colour need to be studied. Recently efforts have been made to trace feeding systems in herbivores in order to implement food safety.

The concentration of carotenoid pigments is high in fresh pastures, such as alfalfa or grass, and low in grains, which means that these compounds can be used as biomarkers or tracers to develop spectrophotometric techniques for the authentication of products from grazing lambs such us discriminant equations developed by Ripoll *et al.* (2008).

The aim of this study was to assess the influence of the lamb feeding system on muscle and fat colour, and to ascertain the capability of reflectance spectrophotometry to discriminate between carcasses from forage-fed and concentrate-fed lambs.

Material and methods

Forty-five Rasa Aragonesa spring single-born male lambs were randomly allocated to four treatments: Grazing (GR): ewes and lambs were continuously stocked on alfalfa pasture. No concentrate was available to dams or lambs. Lambs suckled their mothers and grazed until slaughter. Grazing with supplement for lambs (GR+S): Management was the same as for GR, and additionally the lambs received concentrates *ad libitum* in lamb creep feeders. Rationed grazing (DRL-GRE): ewes grazed on alfalfa pasture for seven hours a day without their lambs and thereafter remained indoors, receiving a mixed ration. Lambs were kept permanently indoors and were fed with ewe milk and concentrate *ad libitum* until weaning (45 days). From this time up to slaughter lambs had free access to concentrate and barley straw. After weaning, ewes were removed from the trial. Indoors (DRL): Ewes and lambs were kept permanently indoors with *ad libitum* access to a total mixed ration and lambs fed with ewe milk and concentrate *ad libitum* until weaning. Thereafter, lambs had access to concentrate only.

When lambs reached 22-24 kg of live weight, they were slaughtered using standard commercial procedures and carcasses were hung by the Achilles tendon and were chilled 24 h at 4 °C in total darkness to preserve carotenoid pigments and weighted. Dressing percentage was calculated and fatness degree was scored from 1 (1−, very low) to 12 (4+, very high) of the scale 1 (low), 2 (slight), 3 (average), 4 (high). A visual appraisal (VA) of the *Musculus rectus abdominis* colour was made according to the Colomer-Rocher scale (Colomer-Rocher, Delfa, and Sierra, 1988) for light carcasses (<13 kg). This scale is divided into pink, pale red and red, expanded to nine points. *Musculus longissimus thoracis* was removed and ultimate pH was measured at the 4th vertebral region with pH-meter equipped with a Crison 507 penetrating electrode. Fat and meat colour were measured at 24 h at 4 °C after slaughter using a Minolta CM-2006 d in the CIELAB space. The lightness (L*), redness (a*) and yellowness (b*) were recorded, and hue angle (H*) and chromaticity (C*) indices were calculated. Caudal subcutaneous fat from the tail root and kidney fat colour values were recorded. Besides trichromatic coordinates, the proportion of reflected light each 10 nm between 450 and 510 nm was collected and the absolute value of the integral of the translated spectrum (SUM) was calculated according to Priolo *et al.* (2002b) in caudal and renal fats. *M. longissimus thoracis* steaks (T11-T12) were vacuum packed, 2 days aged and cooked until 70 °C in steak core. Cooled steaks were cut in 1 cm^2 section samples and sheared with a Warner-Bratzler device to determine stress (kg/cm^2). Meat colour variables were analyzed as repeated measures in time and production system as fixed effect. The other variables were analyzed with a one-way ANOVA.

Results and discussion

The production system had no effect (P>0.05) on the live weight, but GR+S had heavier carcasses than other systems and, consequently, a greater dressing percentage. Carcasses had a fatness degree from slight to average without significant differences (P>0.05) among production systems (Table 1). In a similar way, pH and meat texture (stress) did not show effect of the production system.

In a similar study, Carrasco *et al.* (2007) found that 4 days aged meat from GR group was more tender than the rest, although differences disappeared at 7 days of ageing. Toughness values of this study were tender than those founded by Sañudo *et al.* (2000), Ripoll *et al.* (2006) and Carrasco *et al.* (2007).

The production system had effect (P<0.001) on muscle lightness (L*) and chromaticity (C*), while oxygenation time has effect (P<0.001) on L*, yellowness index (b*), C* and hue angle (H*)

Table 1. Carcass characteristic and Longissimus thoracis *quality.*

System	GR	GR+S	DRL-GRE	DRL	Est. err.	Sig.
n	11	11	12	11		
Live weight (kg)	22.8	23.4	22.5	23.3	0.23	ns
Cold carcass weight (kg)	11.0[ab]	11.6[b]	10.6[a]	10.9[a]	0.21	*
Dressing percentage	48.1[ab]	49.5[a]	47.4[b]	47.0[b]	0.63	*
Fatness degree	4.3	5.3	5.7	6.1	0.59	ns
Visual appraisal	6.4[a]	5.8[ab]	5.2[b]	5.0[b]	0.29	**
Meat quality						
pH_u	5.57	5.57	5.56	5.59	0.01	ns
Meat L* 0 h	39.97[b]	40.26[b]	43.20[a]	42.41[a]	0.42	*
Meat L*24 h	44.46[b]	44.78[b]	46.77[a]	46.03[a]	0.43	*
Meat a* 0 h	10.56[a]	10.47[a]	8.78[b]	9.33[b]	0.49	*
Meat a* 24 h	10.64	10.72	10.54	10.73	0.49	ns
Stress (kg/cm^2)	4.04	4.11	3.73	3.95	0.27	ns
Fat colour						
L* caudal fat	71.67[a]	72.73[a]	70.79[ab]	67.58[b]	1.11	*
C* caudal fat	11.85	11.02	11.37	11.79	0.28	ns
SUM renal fat	218.6[a]	206.3[a]	110.4[b]	101.6[b]	11.19	***
SUM caudal fat	174.0[a]	143.0[a]	73.5[b]	71.6[b]	37.07	***

Values with different superscript letters differ significantly ($P<0.05$).
GR: grazing, GR+S: grazing with supplement, DRL-GRE: rationed grazing, DRL: indoors.

(Figure 1). Time and redness index (a*) had a significative interaction. Lightness increased 4 points in the 4 groups with time. GR and GR+S had greater a* than the others at 0 h ($P<0.005$) but differences disappeared at 24 h, in agreement with Ripoll *et al.* (2005).

Caudal fat L* was affected by production system ($P<0.05$) while it had no effect on C*. GR and GR+S had greater L* than DRL, while DRL-GRE had intermediate values. Ripoll *et al.* (2005) did not find differences on Churra Tensina breed subcutaneous fat lightness. The forage specie could be the responsible of the different results.

Figure 1 shows the translated spectra of the production systems used to calculate SUM that is an indirect measure of carotenoids content. As SUM is the absolute value of the area between spectrum

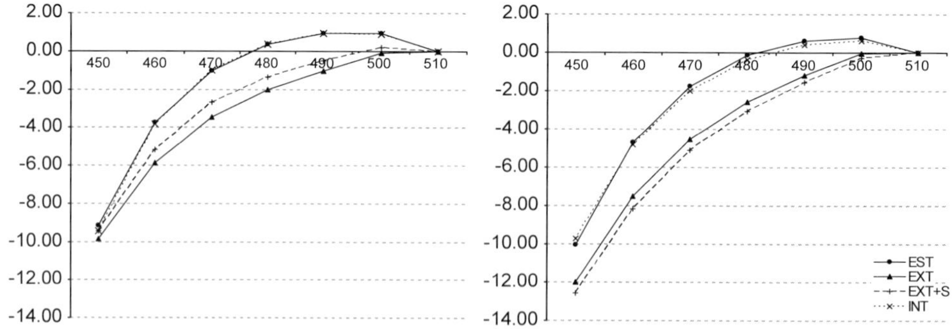

Figure 1. Translated spectra of caudal (left) and renal (right) fats.

and X-axis, spectra with lower values of reflectance gives great SUM values. GR and GR+S had spectra below DRL and DRL-GRE spectra, and renal fat had spectra with lower reflectance values than caudal fat spectra. The differences on spectra were reflected on SUM, hence, extensive groups had greater values than intensive systems ($P<0.001$) and renal fat had greater values than caudal in all the production systems.

Ripoll *et al.* (2008) proposed an equation from a logistic regression with the purpose of classifying lambs from grazing systems. The equation was applied to the present data set and 91.1% of lambs were correctly classified. Three GR+S and 1 DRL were misplaced.

Conclusions

The production system had no effect on fatness degree, meat pH or stress, but rectus abdominis visual appraisal, meat and fat colour and SUM were affected by production systems. Meat from extensive systems were darker than concentrate fed lambs, although differences disappeared at 24 hour of meat oxygenation. Caudal fat of grazing lambs had more lightness. SUM, as an indirect measure of carotenoids content, indicate that fat from extensive systems had more carotenoids content then fat from concentrate fed lambs, and renal more than caudal. SUM and caudal fat colour combined in a regresion equation were useful variables to trace grazing-based production systems.

Acknowledgements

In memory of our colleage and friend Dr. Rafael Delfa. We are grateful to F. Lahoz, F. Gracia, I. Escota and J. Pérez for their collaboration. Study funded by INIA-RTA-03.031.

References

Carrasco, S., B. Panea, G. Ripoll, J. Álvarez-Rodríguez, A. Sanz and M. Joy. 2007. Efecto del sistema de manejo sobre la textura y el análisis sensorial de la carne de corderos de raza Churra Tensina. ITEA vol. Extra. 28: 705-707.

Ripoll G., M. Joy, F. Muñoz and P. Albertí. 2008. Meat and fat colour as a tool to trace grass-feeding systems in light lamb production. Meat Science, 80, 239-248.

Ripoll. G., R. Delfa, M. Joy, A. Sanz. B. Panea, S. Carrasco and P. Albertí. 2006. Evolución del color y de la dureza de la carne de tres tipos de cordero de raza Churra Tensina. XXXI Jornadas Científicas y X Jornadas Internacionales de Ovinotecnia y Caprinotecnia. 20-22/09/2006. Zamora. Spain.

Ripoll, G., A. Sanz, J. Álvarez, M. Joy, R. Delfa and P. Albertí, 2005. Sheep production in Spanish dry mountain areas: 3. The effect of light lambs fattening system on carcass traits. fat and muscle colour and meat texture in Churra Tensina breed. British Society of Animal Science. 04-06/06/2005. York. Inglaterra.

Sañudo, C., M. Alfonso, A. Sánchez, R. Delfa and A. Teixeira. 2000. Carcass and meat quality in light lambs from different fat classes in the EU carcass classification system. Meat Science. 56: 89-94.

The use of subcutaneous fat colour to trace grass-feeding in Parda de Montaña yearling bulls

M. Blanco, I. Casasús, G. Ripoll and M. Joy
Unidad de Tecnología en Producción Animal, CITA, Gobierno de Aragón, Avda. de Montañana 930, 50059 Zaragoza, Spain

Abstract

Grass-feeding in calves can be traced with carotenoid pigments in fat depots. Parda de Montaña calves (227 kg) were fed concentrates (CON), grazed lucerne plus 2 kg/d barley (LUC) or had the same management as LUC calves for 3 months and were finished with concentrates for 2 months (LUC+CON) until a slaughter weight of 450 kg. A spectrophotometer was used to measure subcutaneous fat colour. The proportion of reflected light each 10 nm between 450 and 510 nm was collected to calculate the absolute value of the integral of the translated spectrum (SUM). Lightness (L*) was greater and redness (a*) was lower for CON calves than for their counterparts. On the other hand, yellowness (b*) and SUM were greater in LUC calves than in the rest of the feeding strategies, being both variables highly correlated (r=0.69, P <0.001). Two logistic regressions were obtained to discriminate carcasses according to the fattening strategy, the first one calculated the probability of a carcass to belong to a calf that grazed in lucerne: $P\ (LUC) = (1+e^{18.8-5.6\times L^{*}-36.9\times a^{*}+0.3\times SUM+29.8\times C^{*}})^{-1}$ and the second one calculated the probability of a carcass to belong to a calf that was finished on concentrates after grazing lucerne: $P\ (LUC+CON) = (1+e^{833.7-11.8\times L^{*}+4.7\times a^{*}+0.2\times SUM-2.5\times C^{*}})^{-1}$. These regressions could be used to trace carcasses belonging to calves that had been grazing. Therefore, they could be used in abattoirs and retailers to trace grass-fed carcasses.

Keywords: subcutaneous fat colour, traceability, beef cattle

Introduction

Forage-feeding has a positive image for consumers and could confer specific characteristics to meat (Coulon and Priolo, 2002). However, clear information is demanded regarding production systems, especially animal feeding systems. Forage-feeding in ruminants can be traced with carotenoid pigments in plasma and fat deposits. Prache and Theriez (1999) proposed the use of the reflectance spectrum of the fat in the zone of light absorption by carotenoids to discriminate carcasses of grazing and concentrate-fed lambs. Later, Serrano *et al.* (2006) studied the feasibility of its use in growing cattle concluding that depends on animal age and production system. Thus research in different production systems is needed.

Therefore, the aim of the study was to determine the effect of three fattening strategies involving concentrate feeding, lucerne grazing and lucerne grazing and concentrate finishing on subcutaneous fat colour traits and its suitability to discriminate carcasses.

Material and methods

Twenty-one 5-month old Parda de Montaña purebred male calves were grown from weaning (224 kg) to slaughter at 450 kg in CITA research centre (41° 43' N, 0° 48' W; altitude 225 m), located in the Ebro Valley (Spain). The following day to arrival calves were weighed, blocked by live weight and randomly assigned to one of three fattening strategies.

A group of calves (CON) was allocated in a straw-bedded pen to follow a commercial feeding program with concentrates on *ad libitum* basis. A concentrate was fed from arrival to 350 kg liveweight (11.47 MJ ME/kg DM, 14.9% CP) and a second one from 350 to 450 kg (11.65 MJ ME/kg DM,

13.7% CP). A second group of calves (LUC) rotationally grazed in a 0.9-ha lucerne (*Medicago sativa* L. var. Aragón) plot with 1.8 kg DM barley/day per calf until slaughter. The third group of calves (LUC+CON) had the same management as LUC calves for three months in an adjacent 0.9-ha lucerne plot, and thereafter they were fed the second concentrate for two months on *ad libitum* basis until they reached slaughter weight. All calves received mineral and vitamin supplements, barley straw and water on *ad libitum* basis.

When calves reached 450 kg LW, they were transported to a commercial abattoir, where they were slaughtered and dressed according commercial practices.

Subcutaneous fat colour measurement

Subcutaneous fat colour (CIE L* (lightness), a* (redness), b* (yellowness)) was measured between the 6th and the 11th rib of each left half carcass using a Minolta CM-2600d spectrophotometer (Konica Minolta Holdings, Inc, Osaka, Japan). Reflectance measurements were collected from a 10°-viewing angle with Commission Internationale de l'Eclairage (CIE) illuminant C lighting conditions. Fat colour values were recorded from three locations randomly selected avoiding blood dots, air bubbles and less covered areas. These data were used to calculate Chroma ($C^*= (a^{*2}+b^{*2})^{0.5}$) values (Albertí *et al.*, 2005).

The proportion of reflected light each 10 nm between 450 and 510 nm in subcutaneous fat was collected and the absolute value of the integral of the translated reflectance spectrum (SUM) was calculated according to Priolo *et al.* (2002), who proposed the use of SUM as an estimator of fat carotenoid content. The reflectance spectrum was translated to make reflectance value at 510 nm equal zero (TR). On the translated spectrum, the integral value was calculated as follows:

$$SUM= (TR_{450}/2+TR_{460}+TR_{470}+TR_{480}+TR_{490}+TR_{500}+TR_{510}/2)\times 10 \quad (1)$$

where TR_i was the reflectance value at i nm (for further explanation see Prache and Theriez, 1999).

Subcutaneous fat colour data were tested by Analysis of Variance using the GLM procedure with feeding strategy as fixed effect. Least square means (LS Means) were estimated and differences between LS Means were tested using a t-test. Pearson's correlation coefficients between variables were calculated. For all tests, the level of significance was set at 0.05. A forward stepwise discriminant analysis was carried out for subcutaneous fat colour traits, with L*, a*, b*, C*, H* and SUM as the variables evaluated. Selected variables were included in a multinomial logistic regression to calculate the probability of a carcass to belong to a group that had grazed lucerne (LUC calves) and the probability of a carcass to belong to a group that had grazed lucerne and had been finished on concentrates (LUC+CON calves).

Results

Subcutaneous fat colour and SUM presented differences among fattening strategies (Table 1). Fat of CON calves had the greatest L* and the lowest a* ($P<0.01$). Nevertheless, b* was higher in LUC calves than in the rest of the fattening strategies ($P<0.001$). In the same sense, LUC calves had the highest value for C*, LUC+CON calves intermediate and CON calves the lowest ($P<0.001$). SUM, which is an indicator of the content in carotenoids in subcutaneous fat, was 26.1% and 46.6% higher in LUC calves than in LUC+CON and CON calves, respectively ($P<0.001$). SUM was highly correlated with b* ($r= 0.69$, $P<0.001$) and C* ($r= 0.68$, $P<0.001$).

A discriminant analysis selected L*, a*, C* and SUM as the variables which explained the greatest amount of variation, classifying correctly 100% of carcasses (Figure 1).

Table 1. Subcutaneous fat colour traits according to fattening strategy.

	LUC	LUC+CON	CON	s.e.	*P*-value
L*	73.39^{b}	71.86^{b}	76.05^{a}	1.09	0.001
a*	3.67^{a}	3.53^{a}	1.45^{b}	0.7	0.002
b*	16.8^{a}	12.67^{b}	10.16^{b}	1.3	0.0001
C*	17.22^{a}	13.18^{b}	10.28^{c}	1.35	0.001
SUM, units	324^{a}	257^{b}	221^{b}	25.7	0.0004

Within a row, means with different superscript differ at $P<0.05$.

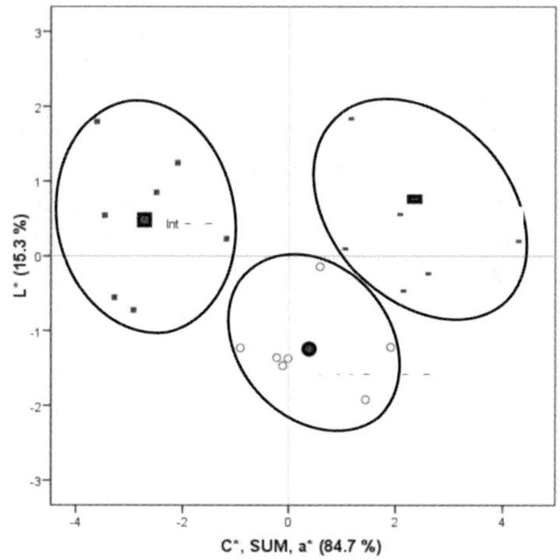

Figure 1. Multivariable analysis of subcutaneous fat colour traits.

The multinomial logistic regression included SUM, L*, a* and C*. The first equation was applied to determine the probability of a carcass to belong to a calf that had grazed lucerne P (LUC) = $(1+e^{18.8-5.6\times L^*-36.9\times a^*+0.3\times SUM+29.8\times C^*})^{-1}$, and the second one to determine the probability of a carcass to belong to a calf that had been finished on concentrates for 2 months P (LUC+CON) = $(1+e^{833.7-11.8\times L^*+4.7\times a^*+0.2\times SUM-2.5\times C^*})^{-1}$.

Discussion

Fat colour results mainly from the accumulation of dietary carotenoids from pastures rich in these pigments (Yang *et al.*, 1992) and therefore forage-fed steers have greater b* than concentrate-fed steers slaughtered at a common backfat thickness (Bidner *et al.*, 1981). Subcutaneous fat b* in LUC calves was 24% and 40% greater than those of LUC+CON and CON calves, respectively. A low carotenoid ration after forage-feeding reduced b* by 9% after 42 days and by 34% after 70 days (Knight *et al.*, 2001), and a concentrate diet by 40% after 70 days (Strachan *et al.*, 1993). Nevertheless the rate of change in fat yellowness depends on the rate of fat deposition, because carotenoids are transported by high density lipoproteins in cattle (Yang *et al.*, 1992) and according to leptin concentrations before slaughter in LUC+CON calves, these animals were depositing fat faster than LUC calves (unpublished data).

The translated reflectance spectrum (SUM) in subcutaneous fat in LUC calves was similar to the value measured in grazing calves by Serrano *et al.* (2006). In the current study, feeding concentrates for 60 days reduced SUM in subcutaneous fat by 23%. Accordingly, Knight *et al.* (2001) reported a reduction in carotenoid concentration of 21% after 42 days on a low carotenoid diet and of 44% after 70 days. These authors concluded that the reduction in fat yellowness was largely caused by a dilution of existing fat with newly deposited whiter fat.

Interestingly, the logistic regression obtained with the data of fat colour and SUM was able to discriminate correctly 100% among the three fattening strategies. When Prache *et al.* (2002) used the reflectance spectrum in steers had an overlap of 33%. To improve the accuracy, Prache *et al.* (2003) included in the model the carotenoid content in blood previous to slaughter, which is more complicated than including fat colour data, as was proposed by Ripoll *et al.* (2008) in lambs.

Finally, recording subcutaneous fat colour traits with a spectrophotometer to obtain logistic regressions is a cheap, easy and reliable method to trace grass-feeding in yearling bulls and could accomplish the consumer demand of information regarding the production system.

References

Albertí, P., B. Panea, G. Ripoll, C. Sañudo, J.L. Olleta, I. Negueruela, M.M. Campo and X. Serra, 2005. Medición del color. In: Estandarización de las metodologías para evaluar la calidad del producto (animal vivo, canal, carne y grasa) en los rumiantes. Monografias INIA: serie ganadera 3: 448.

Bidner, T.D., A.R. Schupp, R.E., Montgomery and J.C. Carpenter, 1981. Acceptability of beef finished on all-forage, forage-plus-grain or high energy diets. J. Anim Sci. 53: 1181-1187.

Coulon, J.-B. and A. Priolo, 2002. La qualité sensorielle des produits laitiers et de la viande dépend des fourrages consommés par les animaux. INRA Prod. Anim. 15: 333-342.

Knight, T.W., A.F. Death, M.G. Lambert and D.B. McDougall, 2001. The rate of reduction in carotenoid concentration in fat of steers fed a low carotenoid ration, and the role of increasing carcass fatness. Aust. J. Agric. Res. 52: 1023-1032.

Prache, S. and M. Theriez, 1999. Traceability of lamb production systems: carotenoids in plasma and adipose tissue. Anim. Sci. 69: 29-36.

Prache, S., A. Priolo, H. Tournadre, R. Jailler, H. Dubroeucq, D. Micol and B. Martin, 2002. Traceability of grass-feeding by quantifying the signature of carotenoid pigments in herbivores meat, milk and cheese. In Proceeding of the 19th General Meeting of the European Grassland Federation La Rochelle, France, 27-30 May. 592-593.

Prache, S., A. Priolo and P. Grolier, 2003. Effect of concentrate finishing on the carotenoid content of perirenal fat in grazing sheep: its significance for discriminating grass-fed, concentrate-fec and concentrate-finished lambs. Anim. Sci. 77: 225-233.

Priolo, A., S. Prache, D. Micol and J. Agabriel, 2002. Reflectance spectrum of adipose tissue to trace grass feeding in sheep: influence of measurement site and shrinkage time after slaughter. J. Anim. Sci. 80: 886-891.

Ripoll, G., M. Joy, F. Muñoz and P. Albertí, 2008. Meat and fat colour as a tool to trace grass-feeding systems in light lamb production. Meat Sci. 80: 239-248.

Serrano, E., S. Prache, B. Chauveau-Duriot, J. Agabriel and D. Micol, 2006. Traceability of grass-feeding in young beef using carotenoid pigments in plasma and adipose tissue. Anim. Sci. 82: 909-918.

Yang, A., T.W. Larsen and R.K. Tume, 1992. Carotenoid and retinol concentrations in serum, adipose tissue and liver and carotenoid transport in sheep, goats and cattle. Aust. J. Agric. Res. 43: 1809-1817.

Wool in Mediterranean regions: a forgotten and neglected product of sheep or a natural and renewable resource for the future?

M.-T. Chaupin[1] and A. Simonpietri[2]
[1]Déléguée aux relations européennes ATELIER, European Wool Group, c/o Filature de Chantemerle, 05330 Saint Chaffrey, France
[2]Lana Corsa, France

Abstract

Concern for the environment and the requirement to save energy have given the opportunity to rediscover a forgotten and neglected product of the sheep breeding, namely, wool. In the Mediterranean regions, sheep breeding is often an important activity and some initiatives are already taken for giving an added value to wool. Examples from Italy, Spain, France and especially in Corsica are given in this paper. Some effects of these initiatives are shown: interest growing for organic wool, local products, creation of jobs in rural areas, links with the traditional cultures combined with innovative and modern applications. Coordination and exchange of know-how between all the initiatives in Europe is necessary for widening all these experiences. The ATELIER-Wools of Europe is active since 1989 at the European level for networking all these activities. It makes a global promotion for the wool as a natural and renewal resource and promotion for the producers as well as the processors using it.

Keywords: wool, Europe, renewable resource

Introduction

Ecological concerns and necessity of energy savings now give an opportunity to rediscover a forgotten and neglected product of the sheep breeding: the wool. Initiatives taken in this sector during recent years by some breeders and processors wishing to upgrade this material despite the lack of interest of most of the European sheep breeding sector will be presented. These experiences take place in France, Italy, Spain and Portugal.

The current situation

Breeding is quite an ancient human activity and the creation of domestic animal breeds is the product of a long and patient work of selection by man in the course of millenniums. Each breed results from an adaptation to its environment, to the soils and the climate. It tends towards optimal use of the natural resources and has developed original specific methods of breeding: transhumance, extensive pastures, mountain pastures… Wool promotion and marketing of wool products could help the different initiatives for the conservation of these breeds.

At the European level, in 2003 the new European regulations *on the protection of geographical indications and designation of origin for agricultural products and foodstuffs* (692/2003) were published. Wool has been included in the list of agricultural products that may profit from this kind of protection.

Positive aspects:
- Wool gives a 'good image' to sheep breeding and it helps the promotion of rare breeds.
- New trends for natural raw material in clothing and home furnishing.
- New (or rediscovered) uses of wool are developed: felt and insulation.

Environmental effects:
- Organic/ethic wool is very much in demand. It will help to give more interest for organic breeding.

- At the moment, wool is considered as a by-product. It is sometimes burnt or put on manure, with environmental consequences.

Social effects:
- Jobs in the wool sector (collecting centres, small processing units) are a contribution for the employment in rural areas.
- Use of traditional, as well as innovative processing, connected with culture and design.

Economical effects:
- Wool processing gives high value products; with a guarantee of quality and traceability if processed in small units.
- Use of wool in rural accommodation, good links with agritourism.

Requirements for the future:
- Training courses for sheep breeders for a better knowledge of wool.
- Promotion of wool 'from the sheep to the product' to the consumers.
- Co-ordination through the whole sector.

A better value of the wool produced could help the breeders at the moment of the diminution of the PAC subsidies.

Some examples of initiatives

Some examples of initiatives are:
- The Piacenza company in Biella (Italy): processing of the wool from the breed 'Sambucana' in Piémont.
- Filature Mazières (a very small spinning mill) in Pyrenees: processing of the fine wool of a rare breed, the Aure et Campan.
- Using the wool of two rare breeds from Italy (Garessina and Sambucana) for the restoration of a masterpiece of Kounellis in the Centre Pompidou in Paris.

The example of Corsica: Lana Corsa[4] initiative

The Corsican sheep is among the last in Europe to conserve a large variety of natural colours: ecru, beige, grey and black. The fleece, which is particularly well-adapted to traversing the Corsican maquis, is long and resistant.

One hundred tonnes are approximately produced each year in Corsica (it represents 1 kg/ewe/year). Lana Corsa collects only 4 tonnes of wool each year from local breeds and merinos breeds. All the remaining wool is wasted. They work with farmers close to their firm, with the association of the breed management and some shearer close from Ajaccio. Each time they collect an average of 15 kg/farm (only young ewe wool) at 1 euro/kg or they barter.

Using wool from younger sheep, blended with the softer wool from the merino sheep, they manufacture a wool thread unique in texture and range of natural colours.

With this wool, they preserve, knit, weave – and dye too – to make quality products. Their wool is prepared by washing with degradable soap only: no chemical treatment, bleaching or carbonisation.

[4] Lana corsa is located at 5 min from Ponte Leccia. For more information, please contact us through www.lana-corsa.com.

All the motifs on their pullovers find their inspiration in the considerable richness of the Corsican patrimony: mosaics, sculptures of the roman churches and of course landscapes, trees, villages, marinas and animals. Products are made to last and their wool bring unequalled comfort: as it is warm, insulating and non inflammable. The representative of Lana Corsa said,'To utilise a local raw material, to transform it into an original product, this is the sense of our work as hand-craftswomen.'

Lana Corsa belongs to 'ATELIER', an association of shepherds and little workshops working local wool in all Europe and having the same objective: quality, re-localisation, environment respect and cultural authenticity.

Besides the workshops, they suggest a permanent exhibition called *Di a pecura à u filu*, which means 'from the sheep to the thread'. This exposition presents a 'circuit' from fleece to the finished product, through spinning, weaving and knitting history and techniques. It also shows diversity in sheep – more than 400 different breeds in Europe – and the thousand ways of dyeing with plants. In their little garden you will recognize the most important plants still used for colouring wool.

The representative of Lana Corsa states: 'In Corsica, we spin wool for at least 8,000 years. Corsican people were shepherds and perfectly knew how to use their own flocks. Women used to weave 'pannu' for clothes and blankets, 'pilone' of goat wool for cloaks and bags; this work was so important that the 'rocca' (distaff) became the engagement gift, decorated by the fiancé himself with a lot of symbolic and traditional patterns…' Today, Lana corsa tries to perpetuate and enrich these traditions.

The example of ATELIER: European Wool Group

ATELIER was founded in 1989 with the aim of developing cooperation between people and wishing to promote and upgrade wool as a natural material and to make it once again an element of local development. Its members – breeders, shearers, processors, designers, scientists – work with wool, mohair, angora and other animal fibres, mostly in less favoured regions. The association was launched in France but has gradually spread to other European countries.

In 1997, ATELIER was one of the founders of the 'European Wool Group' with representatives of the most important sheep breeders' organisations in Europe. Three exhibitions 'From sheep shearing to the finished product', 'Wool types of Europe','Wool' and video films are presented in schools, agricultural shows, craft fairs or museums. ATELIER members also offer a wide range of training courses: sheep shearing, wool grading, dyeing, felting, weaving and general know-how on wool. In a documentation centre they collect books, magazines, technical reviews and other documentation on wool and sheep, as well as wool samples and finished products of more than 100 different European sheep breeds.

A website presents its members and their activities as well as a calendar of events in the fields of sheep breeding, wool and textiles: exhibitions, fairs, conferences and training courses.

Meetings and conferences

Many conferences, tours and visits have been organised. In 2004, the conference 'Wool of Europe, East and West' took place in the Czech Republic. In May 2005, the meeting 'Wool scouring in Europe: regional and ecological projects' was held in Germany. Finally, in September 2007, 'Wool in the Home' was organised in a French centre for training courses and demonstrations of ecological building.

Conclusion

2009 was the 'International Year of the Natural Fibres'. ATELIER organised and participated at several events and exhibitions in order to raise the consumer awareness of the natural fibres.

In 2010, ATELIER set up a new exhibition 'Wools of Europe'. This exhibition presents one hundred European sheep breeds, among them are a lot of breeds from Mediterranean regions in Spain, Portugal, France, Italy, Greece, Croatia, Bosnia-Herzegovina, Albania and Turkey. The exhibition will be held all around Europe in the coming years.

A catalogue of the exhibition was published and helps to spread the knowledge of the biodiversity of the breeds as well as the wide range of the uses of wool, a natural and renewable resource.

Part 2.

Innovations improving the efficiency of livestock systems in the Mediterranean area

D. Animal recording - data identification

Innovations have improved the efficiency of ruminant production systems in Cyprus

C. Papachristoforou and O. Tzamaloukas
Department of Agricultural Sciences, Biotechnology and Food Science, Cyprus University of Technology, P.O. Box 50329, Lemesos 3603, Cyprus

Abstract

Substantial improvements in the efficiency of ruminant production systems in Cyprus resulting from technological and process innovations adopted by farmers have been realised over the last three decades. In the dairy cattle sector, the technological innovations introduced mainly concern mechanical and electronic systems to control the feeding, management and milking of cows. The expansion of artificial insemination has also contributed to the genetic improvement of the animals. All these have driven the national average milk yield from less than 4,800 kg in the mid 1980s, to about 6,300 kg today, while the average yield in recorded herds exceeds 7,000 kg. However, in 1990 there were 535 dairy herds, whereas today there are only 240. In the sheep/goat sector, the most important innovations relate to the adoption of early weaning, artificial rearing, fattening of lambs/kids, changes in the age at slaughter, along with better herd management methods, and the introduction of milking machines. Weaning age has been reduced to 5 weeks for lambs and 7 weeks for kids leading to increases in the quantity of milk sold; further increases have been obtained by artificial rearing of lambs/kids. Fattening from weaning to the slaughter age of 18-20 weeks, exploits the period of optimal efficiency for growth and produces good quality carcasses. The 'two matings per year' system adopted by many farmers is also beneficial, because it extends the marketing period of sheep/goat products. As a consequence of these changes, the sector has moved towards medium-high input systems and larger farms.

Keywords: ruminants, efficiency, intensification

Introduction

The livestock production sector in Cyprus is very important for the agricultural economy of the country, contributing 47.2% to the gross agricultural output. In 2006, this represented €292 million (MANRE, 2007) and from 2003 onwards, the contribution of livestock production was higher than that of crop production (Figure 1). The contribution of ruminant species (cows, sheep, goats) to the total value of animal production was quite substantial, reaching 45%. As shown in Figure 2, dairy cattle products (mainly milk) accounted for 25%, and sheep and goat products (milk and meat) for 20% of the total value of livestock products (Cyprus Dept of Agriculture, 2006). About 6,000 people are employed in ruminant production, and this figure accounts for 20% of all those employed in agriculture. Cyprus is self sufficient in milk and milk products and in goat meat, 90% self sufficient in lamb and mutton and 70% in beef and veal. Depending on the species, increases in production from ruminants realized over the last 20 to 30 years can be attributed to increases in the number of animals raised and to innovations introduced and adopted by producers. There is evidence that the efficiency of production has been improved in the case of cow milk, sheep milk and lamb meat, while in the case of goats, the picture is not quite as clear due to a number of different factors, whose effect is difficult to quantify. The present work examines the driving forces behind the changes that have been observed during the last 25 years in ruminant production in Cyprus, and their consequences for the rural economy of the country.

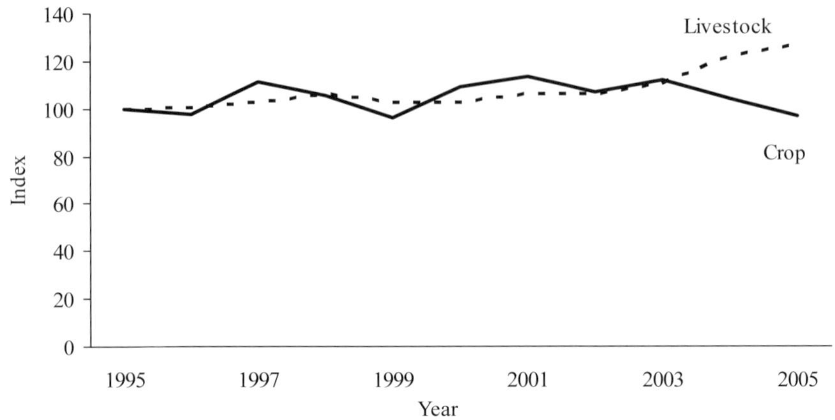

Figure 1. Indices for crop and livestock production of Cyprus for the period between 1995 to 2005 (Indices for 1995= 100).

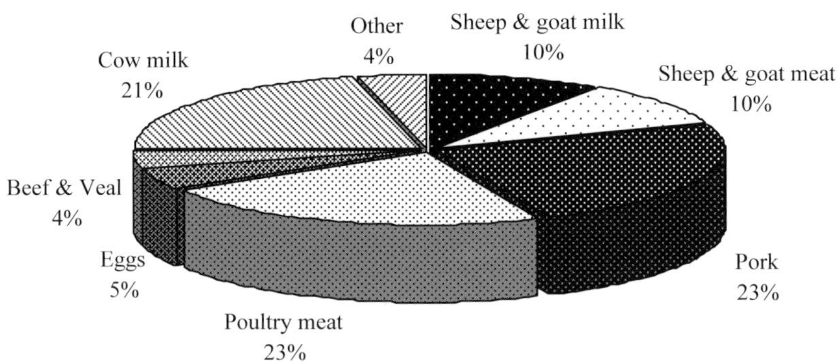

Figure 2. Relative value of the animal products of Cyprus (in 2008 values).

Dairy cattle

General information

Dairy production is an important element of agriculture in Cyprus. Cow milk is part of the daily diet of Cypriots who, on average, consume more than 95 kg of fresh milk annually, while at the same time, cow milk mixed with sheep and/or goat milk is used for making the local 'haloumi' cheese, the main animal product exported. Cow milk is used for other dairy products such as yogurt and other types of cheeses. In 2007, there were about 230 dairy farms in Cyprus producing more than 150,000 tonnes of milk per year and contributing about 21% of the total value of livestock products.

During the 1980s and 1990s, dairy cattle production showed a marked shift from small to larger enterprises with an increasing number of animals per farm unit, while the number of units showed a downwards trend (Figure 3). A feature of dairy cattle units is their concentration around a few villages. The increase in herd size helped producers to increase the efficiency of production and to capitalize on the economy of scale. As shown in Figure 3, over the last 28 years the number of cows per unit has increased from 9 to 109 animals. This development has resulted in the concentration of more cows in fewer farms, and has been accompanied by dramatic increases in milk production per cow

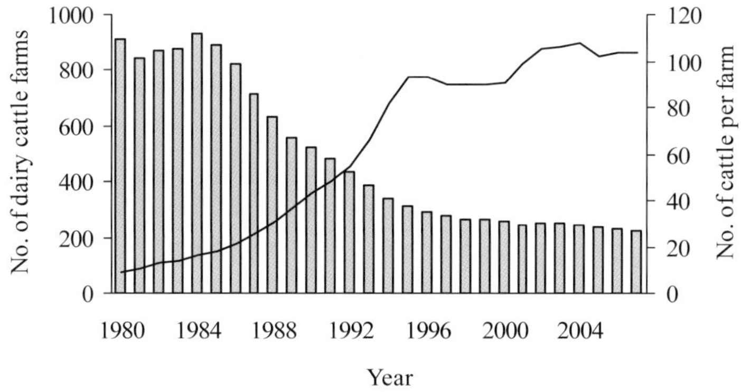

Figure 3. Number of dairy cattle farms in Cyprus (bars) and number of cattle per farm (line) for the period 1980-2007.

(Figure 4). The increased productivity can be attributed to genetic improvement of animals, better feeding, management and health care of cows, and the introduction of technological innovations in all aspects of dairy herd management. Figure 4 shows that over the last three decades, milk production increased by about 50% lifting the national average milk yield per cow annually, from 4,200 in 1980 to 6,300 kg in 2007, while the average annual milk yield of recorded cows exceeds 7,000 kg (ICAR, 2006).

Development of the sector and innovations introduced

Today, modern dairy cattle production in Cyprus is exclusively based on high input systems with animals kept in confinement all year round. Holstein-Friesian is the main breed comprising over 99% of dairy herds in Cyprus. Dairy enterprises are generally well managed and seem to be profitable to farmers (Papachristoforou and Markou, 2006). The level of mechanization of milking and feeding the animals has improved in recent decades and these innovations were taken up by the majority of producers and has had a positive impact on productivity. One example of such innovations is

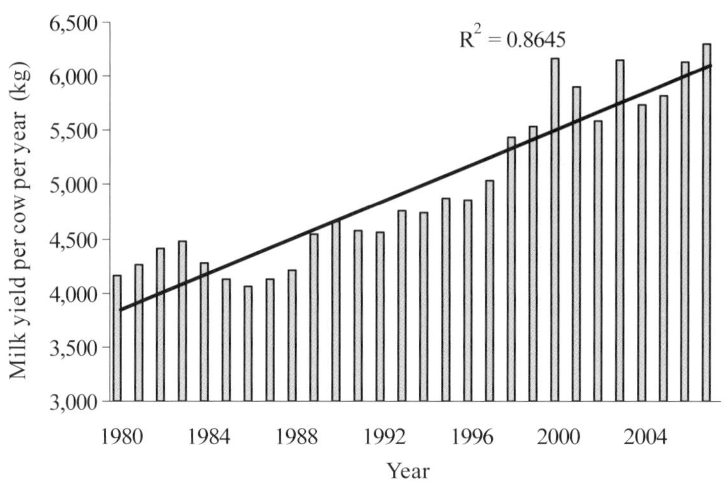

Figure 4. Milk yield per cow per year (bars) over the period 1980-2007 and the regression line of milk yield on time.

the increase in the number of built-in software packages, hardware, and mechanical equipment for ration formulation and automatic feed supply. It is quite characteristic that in 1990, of the 526 dairy units, 16 had a feed mixing wagon and five had a computerized system for individual concentrate feeding, while in 2005, of the 230 units, 123 owned a mixing wagon and 65 had a computerized feeding system. Presently, the number of milking machines in use covers all dairy producers, and at the same time, there is an ongoing process of adopting new technologies for improving milking efficiency and milk quality. Figure 5 shows the percentage of farmers using either the old type of milking machines with milk collected in buckets, or modern equipment with automated control systems and milk collected in tanks with a cooling system.

In order to achieve better conception rates during the hot period when fertility is reduced (Papachristoforou and Hadjicostis, 1996), about 10% of farms have installed animal cooling systems for cows with water spraying and fan cooling. A larger proportion of farms (23%) have devices for the electronic detection of oestrus (pedometers).

Concerning the genetic improvement of animals, the farmers' adoption of artificial insemination as the main tool was a rather slow process that took place over the period under examination. The support provided by state institutions such as the Veterinary Services Department, the Agricultural Research Institute and the extension services of the Cyprus Department of Agriculture, was instrumental in the expansion of artificial insemination. Farmers received training by experts from the Veterinary Services Department in the artificial insemination technique, and by experts from the other two government departments in cow reproduction. This played a key role in the expansion of artificial insemination. As a result, there was a gradual and significant reduction in the number of farms using exclusively natural mating as shown in Figure 6. Today, 58% of farmers use only artificial insemination for impregnating their cows, 25% keep bulls for natural mating, while 17% of farmers use both methods. The semen is imported and comes from proven bulls of superior genetic merit.

Improvements in infrastructure have also played a positive role in the process of transformation of the dairy cattle sector into an intensive, dynamic and efficient agricultural sector. For instance, the transportation of milk to the processing plants was facilitated by the much-improved road system. In addition, the adoption of EU regulations regarding feedstuffs and feed processing units resulted not only in the upgrading of existing equipment and the establishment of new units which provide farmers with better quality feed for animals, but also in substantial improvements in the whole trade system of feedstuffs with respect to storage, quality control and distribution.

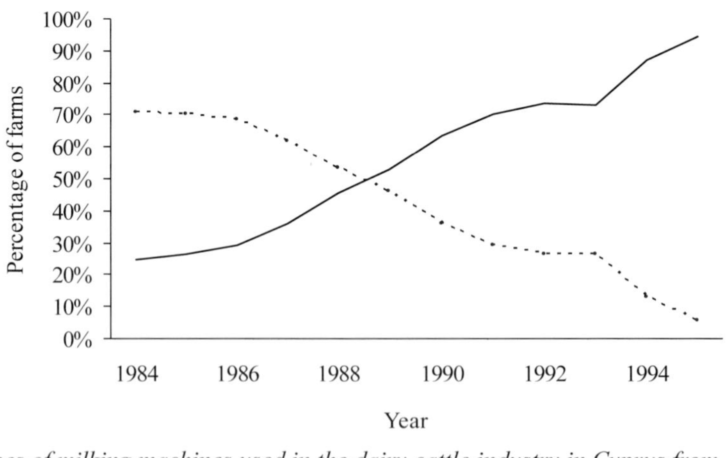

Figure 5. Types of milking machines used in the dairy cattle industry in Cyprus from 1984 to1994. The continuous line indicates modern automatic milking systems and the dotted line, old systems.

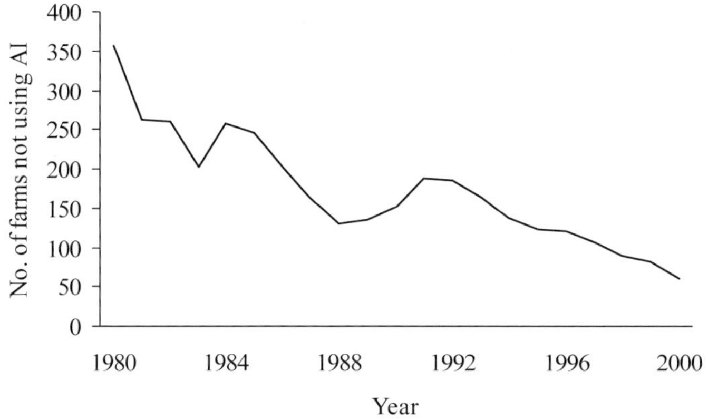

Figure 6. Number of dairy farms in Cyprus not using artificial insemination (AI) over the period 1980-2000.

Two other system innovations adopted by farmers are worth mentioning, the early weaning of calves and cereal hay production. In the early 1980s, calves were weaned at 84 days of age after consuming 400 kg of milk, while today, the weaning age has been reduced to anything between 25 days (with 72 kg of milk) to 56 days (with 200 kg of milk). The reduction in the weaning age resulted in an increase of the quantity of milk sold (Hadjipanayiotou and Kyprianou, 2001). The efficiency of beef production is difficult to assess and, until about 1990, male calves were slaughtered at the age of 18 months which was gradually reduced to 12 months. Regarding cereal hay production, farmers were convinced of the value of feeding hay to their animals after the introduction of hay subsidies almost 30 years ago when barley straw was the main source of roughage. The production of hay increased by 80% (Papachristoforou and Markou, 2006) and presently, with no subsidies paid, cereal hay continues to be produced at even higher levels.

The success story of the dairy cattle sector has not been problem free. Besides its reliance on imported concentrate feeds, the price of which have increased to very high levels, it is quite possible that in the near future, there will be problems with the disposal of manure. According to some estimates, the area under rainfed crops (mainly barley) is just enough for the disposal of manure produced annually by the 56,000 cattle (all ages). This leaves very little area for the disposal of manure from other species, which is currently a matter of concern.

Sheep and goats

General information

Meat and milk products from sheep and goats are regarded as traditional and there is a demand for them in the local market. Milk from sheep and goats was traditionally used as the raw material for haloumi cheese and other local products. During the last 25 years, haloumi has also been made of a mixture of sheep and goat milk with cow milk.

The latest figures (2006) from the Department of Agriculture give the total number of sheep and goat farms as 3,035 of which 50% have goats, 18% sheep, and 32% mixed herds of goats and sheep. From 1996 until 2006, there was a reduction in the number of sheep and goat units and at the same time, an increase in the number of animals per unit (Figure 7). The average size of farms that raise only sheep is 153 animals, that of goats is 110, while mixed herds have on average 198 animals (106 sheep+92 goats).

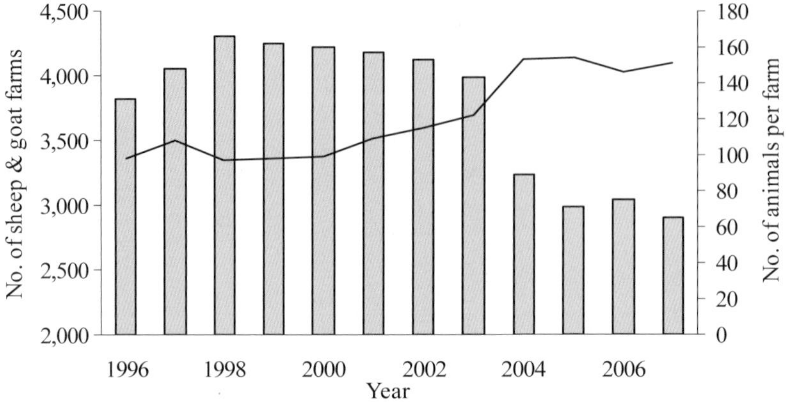

Figure 7. Number of sheep and goat farms in Cyprus (bars) and number of animals per farm (line) from 1980-2007.

Accurate productivity data for sheep and goats is not easy to obtain for many reasons. Compared to cattle farmers, sheep and goat farmers are not very well organized, the majority (58%) are over 50 years of age with no or very basic education and low income. Sheep and goat production appears not to be a profitable business in Cyprus (Markou and Mavrogenis, 2002). In order to obtain a reasonable indication of milk and meat productivity, the total annual milk and meat production for each species was divided by the respective total number of females over 6 months of age that were present in the herd each year.

The period from 1980 to 2005 can be split into two sub-periods, the first from 1980 to 1998 and the second from 1999 to 2005. In 1998, the implementation phase of a national programme aiming at the eradication of sheep genotypes that are sensitive to scrapie started, and the whole sheep production sector was seriously affected as the sole criterion for selection of breeding animals was their genotype. As scrapie spread from sheep to goat herds in the years that followed, goat production was also adversely affected. Approximately 20,000 animals of both species were destroyed every year and this situation caused real problems to producers, as most of the culled animals were productive.

Development of the sector and innovations introduced

Figure 8 shows the average annual milk yield of sheep, Figure 9 that of goats, and Figure 10 shows the average meat production for both species. It is quite clear from Figure 8 that from 1980 to 2005, there was an increase of about 35% in the milk yield per ewe and from 1980 to 1998 an increase of almost 80% in meat per ewe. The milk productivity of goats is not clear and difficult to interpret. There was an upward trend from 1980 to 1988, and then, until 1996 a slight decline which was followed by a further drop in 1997. After a six-year period of constant low productivity, yields increased in 2003 and 2004. The genetic constitution of the national herd if one of the factors that affects goat productivity. Unlike sheep, where there is one dominant breed, (Chios), there are three main goat breeds, the Damascus, the indigenous Machaeras and an ill-defined indigenous population (not really a characterized breed). There is also a large crossbred population (54% of total) of all possible crosses between the breeds. In the early 1990s, the number of indigenous goats increased. Since then, the shift towards improved Damascus or indigenous goats depends on many factors related to market conditions and economic considerations. These changes mostly affect milk and to a lesser extent meat production, because there is a big difference in milk yield between Damascus goats, which produce more than 400 kg of milk per lactation, and indigenous goats, which produce between 100 and 150 kg of milk. There is no difference in prolificacy between Damascus and indigenous goats,

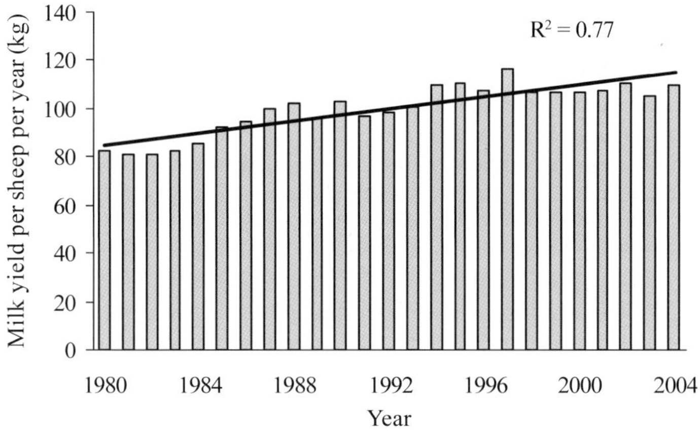

Figure 8. Milk yield per sheep per year over the period 1980-2004 and the regression line of milk yield values on time.

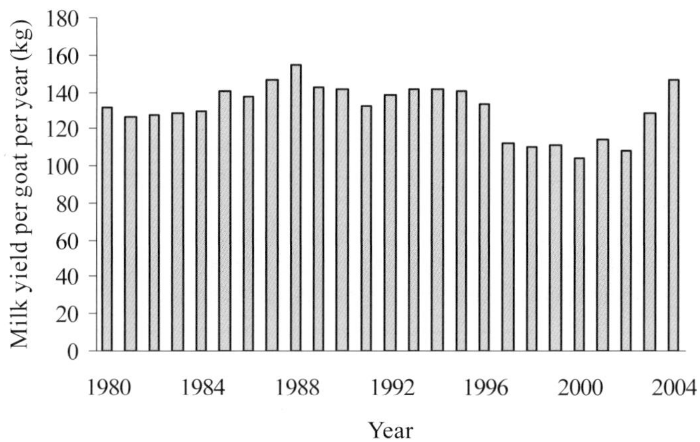

Figure 9. Milk yield per goat per year over the period 1980-2004.

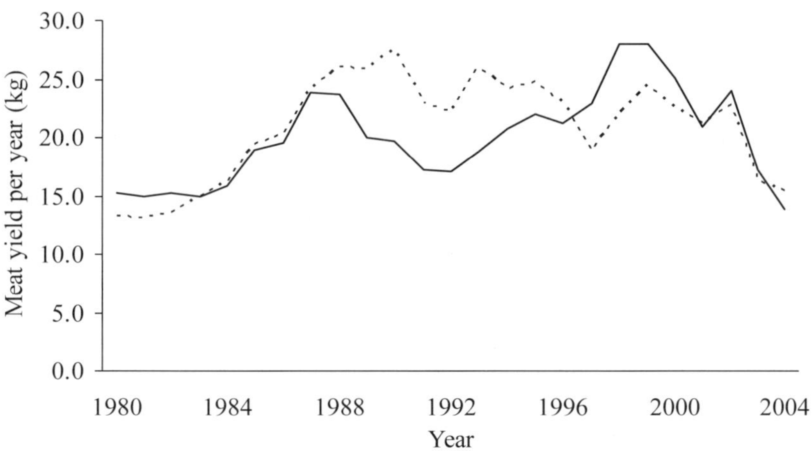

Figure 10. Meat yield per breeding female sheep (continuous line) and goat (dotted line) per year, over the period 1980-2004.

and any effects on meat production can be attributed to the relatively higher growth rate of Damascus compared to indigenous kids.

After 1998, meat produced per breeding female sheep and goat (Figure 10) declined. There is no doubt that the implementation of the scrapie programme had some negative impact on the genetic merit of the sheep population. In addition, before 2000, some lambs and kids used to be slaughtered at the age of 8 to 10 months, because there was a demand among consumers for such carcasses. After 2000, lambs from scrapie infected or suspected herds were slaughtered at younger ages according to precautionary health related measures imposed by the authorities concerned.

As already mentioned, sheep production in Cyprus is based on one breed, Chios and its crosses, and the situation was not very different in the 1980s. The increases in productivity were associated with the genetic improvement in milk and meat production of the Chios breed and with better management and feeding practices adopted by farmers.

Some of the innovations adopted by sheep and goat farmers that had a significant impact on production efficiency were associated with management practices. Lambs and kids used to be weaned at about 12 weeks of age and immediately slaughtered. At present, weaning takes place at the age of 5 to 6 weeks for lambs and 7 to 8 weeks for kids. As a consequence, the commercial milk yield of ewes increased and lambs are fattened on solid feed at a much lower cost and without any adverse effect on their growth. It has been estimated that for one kilogram of lamb or kid carcass, 10 to 12 kg of milk or 8 to 9 kilogram of creep feed is required (Economides, 1984; Economides and Olympios, 1991). Even today, with the very high prices of feedstuffs, the price paid per kilogram of goat and sheep milk, respectively, is two and three times higher than the price of creep feed. A recent development relates to the introduction by a few producers (5.3%) of automatic artificial rearing devices for rearing lambs and kids from birth to weaning on milk substitute. The performance of artificially reared lambs in terms of growth is no different from those on natural suckling, but the commercial milk yield of zero-day weaned dams is significantly increased (Koumas and Papachristoforou, 2008). Similar results were obtained with artificial rearing of kids (Koumas, personal communication).

In addition to early weaning, the slaughter age of lambs and kids was extended to 18 to 20 weeks leading to increased meat production. This system implies fattening lambs and kids from weaning to slaughter, thus taking advantage of the fast growth period and the high conversion efficiency of feed to liveweight (Hadjipanayiotou, 1997). Considering that the breeds kept in Cyprus are dual purpose (milk and meat) animals, carcass quality is considered as quite good at the slaughter ages concerned (Koumas *et al.*, 2005).

Mechanisation of milking has been a long process in the sheep and goat sector. Compared to a handful of producers who owned milking machines back in 1980, today the figure is 41% and continues to rise. Although this proportion may seem low, given that 43% of farmers only own between one and 50 animals with a mean of 24, it is rather unlikely that the majority of these farmers will consider installing milking machines. If these smallholders are excluded, the percentage of those with a milking machine goes up to 71.5, which is a very good figure.

Concerning genetic improvement, the national programme for within breed selection for milk, meat and prolificacy in Chios sheep and Damascus goats, has proved to be successful and genetic gains have been made (Mavrogenis, 2004). Dissemination of genetic progress was achieved through sales of selected males from nucleus government herds to farmers. The programme suffered some drawbacks due to the scrapie problem and the need to transform the nucleus herds within a short period of time (5 to 6 years) into herds that are resistant to scrapie. This was achieved, and nucleus herds played a key role in increasing the genetic resistance of the national sheep herd to scrapie by the dissemination of homozygote (ARR/ARR) rams.

Many farmers having followed a relevant recommendation by the Agricultural Research Institute and adopted two mating periods per year for their herds. Using the 'male effect' adult ewes are mated in April-May and adult goats in June resulting in an autumn drop. Young females of both species and adults which did not conceive at first mating are put to the male in September-October and give birth in February-March. The system ensures milk and meat production almost all year round and therefore better prices. This process also helps reduce the heavy work load and other problems associated with one lambing/kidding period per year.

Sheep and goat farms rely mostly on family labour for their operation. This is very important both in social terms and for the rural economy. Hired labour accounts for only 14% of those working on sheep and goat farms, most being employed in big units. If the trend towards fewer and bigger units continues, it is foreseen that there will be more hired labour. The situation in the dairy cattle industry is indicative of current trends. In 2002, hired labour represented 28.5% of the total employment while in 2007 the figure had gone up to 41%. The real problem associated with these changes, that seem to be inevitable, is the lack of skilled workers to be employed in the ruminant production sector, and this problem appears to be quite difficult to address.

Conclusions

Ruminant production in Cyprus has been moving towards higher input systems, following the current trend in most countries of the world. The number of farms (and producers) is continuously decreasing, which is having an impact on the rural economy. Mountainous and remote areas where small ruminant production is the only economic activity are mostly affected, as these areas are the first to be abandoned by sheep and especially by goat producers. Also, the larger high-input units have to rely almost exclusively on imported grains and high protein feedstuffs to feed their animals and face related risks. For instance, the recent extremely high prices of these commodities on the international market is typical of such risks. On the other hand, larger units offer better opportunities for the introduction of innovations that enhance productivity and increase farmers' incomes. Over the last three decades, dairy cattle producers have introduced many innovations that helped them to significantly improve their performance in terms of production efficiency. However, the high concentration of dairy cattle units in certain areas and the high density of bovine animals in Cyprus are matters of environmental concern. Sheep and goat producers have also introduced many innovations that enhanced their efficiency: most of these innovations are associated with production and management systems. As the ruminant sector, and particularly dairy cattle production, is moving towards intensification, the former 'family' operation is changing into a business and – among other implications – it will rely more and more on hired labour. The availability of skilled labour to be employed in this type of enterprise is one of the problems that the ruminant sector is going to face in the near future.

References

Cyprus Department of Agriculture, 2008. Sheep and Goat Survey for 2006. Ministry of Agriculture, Natural Resources and Environment, Lefkosia, Cyprus, 67 pp.

Economides, S., 1984. The effects of lamb-weaning age on the milk yield of Chios ewes and the growth and carcass gain of lambs. Technical Bulletin 59, Agricultural Research Institute, Lefkosia, Cyprus, 7 pp.

Economides, S. and S. Olympios, 1991. The effect of slaughter weight on carcass merit and conversion of milk or solid feed to meat in Damascus goats. Technical Bulletin 125, Agricultural Research Institute, Lefkosia, Cyprus, 9 pp.

Hadjipanayiotou, M., 1997. Effect of feeding extruded soybeans on the performance of young growing Chios lambs and Damascus kids. Technical Bulletin 180, Agricultural Research Institute, Lefkosia, Cyprus, 8 pp.

Hadjipanayiotou, M. and G. Kyprianou, 2001. The effect of weaning age on the growth of Friesian calves. Miscllaneous Report No. 12, Agricultural Research Institute, Lefkosia, Cyprus, 10 pp.

ICAR, 2006. Available at: http://www.waap.it/enquiry/Cyprus/041_all_breeds_cowscyp_interface/.

Koumas, A. and C. Papachristoforou, 2008. The effect of artificial rearing on lamb growth and ewe milk production in Chios sheep. In: Book of Abstracts No. 14, European Association for Animal Production, Wageningen Academic Publishers, Netherlands, p 193.

Koumas, A., A.P. Mavrogenis, C. Papachristoforou, E. Panopoulou and E. Rogdakis. 2005. Growth potential and carcass characteristics of native Machaeras, Damascus, and MachaerasxDamascus male kids. In: Animal Production and Natural Resources Utilisation in the Mediterranean Mountain Areas. EAAP publication No.115, Ioannina, Greece, 279-282.

MANRE, 2007. Annual Report for 2006 of the Ministry of Agriculture, Natural Resources and Environment, Lefkosia, Cyprus, 176 pp.

Markou, M. and A.P. Mavrogenis, 2002. Agriculrural Economics Report No. 41, Agricultural Research Institute, Lefkosia, Cyprus, 249 pp.

Mavrogenis, A.P., 2004. Genetic improvement of Chios sheep and Damascus goats. In: Review for 2003-2004: 73-77, Agricultural Research Institute, Lefkosia, Cyprus.

Papachristoforou, C. and A. Hadjicostis, 1996. Assessment of the Reproductive status of dairy cattle in commercial herds with the use of milk progesterone concentrations. In: Review for 1996, Agricultural Research Institute, Lefkosia, Cyprus, p 73.

Papachristoforou, C., and M. Markou, 2006. Overview of the economic and social importance of the livestock sector in Cyprus with particular reference to sheep and goats. Small Ruminant Research 62: 193-199.

New morphological and weight measurements by visual image analysis in sheep and goats

P. Negretti[1], G. Bianconi[1], S. Bartocci[2], S. Terramoccia[2] and L. Noè[3]
[1]Department of Animal Production, University of Tuscia, Via C. De Lellis snc, 01100 Viterbo, Italy
[2] Experimental Institute for Zootechnics, C.R.A., Via Salaria 31, 00016 Monterotondo, Roma, Italy
[3]AGER Agriculture and Research, Via Tucidide 56, 20134 Milano, Italy

Abstract

This study presents the application of opto-informatic techniques as Visual Image Analysis (VIA) in the area of sheep and goats production. The main purpose of the research was to bring innovation in terms of both technical and organizational aspects in the field of zootechnic biometry, a discipline where traditional tools are still in use. Such a novelty should not be confined only to research activities, but hopefully also lead to practical applications in farming as well as in production. Morphological traits and weight estimations have been performed on 36 Comisana sheep and 50 Saanen goats. The advanced VIA system has been completely computerised by the implementation of a laser measurement of the distance to allow the automatic calculation of the morphological parameters here reported, and it has become possible to get new angular and surface measurements. Evaluations of animal weight were carried out with the aim to set up regression equations for the indirect determination of live weight. To achieve such a task we measured a surface parameter (Surface Area of the Lateral Profile, SLP) that can only be determined by VIA techniques and has been demonstrated to be highly correlated with live weight (LW) in several zootechnical animals as buffaloes, sports horses, milk- and beef-cattle. In sheep and goats correlations between SLP and LW were 0.95 and 0.96, respectively ($P<0.01$). The regression equation ($LW= -59.25+0.03 \times SLP$) has been utilised in sheep, while ($LW= -3.5+0.014 \times SLP$) has been utilised to estimate live weight in goats.

Keywords: sheep, goat, visual image analysis, weight estimation, linear evaluation

Introduction

Visual Image Analysis techniques are well known and applied since two decades, thanks to technological developments and cost reduction, and they were really useful for zootechnical researches.

The application of VIA techniques to sheep and goats breeding is useful for both morphological measurements and weight determination through photographic and video images. The main advantage is to avoid direct contacts between the animal and the operator, with significant benefits for the animal welfare, mostly due to the minor stress caused by this innovative system compared to traditional tools. In this way the animal keeps its natural posture with a much better matching between biometric measurements and real body dimensions. This improvement is even more important for small size zootechnical species that usually, when exposed to operator manipulation, behave in a scared or panicked mode (crouching, flattening to the ground, jumping, pushing, etc.). The applied research work carried out by us concerned the most interesting zootechnical species in Mediterranean areas (buffaloes, milk- and beef-cattle, sports horses, rabbits, goats and sheep), and involved research institutes, SMEs, as well as local producers. These partners have always collaborated in order to supply consumers with traceable and quality products, possibly derived from production processes having low environmental impact (Negretti *et al.*, 1995; Negretti and Bianconi, 2005; Negretti *et al.*, 2008; White *et al.*, 2004; Zehender, *et al.*, 1996). Since 1994, when we began the conceptual development of image analysis for applied zootechnics, the first aim was to achieve a very versatile and absolutely innovative methodology; the second task was to develop such an application at a reasonable cost and suitable to fit the diverse characteristics (environmental, social, economical, etc.) of farms. Once its reliability and effectiveness was assured, the results were shared and elaborated by interdisciplinary working groups; the aim of this final step was to point out priorities, best applications and optimal tools in order to have a beneficial impact on rural communities. One of the greatest advantage for farmers will be the achievement of a more rational and practical work, that will be also faster and safer for the operator. A further positive aspect is the professional updating for working people, being involved directly in the computerised management of measurements, databases and genetic selection of animals. Perspectives are very promising but still much work has to be done, because at first such innovation in the field was not always well received, since it required significant changes at different levels, including cultural and operational new approaches.

Material and methods

Thirty-six adult and female Comisana sheep were examined using 20 individuals to calibrate the equations and 16 to test it. In order to achieve useful images for the indirect determination of weight by VIA techniques, animals were previously sheared.

Fifty adult and female Saanen goats were examined using 20 individuals to calibrate the equation and 30 to test it. Thanks to fruitful collaborations among research institutes, farmers associations, and local producers, the linear and morphological survey was performed by official evaluators belonging to ASSONAPA (National Association of Sheep-Rearing) simultaneously to our determinations, carried out by VIA techniques VIA (Table 1, Figures 1-3). For both species weight measurements were obtained by the electronic scale, as well as by the opto-informatic methodologies previously set up (Negretti *et al.*, 2004).

Results and discussion

To estimate live weight in sheep the regression equation (LW= -59.25+0.03 × SLP) has been utilised. Results showed a correlation of 0.95 between the live weight measured by scale and that estimated by VIA. The difference between these values was in the range of 0-3% for 55% of sample, and in the range of 4-5% for 45% of sample; a maximal difference of 4 kg was determined.

To estimate live weight in goats the regression equation (LW= -3.5+0.014 × SLP) has been utilised. Results showed a correlation of 0.96 between the live weight measured by scale and that estimated by VIA. The difference between these values was in the range of 0-5% for 74% of sample, and in the range of 6-11% for 26% of sample; a maximal difference of 6 kg was determined. The higher difference in percentage recorded with goats is due to the fact that, because of farm management needs, the animals had already been fed when measurements were performed; this lead to an alteration of values for about 1.2-1.8 kg. Considering the linear morphological survey, it was possible to establish that in goat farming many parameters presently estimated by eye can be measured by VIA techniques, leading to a more reliable and scientific evaluation (Table 2).

Table 1. Codes and main morphological parameters evaluated by eye (italic) and measured by VIA techniques (regular) in Saanen goats.

Code	Morphological parameter
A1	*Back line*
A1	Measurement of the back line inclination (Figure 1)
A2	*Rump angle*
A2	Measurement of the rump inclination (Figure 1)
B1	*Perpendicularity of limbs and hock distance*
B1	Measurement of limbs perpendicularity (Figure 3)
B3	*Perpendicularity of limbs from the side view*
B3	Measurement of limbs perpendicularity from the side view (Figure 1)
C1	*Fore udder attachment*
C1	Measurement of the fore udder attachment inclination (Figure 2)
C2	*Udder outline*
C2	Measurement of the udder inclination in respect to a parallel line to the ground (Figure 2)
C3	*Udder depth*
C3	Measurement of the udder depth in respect to the hock perpendicular line (Figure 2)
C5	*Rear udder attachment*
C5	Measurement of the rear udder attachment width (Figure 3)
D1	*Teats length*
D1	Measurement of the teats length (Figure 2)
D3	*Teats angle*
D3	Measurement of the teats angle (Figure 2)
D4	*Teats orientation*
D4	Measurement of the teats orientation (Figure 3)

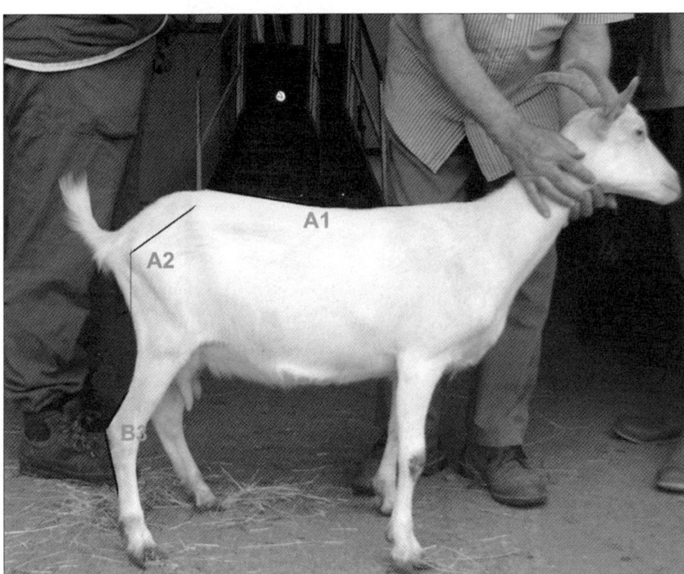

Figure 1. Side view.
A1: measurement of the back line inclination, A2: measurement of the rump inclination, B3: measurement of limbs perpendicularity from the side view.

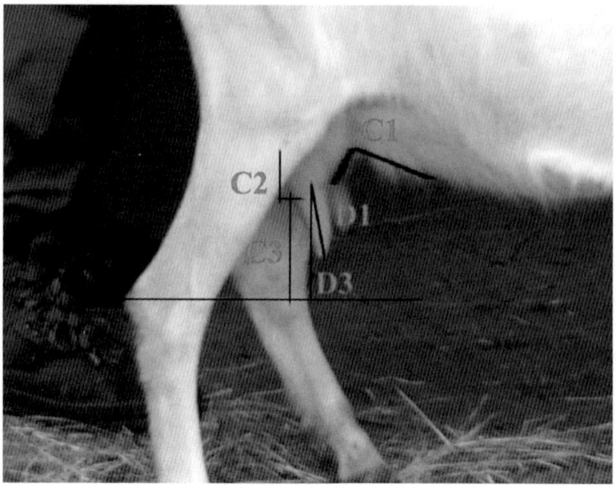

Figure 2. Udder apparatus.
C1: measurement of the fore udder attachment inclination, C2: measurement of the udder inclination in respect to a parallel line to the ground, C3: measurement of the udder depth in respect to the hock perpendicular line.

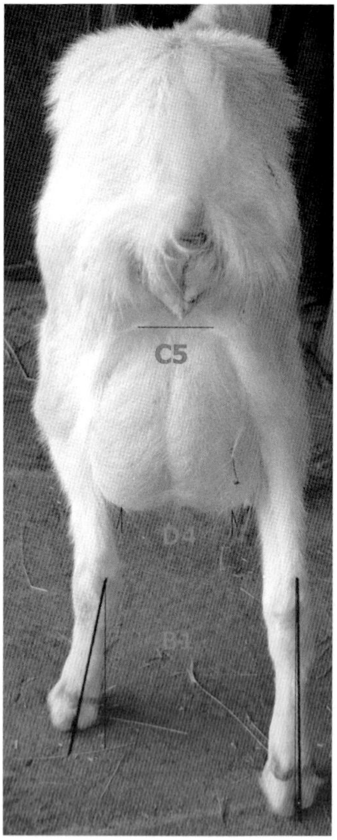

Figure 3. Rear view.
B1: measurement of limbs perpendicularity, C5: measurement of the rear udder attachment width, D4: measurement of the teats orientation.

Table 2. Determination of the main morphological parameters estimated by the Official Morphological Evaluation and those measured by the opto-informatic techniques in Saanen goats.

	Official and linear morphological evaluation											Measurements by visual image analysis												
												La	La	Re		La	La	La	La	Re	La	La	Re	
Subjects	A1	A2	B1	B3	C1	C2	C3	C5	D1	D3	D4	A1	A2	B1		B3	C1	C2	C3	C5	D1	D3	D4	
												(d)	(d)	L (d)	R (d)	(d)	(d)	(d)	(cm)	(cm)	(cm)	(d)	L (d)	R (d)
1	6	4	7	5	2	9	9	5	8	9	5	170	132	4	6	137	123	-	-	5	5.1	18	0	-
2	5	5	6	4	1	5	6	4	7	3	5	164	127	10	1.6	143	119	-	-	5	6.5	16	-	-
3	5	5	6	3	1	7	6	2	6	4	4	165	127	3	2	134	116	90	4.7	-	5.2	23	-	-
4	5	4	6	5	1	7	7	4	5	7	5	166	133	6	1.5	141	122	90	6.4	-	4.2	17	-	-
5	5	5	7	4	2	5	9	4	4	6	4	169	127	-	-	138	126	-	-	4	4	10	-	-
6	5	4	6	5	2	4	7	7	6	4	4	172	130	12	2	133	105	-	-	6	5.4	14	21	-
7	5	5	7	4	2	4	7	8	5	4	5	169	123	9	0	129	-	-	-	7	-	-	4	-
8	5	4	6	5	1	9	5	6	10	9	5	164	134	8	0	148	142	-	-	-	-	-	57	23
9	5	5	8	5	7	7	8	7	5	5	4	169	117	13	10	138	127	-	7.6	6	3.7	4.4	31	30
10	5	5	7	4	2	4	7	8	5	4	5	166	129	8	21	142	122	-	-	8	4.5	0	-	-
11	5	5	8	5	6	8	7	3	7	4	3	169	119	13	11	143	118	139	6.6	3	4.8	0	-	-
12	4	4	7	5	5	9	6	5	6	5	5	164	121	11	13	141	127	-	-	5	-	-	-	-
13	5	5	9	5	6	9	9	3	5	7	4	165	130	-	-	141	147	124	6.7	-	3.8	0	-	-
14	4	5	8	5	1	4	7	1	5	6	4	168	125	-	-	135	111	109	8.9	-	5.5	8.9	-	-
15	5	5	7	5	3	9	8	4	6	5	5	151	118	-	-	144	124	116	7.6	-	5	3.5	-	-
16	5	5	7	5	4	9	8	4	5	8	5	162	110	17	5	135	120	135	4	5	4.3	14	25	16
17	5	5	7	4	4	8	8	3	5	8	4	151	123	24	-	145	136	109	8.6	4	4.5	22	33	17
18	5	5	7	3	1	5	7	2	5	5	4	164	127	14	14	131	114	-	-	-	5.8	5	0	-
19	5	5	6	5	2	6	5	4	6	7	4	163	126	-	-	126	124	-	-	-	5.7	2	-	-
20	5	5	5	4	1	6	5	1	5	7	5	172	125	-	-	128	120	123	-	-	3.4	0	-	-
21	3	3	5	7	1	4	9	4	6	2	5	170	108	-	-	140	112	127	9.6	-	5.3	26	-	-
22	5	5	5	5	2	7	9	6	6	7	4	168	127	-	-	146	96	104	7.6	-	5.5	19	-	-
23	5	5	7	5	1	8	8	1	7	4	4	173	118	-	-	141	117	112	8	-	5.1	33	-	-
24	5	5	8	5	3	8	8	4	5	5	4	176	127	-	-	134	116	108	7	-	4	15	-	-
25	5	5	7	7	5	5	8	4	4	5	4	166	126	-	-	142	125	-	7	-	3	-	-	-
26	5	5	9	9	1	6	9	4	7	5	5	169	132	-	-	139	109	-	-	-	-	-	-	-

La = lateral picture; Re = rear picture; L = left; R = right; d = degrees; cm = centimetres.

References

Negretti, P., G. Filippi Balestra, G. Bianconi and R. Tonielli, 1995. Analisi d'immagine e identificazione dei parametri morfologici del modello ideale della bovina da latte. In: Atti Conv. Naz. Società Italiana delle Scienze Veterinarie 49: 955-956.

Negretti, P., G. Bianconi, A. D'Angelo, A. Gaviraghi and L. Noè, 2004. Application of the opto-informatic system to the morpho-weighted evaluation of goats: preliminary communication. 39° Simposio Internazionale di Zootecnia 'Meat Science and Research' Roma, Italy, 433-440.

Negretti, P. and G. Bianconi, 2005. Indirect determination of live weight, dead weight, and yield at slaughtering of Chianini beef by means of opto-informatic evaluation methods. In: Proc. 4th World It. Beef Cattle Congr. 229-234.

Negretti, P., G. Bianconi, S. Bartocci, S. Terramoccia and M. Verna, 2008. Evaluation of live weight and body condition score on Mediterranean Buffalo during lactation by the Visual Image Analysis. Livestock Science, 113: 1-7.

White, R.P., C.P. Schofield, D.M. Green, D. J. Parsone and C.T. Whittemore, 2004. The effectiveness of a visual image analysis (VIA) system for monitoring the performance of growing/finishing pigs. In: Animal Science, 78: 409-418.

Zehender, G., L.P., Cordella, A. Chianese, A. Lino Ferrara, A. del Pozzo, S. Barbera, A. Bosticco, P. Negretti, G. Bianconi, G. Filippi Balestra and R. Tonielli, 1996. Image analysis in morphological animal evaluation: a group for the development of new techniques in zoometry. Animal Genetic Resources Information (AGRI), 20: 71-7.

An information system for the assessment of the threats to the heritage sheep breeds

C. Ligda[1], Ch. Mizeli[2], A. Carson[3], D. Bowles[4] and A. Georgoudis[5]
[1]Centre for Agricultural Research of Northern Greece, NAGREF, Thessaloniki, Greece
[2]Holstein Association of Hellas, Derveni, Thessaloniki, Greece
[3]The Sheep Trust, Cumbria, United Kingdom
[4]Departement of Biology, University of York, York, United Kingdom
[5]Department of Animal Production, School of Agriculture, Aristotle University of Thessaloniki, Greece

Abstract

The heritage sheep breeds are geographically concentrated, environmentally adapted, genetically distinct and commercially farmed breeds to economically support the local communities. In the frame of the HERITAGESHEEP project funded under the EU Regulation 870/04, a database has been developed to combine all (the) available information for the assessment of threats for the heritage sheep breeds. This information is grouped in three parts, where the first refers to the population numbers and trends, the second refers to the potential threats, which include social, political, disease and environmental pressures, and the third part focuses on the determination of the value of the breed to the region where it is raised, through its contribution to the environment, adaptation and grazing habits. The system aims to serve as a tool to promote the sustainable use of the heritage sheep breeds, integrating information on the characteristics of the breeds, their use, the threats they are exposed to and the values attached to the breeds and their products.

Keywords: AnGR databases, sheep breeds, information system

Introduction

Heritage sheep breeds (HSBs) are defined as genetically distinct, geographically concentrated and adapted to their environments. Typically, these sheep breeds are 'local' breeds, traditionally farmed for commercial use and play an important role in the culture and rural economy of the regions in which they are managed (Bowles *et al.*, 2003).

The aim of the HERITAGESHEEP project was to establish a European-wide conservation programme of HSB genetic resources for the diversification of production in livestock agriculture and for their central importance in the long-term sustainability of medium to low input farming systems. The outcomes of this programme has positive impact: (1) on the environment, reducing the negative impact of intensive agricultural practices; (2) on the regional communities by using natural resources in a sustainable way to manufacture products; (3) on the rural economies developing strategies to enhance the profitability of local breeds.

Through a European Regional Focal Point (ERFP) scoping study, it has become clear that all member states across the European Union (EU) can identify HSBs. In each case, the breeds are confined in specific regions of the member states, and through their adaptation to local environmental conditions, are used successfully in extensive, medium to low-input farming systems. Despite the value of these breeds and their genetic resources for both environmental and economic sustainability of local communities throughout the EU, they are only beginning to be recognised as 'breeds at risk' within National Action Plans of Farm Animal Genetic Resources (FAnGR) (ERFP, 2006).

The information system

Within the project, a questionnaire was developed that aimed to collect information on the threats to the Heritage Sheep breeds, the values of the breeds according to the Breeders Societies, the current situation on *in situ* and *ex situ* conservation activities and the future trends of the breeds.

The questionnaire was divided into the following sections:
- section 1: numbers and trends;
- section 2: threats;
- section 3. values;
- section 4: current situation and future trends.

All data collected are provided by Breed Societies. According to the societies, the main causes of the reduction of the number of heads within a breed were: the decrease in public funding, the lack of political will to support rural communities, policies and legislation, diseases, predators, urbanisation, poor return on product, ageing of farmers, lack of marketing support, inbreeding and loss of skills (Carson *et al.,* 2008). All information collected within the project is stored in a database in order to assess the degree of threat of each breed and visualize the results. The aims of the database are to serve as a tool to promote the sustainable use of the heritage sheep breeds, integrating information on the characteristics of the breeds, their use, the threats they are exposed to and the values attached to the breeds and their products. A permanent and widely accessible European network of national inventories of HSBs is established that informs on the activities concerning these breeds including related *in situ* and *ex situ* conservation programmes.

The portal development consists of two parts:
1. Database development. The database was developed by using Microsoft Access. The tables are connected to the breed and the country fields. The tables are generated under the different groups of information gathered (data collected from WP1, as described earlier). A user interface using graphical forms and list boxes was created in order that all the partners enter the collected data per breed. Finally, several queries are generated for to develop reports on breed or country level. The tables' relationships are presented in Figure 1.
2. Portal (website) development. The website runs under the http://www.heritagesheep.eu URL. The portal consists of two parts, static and dynamic content. In the static part of the site there is information about the description and the objectives of the program, work-packages, sheep breeds and all partners. Also there is a part to view or download all reports and presentations of the different meetings.

The dynamic part of the web page uses ASP technology. It has developed by using Microsoft Visual Web Developer and Microsoft Visual studio. There is a distinct environment separated under the different thematic fields of the data collected and queries generated. The user is able to select the country and then the breed and the related database information are presented in the website. An overview of two of the forms as they are viewed on the website is presented in Figures 2 and 3.

Further development and exploitation of the information system

All parts of the website and the database are in open access for all visitors. Any stakeholder can find information on the implementation of conservation strategies of HSB genetic resources in national initiatives, as local breeding and farming groups, the general public etc. The website aims to raise the public awareness on the issue of the conservation of FAnGR and also to provide information to the policy makers and the relevant authorities at national and EU level on the significance of the genetic resources of the Heritage Sheep Breeds and the urgent need for action to conserve these resources.

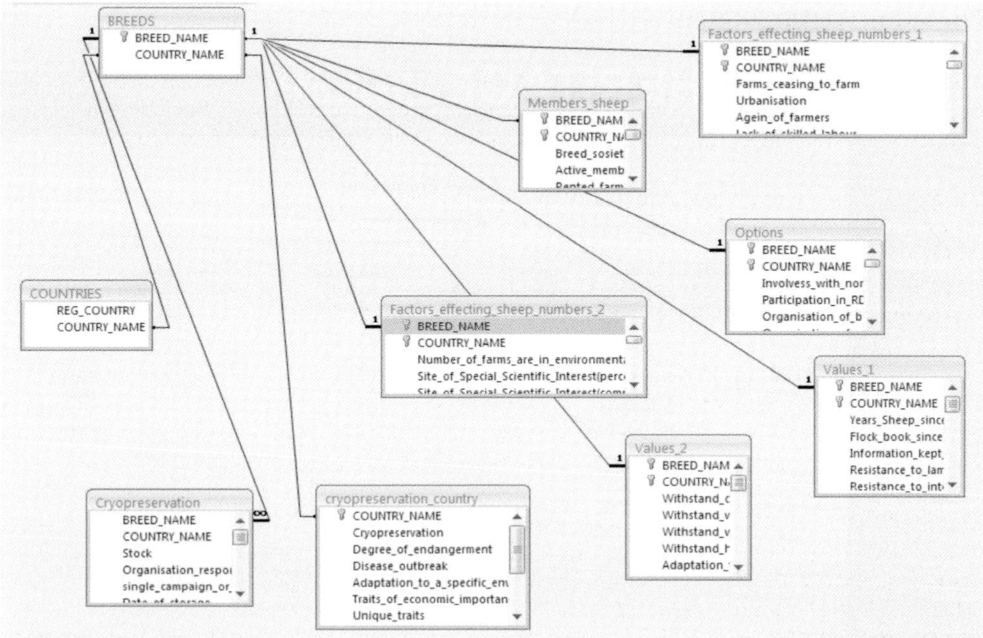

Figure 1. Table relationships.

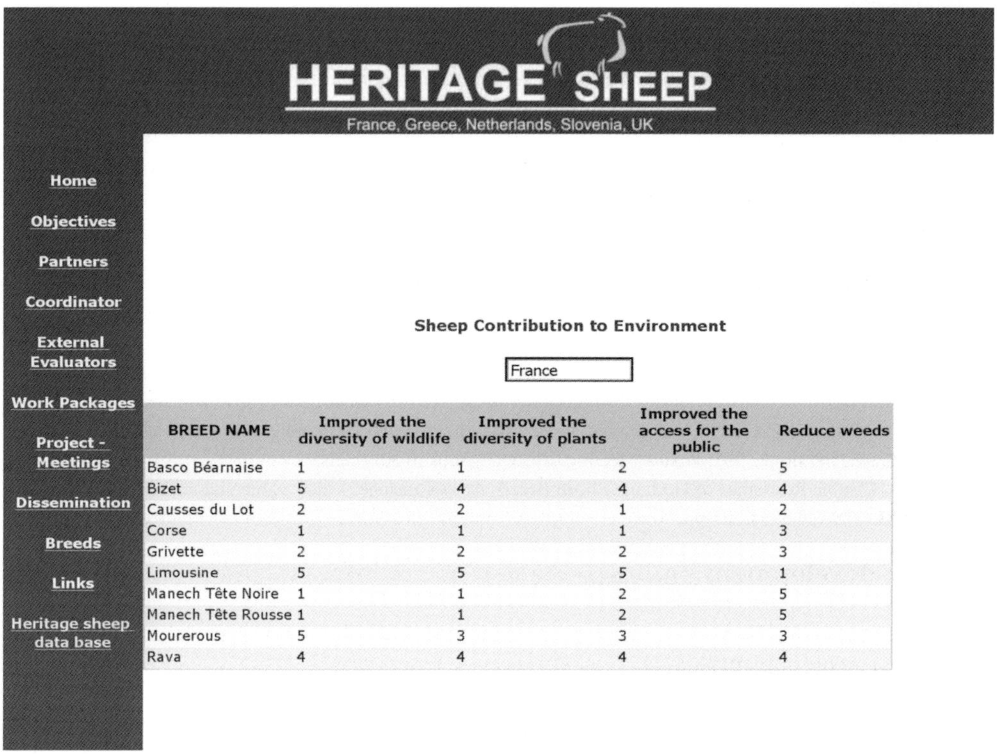

BREED NAME	Improved the diversity of wildlife	Improved the diversity of plants	Improved the access for the public	Reduce weeds
Basco Béarnaise	1	1	2	5
Bizet	5	4	4	4
Causses du Lot	2	2	1	2
Corse	1	1	1	3
Grivette	2	2	2	3
Limousine	5	5	5	1
Manech Tête Noire	1	1	2	5
Manech Tête Rousse	1	1	2	5
Mourerous	5	3	3	3
Rava	4	4	4	4

Figure 2. The database on the web, table: sheep contribution to the environment.

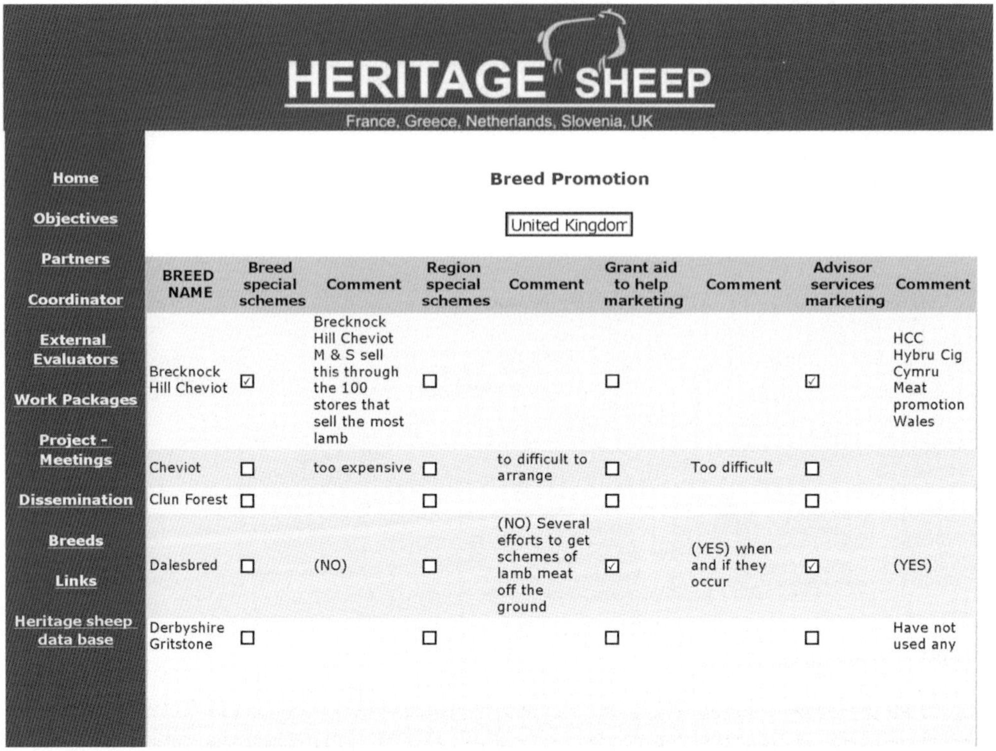

HERITAGE SHEEP

France, Greece, Netherlands, Slovenia, UK

Home
Objectives
Partners
Coordinator
External Evaluators
Work Packages
Project - Meetings
Dissemination
Breeds
Links
Heritage sheep data base

Breed Promotion

United Kingdom

BREED NAME	Breed special schemes	Comment	Region special schemes	Comment	Grant aid to help marketing	Comment	Advisor services marketing	Comment
Brecknock Hill Cheviot	☑	Brecknock Hill Cheviot M & S sell this through the 100 stores that sell the most lamb	☐		☐		☑	HCC Hybru Cig Cymru Meat promotion Wales
Cheviot	☐	too expensive	☐	to difficult to arrange	☐	Too difficult	☐	
Clun Forest	☐		☐		☐		☐	
Dalesbred	☐	(NO)	☐	(NO) Several efforts to get schemes of lamb meat off the ground	☑	(YES) when and if they occur	☑	(YES)
Derbyshire Gritstone	☐		☐		☐		☐	Have not used any

Figure 3. The database on the web, table: breed promotion activities.

The geographical representation of the results is under development, where it is investigated the feasibility of an early warning system to predict which breeds are likely to become threatened, or change their current status.

References

Bowles, D., P. Gilmartin, W. Holt, H. Leese, J. Mylne, H. Picton, J. Robinson and G. Simm, 2003. The emergency of 2001: cryopreservation of sheep germplasm during the UK crisis of foot and mouth disease. Workshop on Cryopreservation of Animals Genetic Resources in Europe, (ed. D. Planchenault), p. 117-122.

Carson, A., Ch. Ligda, C. Danchin, D. Kompan, L. Kaal-Lansbergen, A. Georgoudis and D. Bowles, 2008. Heritage Sheep breeds in Europe, 59th Annual Meeting of EAAP, Book of Abstracts.

ERFP, 2006. Genetic Resources of Heritage Sheep Breeds Across Europe: - their value and conservation. Jones, S. and D. Bowles (editors). 221 pp.

Permanence of electronic ear tags in extensive rearing of Majorcan black pig

J. Jaume[1], J. Tibau[2] and J. Gonzalez[3]
[1]IBABSA Balearic Institute for Animal Biology, Spain
[2]IRTA Food Technology, Spain
[3]IRTA Animal Production, Spain

Abstract

The aim of this study was to evaluate the performance of electronic ear tags in Majorcan black pigs in extensive conditions. To achieve this objective, a total of 1,221 sows and 94 boars from 66 farms of Majorcan black pig were identified with an electronic tag. After one year, failures, losses and permanence in dynamic reading were observed. No negative effects on animal health were observed after application and during the trial. Perimeter wall effect was statistically significant. Losses in grid perimeter wall (24.4%) were greater than in stone (3%). Despite losses, results in the Majorcan Black Pig breed under extensive conditions surpassed the acceptable level of permanence. In addition, electronic ear tags identification proved to be an acceptable technology for dynamic reading in extensive conditions.

Keywords: electronic identification, traceability, tags, transponders, Majorcan black pig

Introduction

The Mallorcan black pig is an endangered breed native to Mallorca in Spain. Preservation of Majorcan black pig is mainly sustained by the utilization of pure breed animals in the elaboration of the *Sobrassada de Mallorca de Porc Negre Mallorqui*, a kind of cured sausage with paprika which enjoys a Protected Geographical Indication, and also by the commercialization of the *Porcella*, the pig slaughtered at 7-12 kg live weight approximately, between 1-2 months, consumed as roasted meat (Jaume, 2000). This preservation is based on the existence of the herd book, officially approved in 1997, which is administered by the Majorcan black pig producer's association, and currently controls 1,300 sows and 120 boars. This breed is adapted to the Mediterranean conditions of production, and it is reared in an extensive regime, taking advantage of the limited local resources spontaneous and cultivated. Grazing occurs on land bounded by stone walls or grid. Food through grazing is seasonal and depends on the farm and the type of animal and is complemented with cereal grains, legumes, carobs and figs (Jaume *et al.*, 2007).

The use of the electronic identification in swine has been investigated (Lambooij and Merks, 1989; Stärk *et al.*, 1998; Caja *et al.*, 2005; Babot *et al.*, 2006) because of the necessity to develop a permanent and tamperproof identification system for improving farm management and animal control. The uUse of subcutaneous microchips in animals intended for consumption presents problems in his recovery. Different studies in pigs have been done to gauge interest in the use ear tag with a microchip, and have proved their efficiency under conditions of commercial pig farms (Santamarina *et al.*, 2007). There are not available studies on electronic identification in Majorcan Black Pig.

The aim of this study was to evaluate in extensive conditions the performance of electronic ear tags in sows and boars included in the herd book of Majorcan black pig.

Material and methods

Identification and reading system

One type of ear tag plastic button, 25 mm in diameter, is used and numbered individually for each animal, with a transponder (LW FDX) incorporated. The electronic transponders contained a single code with the coding standard established by the ISO 11784 mark ALFLEX, supplied by AZASA. There are unequivocal correlations between the number visual individual and the transponder.

All the sows and boars in the herd book of Majorcan Black Pig from each farm were identified whit an ear tag plastic button with a control of identification on the day of placement during February and March of 2007 and a control of identification, retention, cracks and faults of reading after one year, between February 18 and March 12, 2008.

A total of 1,221 sows and 94 boars from 66 farms of Majorcan black pig were used. The animals have been kept in extensive conditions, and information regarding the type of enclosure of the holding and impact health in the ear was collected.

The operation of all the transponders and verification of the correspondence between numbers and visual electronics have been tested before application. The implementation of the ear tags has been made on the same day for all animals in each farm.

Statistical analysis

Statistical analysis was performed using the computer software Statgraphics Plus. Losses, electronic failures, and recovery were analyzed using the GLM procedure. Significance was fixed at $P<0.05$.

Results and discussion

Descriptive analysis of data is shown in Table 1. The study was conducted on 66 farms and about 1,318 animals; 126 animals from 35 farms lost the ear tag with the microchip. This represents a 9.56% loss in animals and a durability of 90.21%. There were only 3 failures reading. It should be stressed that on 42.42% of the farms the identification of all animals remained in place.

The use of individual electronic ear tags to identify animals included in the herd book of Majorcan black pig in extensive conditions has shown recoveries of 90% after one year. Ear tag losses were similar to values reported by Caja *et al.* (2005) in intensive conditions.

Both the placement as the reading dynamics in these conditions have been readily achievable. No apparent negative effects on animal welfare were observed after the application of the ear tags or at the reading control. This fact agrees with previous results reported by Babot *et al.* (2006) using

Table 1. Descriptive analysis of data.

	Animals		Farms	
	n	%	n	Farms
Failures	3	0.23	3	4.55
Losses	126	9.56	35	53.03
Durability	1,189	90.21	28	42.42
Total	1,318		66	

similar ear tags in pigs in intensive conditions. This results possibility their use as an electronic key for automatic data recording and farm management, in estensive conditions, which should reduce labor costs.

Table 2 shows the losses depending on the type of enclosure of the plots. The losses were minor ($P<0.05$) in those farms with stone walls compared to those with grid enclosures (3.0 vs. 24.4). The main reason which explains these differences is that ear tags are easily removed in farms with grid enclosures, due to the fact that they are often stuck in the grids.

Implications

Electronic ear tags in extensive rearing of Majorcan black pig proved to be a good method for the identification and the automatic reading of the pigs belonging to the herd book.

A strong farm effect due to enclosure was observed on losses of electronic ear tags. In this sense, the combination of visual and other electronic ear tag in each ear seems to be a higher efficient system to achieve complete identification. We also need to develop the ability to make copies of the electronic ear tag for its replacement in case of loss.

Table 2. Comparing results depending on the type of enclosure of the plots.

Enclosure	Animals			Farms		
	n	Losses %	Std dev	n	Losses	% losses
Stone	909	3.0	2.7	45	15	33
Grid	406	24.4	15.6	21	20	95
Total	1,315	9.56		66	35	53.03

References

Babot, D., M. Hernández-Jover, C. Caja, C. Santamarina and J.J. Ghirardi, 2006. Comparison of visual and electronic identification devices in pigs: On-farm performances. J. Anim. Sci. 84:2575-2581.

Caja, G., M. Hernández-Jover, D. Garı´n, C. Conill, X. Alabern, B. Farriol, and J. Ghirardi. 2005. Use of ear tags and injectable transponders for the identification and traceability of pigs from birth to the end of the slaughter line. J. Anim. Sci. 83: 2215-2224.

Jaume, J., A. Rigo, E. Fábreg, J. Tibau, 2007. Majorcan Black Pig Production System: Management in extensive conditions and performances. 6th International Symposium on the Mediterranean Pig. 11-13 October 2007 Capo d'Orlando – Messina (Italy).

Jaume, J. and L. Alfonso, 2000. The Majorcan Black pig. Animal Genetic Resources Information 27: 53-58.

Lambooij, E. and J.W.M. Merks, 1989. Technique and injection place of electronic identification numbers in pigs. Pages 5-14 in Rapport B-335. Res. Inst. Anim. Prod. 'Schoonoord,' Zeist, Utrecht, the Netherlands.

Stärk, K.D.C., R.S. Morris and D.U. Pfeiffer, 1998. Comparison of electronic and visual identification systems in pigs. Livest. Prod. Sci. 53: 143-152.

Part 3.

Innovations improving the socio-cultural basis or increasing ethical issues in livestock systems from the Mediterranean area

Mediterranean pastorality[5]: sustainable know-how in search of development

R. Bouche
INRA SAD LRDE, Quartier Grossetti, 20250 Corte, France

Abstract

While the new societal expectations, focused on sustainable development, endow Mediterranean pastoral systems with tangible advantages (environmentally-friendly breeding, quality products, keeper of traditions), they seem mired in a strange paradox where the robustness forged over thousands of years as being the only way of maintaining them in a extensive environment, actually makes them highly vulnerable to the technological innovations and developments of the current socio-economic environment. More importantly, as the only ones to possess original assets (local breeds, know-how etc.) and able to differentiate them and add value to them in the current context of globalization, these systems probably make the least effort to preserve such resources and such a heritage. Referring to a number of examples of recent innovations and based on the paradigm of complexity and systemic analysis, the author reflects on some of the reasons for this paradox. In his analysis he suggests combining the field of technical innovation with cognitive, organizational and cultural aspects to form 'an anthropological whole' which is necessary to maintain a balance between the system and its environment and, by the same token, to explain the conditions for ownership and for learning innovation.

Keywords: know-how, pastoralism

(Caricatural and) contextual iconography

In the distance, the dairy inspection vehicle speeds away in a cloud of dust. Nostalgically, Charles the goatherd begins thinking of the same track, recently repaired by the council in an attempt to mollycoddle the voters. He can remember riding down this stony, winding track sitting on Fajanu the mule. He would join his grandfather Charles – proud to have the same first name – as he brought the white cheeses up from the sheepfold on the plain to the women in his village. They would work these young cheeses all day long, turning them, washing them with a piece of damp cloth or a fern: patience would melt into a creaminess which is mystified in the present day. In those days, his father Auguste would repeat a seemingly unchanging ritual. After the five o'clock milking that he would perform squatting among his goats, he would whistle stridently through his teeth to wake 'that lazy dog' he would grumble, referring to his four-legged friend who was already watchful, tail wagging in the middle of the flock. Showing him first his goatherd's haversack then his shoes, he would say: 'Never forget, son, these are what make milk here. There is no other secret and there never will be'. Just like every other day, the 80 goats would shake themselves as one, a shoal of fish, a flock of starlings, a swarm of bees! A herd, yes, scattered but calmly moulting at a peaceful rhythm and setting off for a long day's grazing. Today, there are almost 200 goats, Ziu Carlu is resting while August is still there, growing vegetables to supplement the meagre pension paid by his agricultural mutual insurance but too worn out to set out with the herd. He simply tells his goat stories to passing tourists, sitting on his stone bench which looks over the church square and the entire valley beyond. The cheese is more salty, ferments and cold rooms have replaced the women. He asks the younger man no questions. He smiles knowingly, because from his look-out he has seen the herd cross the ridge named 'a sulana'; he knows that there will be 15 litres extra this evening after milking. Strange knowledge which is of great interest to the people at the INRA, who believe that this empirical know-how is essential to decipher the complexity of the phenomena. They arrive at 10 a.m. with the GPS

[5] This neologism is used here by the author to emphasise and reify the quality and cleanliness of the pastoral world.

systems that they attach to the goats' bells to monitor their movements on a computer. Charles will be there too to talk to them ....They will discuss the hows and whys ... and the how of the why. He will answer willingly, even if he does not understand the meaning of what is obvious to him: 'How do you choose the next generation? Based on what criteria?'; 'Well we do what we have always done, we choose the best one, the most adaptable.'; 'Which ones are the best? Those which give the most milk to the dairy inspector?'; 'No, we don't pay much attention to the dairy inspector; I know my goats,.... But nowadays it's easier, we let the goats go and we distribute the food – it's expensive but there's no other way, otherwise we'd have to follow them all day like Auguste senior did from the age of 14. I open the gate in the morning, I walk with them at the start of the circuit and depending on where we start, they know where they have to go until the evening'. Sitting in the 4-wheel drive, Kevin, the latest addition to the family and named, to the great displeasure of his grandfather, after a TV hero that no-one knows any more, plays on his games console.... Times have certainly changed.

Context and scope of the analysis

This insight into Mediterranean animal husbandry systems firstly encourages us to explain the specific context chosen for our analysis, even if it could be broadened – at the request of the reader – to other 'villages'. What we feel characterises these breeding systems first and foremost cannot simply be explained by geographic location but by the influence of all the facets of this geography (climatic, human, social etc.) on the systems studied. In other words, our study does not concern the breeding systems in general but more specifically those which, in this part of the world, are influenced by this particular milieu and this particular environment. It refers to systems which have been shaped over the centuries both by and for the specific conditions of the environment which retain a memory from both the point of view of biophysical resources (e.g. breed) and that of the human components capable of activating them; in other words systems of knowledge and know-how which have developed gradually by interacting with this particular environment. Our observation framework concerns more specifically the pastoral systems of small, extensive dairy ruminants, in particular those which have been widely studied in Corsica as the archetype of a certain number of situations in the non-arid Mediterranean perimeter. Furthermore, avoiding the economic aspects of innovation although not ignoring them, this analysis more specifically concerns the informational and organisational innovations likely to herald profound changes which may call the viability of these systems into question.

Pastoralism, a question of know-how

In Corsica, with its strong cultural ties, pastoralism, and more specifically extensive dairy sheep and goat breeding[6], remains a distinctive characteristic of regional identity, despite a marked decline. This breeding remains one of the only 'productive' activities capable of maintaining the mountainous areas at the heart of the island. As such, it provides the community with a certain number of advantages for sustainable development (environmentally-friendly breeding, quality products, preservation of traditions) which tour operators put to great use whereas, paradoxically, it is rooted in very robust systems at an individual level which in theory are very difficult to reproduce for future generations (low productivity, time-consuming work, social isolation).

In the traditional systems, the breeders tried to adapt to a restrictive environment and to manage uncertainty over long periods of time (Ravis-Giordani, 1983). They therefore developed specific practices and know-how to guide their herds over vast areas (Pernet and Lenclud, 1977). The invented indescribable signs allowing them to communicate (using forms which are not necessarily taken from everyday language) with pastoral socio-technical networks representing essential human

[6] It should be noted that while traditional pastoral systems use extensive areas, they have nevertheless become highly intensified with regard to the mobilisation of human capital and in particular of know-how.

relays for this type of extensive herding (Bouche, 2001). Extensive breeding can therefore be clearly distinguished from intensive breeding systems in terms of its relationship with the natural and cultural environment and in its aims and decision-making processes. An extensive breeder is obliged to plan production so that it takes unforeseen events into account. The events might occur (1) daily, depending on bad weather, (2) with seasonal fluctuations combining periods of scarce resources in winter and summer and more plentiful periods with more than enough food for the animals, and (3) from one year to the next, with a succession of years with completely different profiles. The practices are not thus aimed at maximising the expression of the animal's individual productive potential by nullifying the effects of the environment, but rather at combining a number of individual performances within the 'herd' (Santucci, 1991) over long periods of time when decision-making cycles are more concerned by biological maturations and learning the behaviour of the animals than by annual accounting objectives.

We see here the elements of a fundamental identity characterised by complex interactions between a system and its environment. Hence, whereas intensive systems secure and balance their dependence with the ups and downs of the natural environment by increasing the level of inputs (provision of food, regulated temperature etc.), Mediterranean pastoral breeders have developed specific know-how in order to survive in a climate of great uncertainty, thereby increasing the complexity of their systems. This complexity can be seen both in the internal elements of the system (breeder, herd, territory etc.) and in the social relations capable of supporting it at higher levels of organisation.

Driving herds over large areas (sometimes >300 ha) has allowed breeders to developed *abductive* (the capacity implemented by an individual to spontaneously favour one scenario over a large number of other potential scenarios) mechanisms and refined *tacit* knowledge (tacit knowledge is linked to the experience of those who have it: 'It might be a means or specific talent of individuals or an interiorisation of explicit past knowledge which has gradually become subconscious and which is reflected by automatisms', Nonaka and Takeuchi, 1995) of their animals and the land on which they graze. On the one hand, in addition to an excellent capacity for adaptation to feeding in a difficult environment (walking, agility, resistance, toughness), the herd must demonstrate intrinsic 'social' qualities (gregariousness, existence of docile leaders). It must also be possible to identify it from a distance (diversity of colours). These qualities are desired, expected, perceived and then perpetuated by the breeder by means of a specific practice of animal selection. At the same time, they require an excellent knowledge of the behaviour of each animal in order to anticipate its reactions with regard to unforeseen events and then react accordingly. This knowledge is developed and accumulated in a dynamic process over generations throughout a pastoral and social history which organises and regulates the forms and objects of intra- and inter-generational transmission.

These numerous phenomena can be seen both in driving the herd and in identifying the vegetation which will feed the herd or damage the desired products: one type of pasture is potentially unfavourable to the volume ('if I drive the herd under that ridge tomorrow, milk output will drop by 20 litres') or quality ('if they walk under the olive trees, the cheese will run') of milk production. This complex knowledge, rooted in place names (like so many signs expressing the collective memory), relayed culturally by social-technical networks, is conveyed and even ritualised throughout society by sayings, proverbs or events which constantly refer to it.

Technical acts and cognitive priorities primarily focus on the animal's relationship with its environment. The activities of transforming milk into cheese and marketing the final product are of secondary importance and may be delegated to a third party. If the breeder is also a processor, he may establish priorities between the different tasks considered to be essential to maintaining the global coherence of his system. For example, it was not a rare sight to see a breeder leave a vat during the curdling process to go and find an animal that was lost. Even more strange, the processing vat might be left unattended when a neighbour arrives with a piece of information (concerning an animal separated from the herd). However, it is likely that of these three options, the third, i.e. information and even more so linguistic communication, will be preferred to searching for an animal alone.

In this context, even if real cheese-making know-how exists, the system works as if, from the point of view of processing, the operating chain aimed at stabilising the milk proteins quickly to ensure their conservation can at any time be sequenced, interrupted or delegated to a third party. The maturing process, a sedentary activity which in the past was entrusted to women[7] or old people, is relegated to a position of secondary importance by the goatherd ranking after watchful surveillance of the herd and the land and the search for information which may facilitate his task.

An overview of some more recent innovations

Naturally, this reality changes, and many changes (rural exodus, breakup of the pastoral society, etc.) erase past balances. Herds grow, become more sedentary and the animals are given supplementary feed. However, in the light of their lack of competitiveness in terms of productivity, breeding activities in these regions are again confronted by the inefficiency of technical innovations designed for intensive breeding. The supposedly age-old know-how is no longer transmitted by traditional methods. We are currently witnessing a pastoral crisis, primarily resulting from the breakdown of know-how and the associated cultural aspects. Should we then today speak of a *genuine extensification* focusing on an unsound hybridisation, clumsily combining the stereotypes of two antinomic worlds? The younger generation is trapped between disappearing traditional practices and technical tools unsuitable for pastoral breeding. They do not have the necessary technical models and referentials. Through the phenomena of self-education, the remaining breeders gradually adopt an intrinsic complexification of their practices which, individually, often makes them too robust. Paradoxically, these practices and this new know-how are difficult to hand down to subsequent generations. In this context, the capacity to increment this know-how is becoming a key element in optimising and reproducing the pastoral system.

To illustrate our remarks, we have taken a number of present-day innovations which are in themselves unimportant but which give us food for thought concerning the changes they bring about.

Imperative memorisation of mother-kid pairs at birth

More than maximising performances, the breeder wishes to optimise the system as a whole. As well as the choice of animals which are individually adapted to the ecosystem, optimisation involves global herding which includes managing lineage, maintaining a gregarious herd, culling disruptive goats, etc. For example, the homogeneity of a herd should be established as early as parturition. In addition to selecting the optimum renewal rate (20%), monitoring parturition enables the kids to be identified and paired with their mothers. This requires a constant presence and attention during the entire period to observe and memorise the relationships: 'personally, if I don't witness the birth, I don't keep the kid, as it might have been adopted'. Breeders with small herds applied similar practices (observation of the herd while tending it on a daily basis) to memorise[8] the mother-kid

[7] In this context, the know-how related to developing taste is recorded in a housekeeping book passed on from mother to daughter in the same way as sewing or cooking procedures.

[8] Some talk about a gift ('if you don't have the gift at parturition, when all the kids arrive at the same time, you're lost'), others refer to the need 'to be born into it' to learn these skills.

pairs. As they are now obliged to increase the size of their herds (the average size is currently about 180 head, although some herds number 500 to 600 head) for economic reasons, breeders must adapt the cognitive and organisational artefacts that enabled them to build a productive herd for which regulatory individual identification (tip-tag) would appear unsuitable and, more particularly, far removed from this essential identification and memorisation work. This is hindered by a natural grouping of births over a maximum of about 10 days: 'Sometimes 50 to 60 kids are born in the space of one morning. You have to be very alert to see them'. Breeders adopt innovative strategies which, for technical and economic results which are often very similar in the short-term, have very different results in the long term. For example, to avoid the risks of adoption or confusion in relationships, (1) some breeders isolate parturient goats by placing them in separate pens. While this 'innovative' method would appear to ensure the creation of a mother-kid pair without any particular cognitive effort, handling becomes prohibitive as it requires a large number of personnel. It does not exclude the presence of the goatherd, who speaks of a 'mental impression' (sic) of the pairs at the moment of birth. Although this 'impression' is not unique and crucial at the actual birth, its absence can subsequently cause major problems in recognising the animals and managing them when the animals are on the move. Inversely, following ovine models, (2) other breeders let the birthing take place in vast open flat areas which they watch from a distance, sometimes using binoculars, to memorise the pairs. This technique would appear to be positive as it does not affect the '1st essential view when the kid arrives' (sic). However, it requires large enclosed areas. Finally, we even met a breeder who, in order not to lose 'this instant memory' (sic) while avoiding the need for long periods of surveillance, has his goatherd film the births using a camcorder so that, 'if necessary', he can watch the event at a later date in order to memorise it.

In this example of memorising genealogies, we are confronted by animal husbandry and economic criteria which guarantee the breeder's income but which also require assessment of the organisational and cognitive room for manoeuvre available to him to remain viable. Knowing how to identify the mother-kid pairs remains a key to success for pastoral goat breeders. Few innovative solutions would appear to be genuinely adapted to simplifying the cognitive effort focussed on this event.

Surveillance, mobile telephones and the socio-technical network

To drive their herds over vast territories, the goatherds rely on other operators, often activated subconsciously, in particular with regard to transmitting information. Extensive breeding relies on distributed guidance, an essential system of 'instinctive' mutual help which we will refer to as pastoral echolocation[9]: using a specific code (marks on the ears, colour of the animals, etc.) and a detailed knowledge of the toponymy of the land, the breeder can find a stray animal or find out about the condition of a distant pasture by means of social information relays. Echolocation therefore relies on the existence of a network, conscious or otherwise, which allows the different users of the space to situate themselves. This communication technique requires a common vocabulary concerning the animals and the environment. Tending a herd in an extensive system therefore requires the implementation of specific listening mechanisms (which may sometimes lead to cognitive saturation) in order to connect to the pastoral networks and receive or emit useful information. This use of the land still exists today, although the mobile phone is gradually replacing the presence of echolocation agents as the new communication media in pastoral areas.

The mobile phone, often perceived as an 'incremental innovation' which frees landline telephones of their bonds, can be seen here as a 'procedural innovation' in that, like fences, it profoundly changes the system of echolocational territorial surveillance. However, it is also a 'breakthrough innovation' which radically changes the relationship of breeding with its environment and with the society that has accompanied it up to now by specifying, specialising and reducing communication at the centre

[9] Bouche and Hugot (1999) suggest this term as an analogy to the guidance system observed in bats.

and at the extremities of a professional network. The goatherd no longer relies on travellers to collect information concerning a stray animal; he now obtains information from a network of initiates (in his phone directory), thereby relieving the other land users of their contribution to the system while at the same time rendering obsolete the codes and specific vocabularies familiar to all and essential to this method of using the land. The network of information which extended to any place where a human might be present (hunters, foresters, farmers etc.) has been restricted to areas occupied by the members of the network, leaving huge zones without surveillance, where the tasks of watching and 'listening' remain highly demanding.

GPS, territory and new consulting tools

What is particularly striking to the observer in these situations, or in pastoral dairy breeding, are the difficulties encountered with regard to providing advice on how to tackle the problem of food and what the land provides. For some time, the science of animal husbandry has been capable of providing extension agents and breeders with methods and tools for rationing animals in stalls or in reasoned grazing pasture. However, major difficulties remain in estimating accurately what these extensive territories provide and in identifying the supplementary feeding adapted to the animals' needs. *De facto*, in the absence of this knowledge, we increasingly see over-distribution of supplementary feed to avoid the risk of a negative inflection of the winter production curve. Among traditional breeders, this concern makes very little sense as they succeed in compensating for a drop in winter production 'without drying up' by appropriate adjustments to prepare for an upturn in production in spring when resources are plentiful. At the same time, they accept that certain animals may remain unproductive for one season in order to renew their reserves by using the herd as a regulated whole. For them, goat herding is not simply a question of surveillance to prevent their animals from grazing in gardens and on crops. It is also a question of stimulating ruminants which are naturally lazy and providing supplementary feed in the form of branches when grass is in short supply. With the gradual rejection of this type of herding and the relatively high price of milk which enabled them to purchase concentrated feed (but for how long?), the new generation of Corsican breeders for the most part lose the advantages of pastoral resources and, by the same token, the related know-how. In the absence of appropriate methods and tools, consultants attempt to identify the appropriate rations which they naturally tend to overestimate out of simple precaution. For some years, Geographic Information Systems have allowed technicians to obtain more accurate illustrations of the paths actually used by the animals (Figure 1). This innovation is nevertheless used more for tax declarations concerning the area of the land than in the context of scientific animal husbandry. Recent developments include location sensors (GPS) which provide a reasonably precise indication of the animals' behaviour within the territory. This approach, which facilitates the precise analysis of land use, will also facilitate the acquisition of new competences in terms of providing support and advice.

Single milking: a simplifying old myth

Let us consider the example of a very old simplification process concerning a single milking session per day. Changing from two to one milking session per day represents a considerable reduction in a time-consuming workload. Although it has existed for a very long time, nobody really dared to test it except at the end of lactation, relying on ancient prescriptions dating from the time when the least litre of milk (without cost, as supplementary feeding did not yet exist) had a certain value. At that time, if the goatherd faced the slightest problem, a system of mutual local help would be immediately mobilized to prevent the animals drying up prematurely even if, as we can observe in the individual behaviour of goats, certain animals do not come in for milking in the evening. During the 2008 season, in the absence of a tried and tested experimental measure, at least two experiments (ovine and caprine) were conducted by the breeders themselves. In both cases, the drop in annual production in this single-milking configuration did not exceed 10%. Both innovators, closely observed by the network, consider the experiment to be a success. This practice has since become routine. It is the

Figure 1. Modelling the paths used by the animals in the pastoral land.

subject of intense discussion among the other breeders. Some believe that a particularly favourable year cannot be used as a reference and, in the name of tradition, denounce an innovation which breaks with the pastoral practices based on long and circular timeframes. Others, who feel that the loss of milk is largely offset by the savings in terms of the cost of food distributed during milking, work time, time spend washing the machine etc, are willing to embark on this adventure immediately. In this context, the technical world of research and transfer, hampered by the necessary hindsight concerning experimentation and the absence of references, is unable to fulfil its role concerning such rapid implementation.

Discussion: pastoral systems and innovation

In addition to the usual categories opposing radical innovation and incremental innovation, innovation also develops in a strange temporal paradox: that of an innovation which combines with the past (need for proof) with the utopia of a future built on models which are only partially accessible at the present time.

Innovation is only meaningful once it is installed, tried and tested, cunningly forcing us to detect, *in the course of change*,

- Both what is linked to the global development of a collective which has gradually adapted to the changing environment and which, *de facto*, becomes the norm (for example in the case of pastoral systems, the sedentarisation or dropping of surveillance). This may involve profound changes which evolve with the development of guidance by means of the mobile phone. Innovation is less common, less frequent, but in the modification of the information system which shifts towards socio-technical networks due to a relationship between society and breeding. Telephone guidance is more specialised and has a broader scope (between micro-regions), and is safer and quicker although, paradoxically, it is often less efficient due to the reduced number of observers and the extent of the zones observed.
- It is the result of an individual change that the breeder introduces to simplify his life, but which does not necessarily involve a profound modification of the socio-technical environment and

the codes conveyed within it. Here we can cite the example of the camcorder used to identify the mother-kid pairs.

Should we then consider innovation in these systems as a (neo-)Darwinian *process* (variation-stabilisation) with selected phenotypcial protrusions (individual changes), adopted as being inscribable in the heritage (genetic potential) of the collective for the future (Heudin, 1998) (Figure 2) with all that this implies in terms of testing (natural selection) and codes and learning (inscription in the norm)?

In this vein, the distance to the exogenous model which contributes to the construction of the illustration to which the system refers is decisive. For example, in the Mediterranean region, it is easier for dairy ovine chains to develop a representation of the future based on observation of what has been achieved in Sardinia or the Aveyron, whereas it appears to be much more complicated for the caprine chains to identify a reassuring archetype between the intensified caprine situations observed in western France and the neighbouring ovine situations, too often amalgamated by public policies in a 'small ruminant' group despite the significant differences between them. This calls into question the role of consulting and extension services.

Consequently, we adopt the notion of *process* and the idea of a gradual spiral integration of innovation reflecting the knowledge dynamics between the four states put forward by Nonaka and Takeuchi (1997) whereby specialisation, exteriorisation, combination and interiorisation tightly and successively combine tension between tacit/implicit and individual/collective knowledge (Figure 3).

'*Nowadays, the real innovation is lasts*'.This remark[10], which can easily be imagined in the mouth of a traditional breeder, indicates the main purpose of pastoral systems: it is not a question of relentless *growth* but simply possessing the capacities needed to at least go on as before. Augie-Laribe (1955), referring to the second agricultural revolution, defines intensive agriculture as the expression of a process of simplifying agri-systems. *De facto*, as we have seen, intensive breeding will concentrate its energy on techniques likely to dissociate it from its environment while pastoral

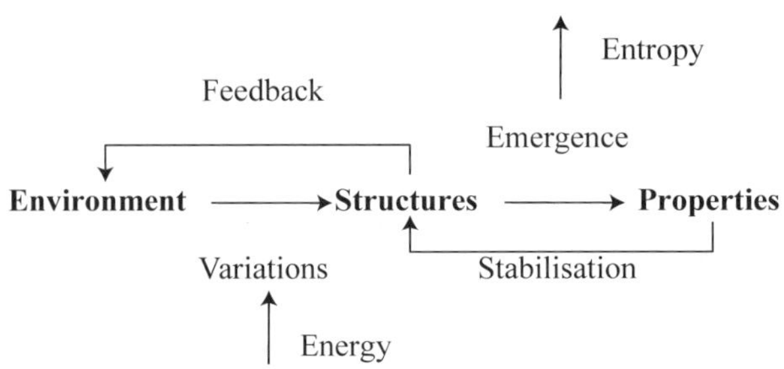

Figure 2. Principle of variation-stabilisation.

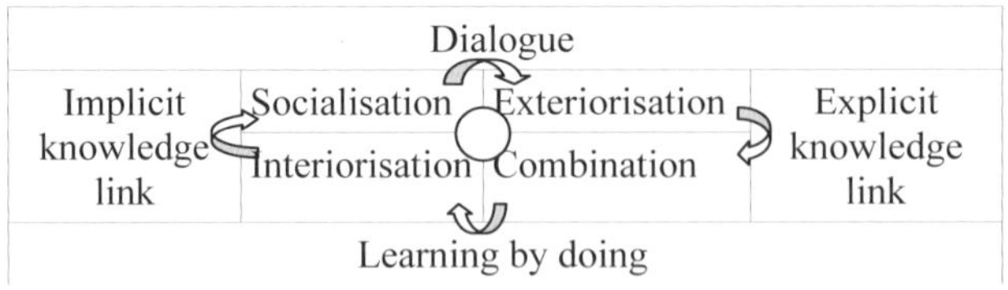

Figure 3. The knowledge spiral.

[10] Invented for the occasion but referring to real remarks made by breeders collected in the field (Bouche *et al.*, 2004).

breeding will concentrate its efforts on adapting to this environment, to the point of deploying a certain number of strategies which, strange as it might seem, make its operations more complicated (e.g. Santucci's PTI model, 1991). As already mentioned, the communication theory postulates that the two information flows (from the system towards the environment and from the environment towards the system) still have the same value. In other words, the information that the system provides concerning the behaviour of the environment always coincides exactly with the information provided by the environment concerning the behaviour of the system according to its fundamental identity (Wagensberg, 1997) which is: 'The complexity of a system, minus its capacity for anticipation vis-à-vis its environment equals the uncertainty of the environment minus its sensitivity vis-à-vis the system.' Consequently, to obey this law and to continue to exist, if the uncertainty of the environment increases, the system will increase its complexity and its capacity for anticipation, and decrease its effect on the environment. Pastoral systems are subject to a strong increase in:

1. endogenous uncertainty through the impoverishment of internal references for guidance (isolation and a drop in the number of similar farms, abandoned exchange networks, loss of the transmission of know-how, etc.);
2. exogenous uncertainty relating to changing standards and economic parameters (produce more, find new markets etc.).

Consequently, surviving in the face of an increase in uncertainty in the field of extensive breeding implies an increase in the complexity of the system, often to the detriment of its capacity for anticipation. This increased complexity, which is a source of flexibility and robustness in the face of uncertainties, is also accompanied by a certain number of handicaps and weaknesses and a certain inertia:

1. Concentration of information functions 'listening' to the signals of the physical (weather, animal behaviour, vegetation) and human environment (informers) making it relatively unreceptive to other signals and other emission zones.
2. Extreme flexibility of the productive system (decision-making entity, the herd is greater than the number of individual animals, maintained diversity) within the framework of fixed empirical knowledge and beliefs. We can observe the difficulties which exist today in hybrid systems when intensive approaches contradict this adjustment by increasing the level of complexity and creating new weaknesses.

Today, the paradox lies in the evolution of the socio-economic environment which is rather predisposed to favouring such pastoral systems, primarily for their natural environmentally-friendly nature and the quality of their products. Unfortunately, encysted and focussing on its end purpose (survival), the system has difficulty in adapting to the new configuration. For example, it is only with great difficulty that it can adopt an 'organic agriculture' certification procedure, despite the obvious absence of pollutants in its products. Even though they are naturally based on the use of local breeds (which are now highly valued), pastoral systems are today almost incapable of generating a differential rent. As a result, despite their many advantages, these systems are incapable of integrating appropriate valorisation circuits (sufficient production in terms of quantity and regularity etc.).

The importance of know-how

In the above description, we noted the importance of specific know-how in pastoral systems which has to be combined, in addition to its technical dimension, with anticipation, communication and coordination requirements. Knowledge of extensive breeding, which can in no way be made artificial or mechanised (unlike cheese processing, for example), depends on an indispensable eco-systemic optimisation of the natural and human environment which we echo Edgard Morin (1973) in considering it as the heart of an 'anthropological whole'. In its traditional form, Corsican pastoralism, forged over centuries in response to difficult production conditions, can be analysed here as an archetypal model (Figure 4) of territorially-anchored collective know-how (TACKH)

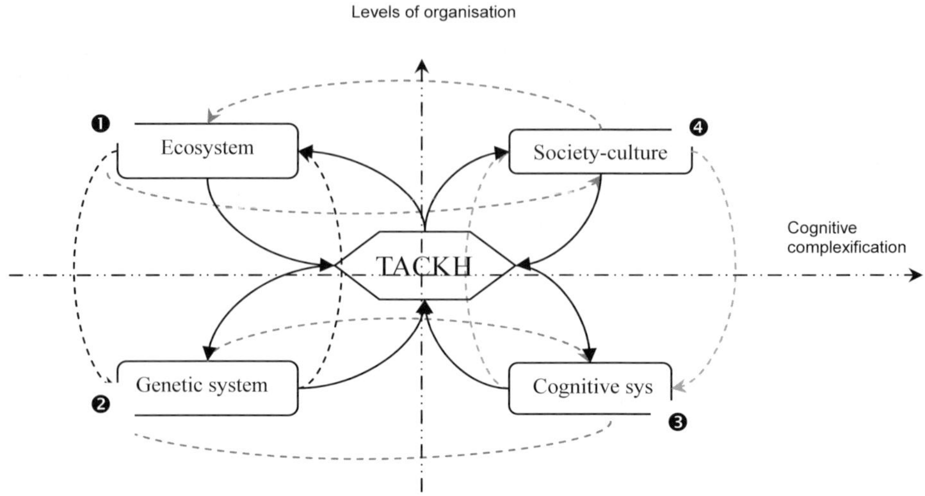

Figure 4. Collective pastoral know-how at the heart of a complex process.

(Bouche and Moity-Maïzi, 2008). This model enables us to classify and relate in a single entity, without opposing them:

- 'phenotypical' competencies tested on a daily basis which are included in this know-how (for example anticipation, creating relations);
- the 'genetic' attributes of these competencies which make it possible to reproduce them locally from one generation to the next, but without being able to generalise them in ubiquitous dimensions which would lead to their becoming commonplace, to the detriment of income based on originality.

We felt that it would be interesting to define and present this archetype as being the heart of a process of multidimensional differentiation where organisational phenomena are inevitably part of individual-collective, natural-artificial and physical-cognitive dimensions. In this systemic approach, we define TACKH as a 'computational' node that connects four sub-systems in which the significance and quality of the interactions are decisive in establishing its specificity and potential for patrimonialisation.

By assuming an increase in the levels of organisation between physical phenomena ((1) → (2) from genetic coding to the ecosystem) and cognitive phenomena ((3)→(4) from the individual cognitive system to social organisation), TACKH thus depends on and is an actor in:

1. the particularities of the natural environment;
2. the significance of a genetic system which is both productive (technical) and memory-based (heritage);
3. a cognitive system specifically dedicated to understanding, memorisation and communication between sub-systems;
4. social and cultural environment, a vector of distribution and transmission and an engine of evolutions and innovations. It is in this cultural environment that the 'epidemiology of representations' would occur' (Sperber, 1996).

This notion is all the more important as it cannot be evenly applied to the measure concerned. The observed innovation impacts the cognitive system of its user to a greater or lesser extent. Throughout the making of cheese, its processing and marketing, we distinguish areas that are differentiated by the cognitive mechanisms that are implemented. For example, knowledge of breeding is completely different from knowledge of processing. The relation with the artefacts which are increasingly present in the dairy sector (even traditional) such as the use of ferments, cold rooms etc. as part of the formal knowledge and procedures, generate operator postures which are completely different from those deployed in herding animals across a large area of land. Whenever implementation and learning situations are viewed in a totally different manner, even if they concern the same individual located in the same social environment (inertia of a herd vs. wasted cheese products), the question of territorial anchoring in a specific/generic context requiring ubiquitous techniques and knowledge arises. Indeed, as we have seen, know-how anchored in the territory due to knowledge of breeds and ecosystems are monopolised by the necessary learning, monitoring and collection of information; the greater the distance between the production function (processing, sales) and the territory, the more the know-how appears to have become formalised, ubiquitous, interchangeable and protected by the artefacts or the operators.

Today, innovations in 'cheese-making' naturally become part of a top-down process of transfer from the field of design and consulting services towards the user via the medium of technology and its artefacts and is relatively easy to understand for a large number of producers. Standardised and normative recipes exist which are generally ubiquitous and reassuring. They are easy to apply, even if this is damaging to the original diversity which is today becoming an advantage in new markets. However, innovation in the pastoral sector is rarely of this type. Long individual attempts by trial and error over long timeframes (one generation of animals) are required to develop learning, to such an extent that breeders often forge their own experience without resorting to the world of consulting which is becoming very rigid.

We should now return to our postulate of balancing a system with its environment to propose new lines to be explored as an innovative alternative to encystment, in particular the capacity for anticipation

Innovation: increasing the capacity for anticipation

Beyond economic improvements, which in this context often depend on agricultural support policies, and on social conditions (work conditions, *cf.* single milking session, bachelorhood) which depend on initiatives and individual parameters, innovation will be all the more relevant as it will be easy to adopt and likely to *increase the capacity for anticipation* of the parties involved with regard to uncertainty. Consequently, in these ultra-complex systems, two possibilities should be explored with regard to both their individual and collective dimensions:

- The first to shore up the crumbling traditional networks which contribute to the encystment of the pastoral system in a robustness which is difficult to reproduce. The question of organisation arises here as a meta level which may relieve the breeder of a certain number of cognitive functions which saturate his capacity for anticipation. In addition to the services traditionally provided by supply and marketing, we can easily see the need for information services formerly ensured within the cultural sphere of the surrounding society. We might also imagine measures which provide information on the management of extensive areas for pastoral guidance (directory). What of the organisations of this type and their institutionalisation around the meta-techniques? This type of organisation must above all allow the complexity to be reduced to level n-1 by providing the capacity for collective anticipation while rebuilding the collective choice functions formerly performed at the level of whole pastoral communities.
- The second concerns the individual in his capacity to adopt new techniques as seen in the first three examples presented here in which each innovation is based on a tool that costs less and is quite ordinary (even if their design and environment required enormous resources). For the most part, these objects are already used by the breeder in his family environment (camcorder, telephone, vehicle GPS). These three tools are directly linked to information systems and can consequently help provide the decision maker in the system with information at several different levels.

Conclusion

This brief overview of some of the innovations observed in the pastoral system encourages us to give real thought to the real modifications which are currently appearing concerning the knowledge procedures within these systems. It is a lesson we should learn from the analysis of the current systems which are an incongruous hybridisation of two antinomic systems. Innovation cannot and should not be restricted to a simple review of their technical and technological aspects. It should also explore the *cognitive and organisational* changes that it causes. It is, *de facto*, a *cognitive and organisational change.* It can no longer be seen as a static element, an article from a catalogue ready for use; rather it should be analysed as a process, a spiral which, like knowledge, is the result of the interactions between its interiorisation/exteriorisation phases at different organisational levels that are both individual and collective.

References

Auge-Laribe, M., 1955, La Révolution agricole, Ed Albin Michel, 479 pp.

Bouche, R. and S. Hugot, 1999. L'île des bergers: paradoxe d'une évolution figée entre reproductibilité et progrès. In: Alain Piezzola Ed. L'île laboratoire, Colloque de l'Université de Corse, 19-21 juin, textes réunis par Anne Meistersheim, p. 175- 189.

Bouche, R., 2001. Le nom de la chèvre. Dans: Congrès 'Environnement et Identités en Méditerranée', Vol. 3, Agro-Ressources. Università di Corsica, pp. 163-176.

Bouche, R., J.Y. Gambotti and O. Maestrini, 2004. L'avenir c'est de durer. Quand le changement en élevage extensif nécessite le maintien ou la réinvention de systèmes collectifs. Cas des caprins corses. In: L'évolution des systèmes de production ovine et caprine: avenir des systèmes extensifs face aux changements de la société Options Méditerranéennes: Série A. CIHEAM IAMZ, 2004. p. 319-327.

Bouche, R. and P. Moity-Maïzi, 2008. Technique, ecology and culture: the territorial anchorage of Corsican cheese-producers' knowledge, 8th European IFSA Symposium, France, 6-10 july 2008.

Heudin, J.C., 1998. L'évolution au bord du chaos, Ed Hermès, 185 pp.

Morin, E., 1973. Le paradigme perdu: la nature humaine. ed du Seuil Paris, 246 pp.

Nonaka, I. and H. Takeuchi, 1995, The Knowledge-Creating Company: How Japanese Companies Create the Dynamics of Innovation, Oxford University Press, 284 pp.

Pernet, F. and G. Lenclud, 1977. Berger en Corse: Essai sur la Question Pastorale. Presses Universitaires de Grenoble, 190 pp.

Ravis-Giordanni G., 1983, Bergers Corses, les communautés villageoises du Niolu.EDISUD, 505 p.

Santucci, P.M., 1991. Le troupeau et ces propriétés régulatrices, bases de l'élevage caprin extensif. Thèse de doctorat, académie de Montpellier, Université de Montpellier II.85 pp.

Sperber, D., 1996. La contagion des idées, ed Odile Jacob, Paris, 243 pp.

Wagensberg, J., 1997. L'âme de la méduse: idées sur la complexité du monde, Seuil, 169 pp.

The adoption of technical and organizational innovations and their impacts on dairy sheep breeding in Tunisia

A. Mohamed[1], R. Khaldi[2] and G. Khaldi[1]
[1]Institut National Agronomique de Tunisie, 43 Avenue Charles Nicolle, 1082 Tunis Mahrajène, Tunisie
[2]Institut National de la Recherche Agronomique de Tunisie, Rue Hédi Karray, 2080 Ariana, Tunisie

Abstract

In Tunisia, the size of the dairy Sicilo-Sarde sheep population has decreased by 90%. Its productivity has stagnated at around 70 liters/ewe/lactation over the last decade. The entire milk yield is used for cheese manufacture. The objective of this work was to identify the innovations that have been adopted and their impacts on flock performance. Two types of innovation are concerned: technical innovations (food supplementation, early weaning and crossing with a Sardinian breed) and organizational innovations (membership of breeders' associations). The method used was based on an exhaustive survey. Two milk production levels were selected for analysis 'class 1' with a production level >70 liters/ewe/lactation and 'class 2': production level ≤70 liters/ewe/lactation. The results showed that, whatever the milk production level, the rates of adoption of innovations such as food supplementation and early weaning are still limited, but were higher in 'class 1' than in 'class 2'. The same was true for organizational innovation which appeared to slightly improve the herds' performance. The farmers' membership of the breeders' association led to an increase in the sale price of milk and reduced the monopoly imposed on the sector. Overall, the innovation implementation rate did not exceed 52% in 'class 1' and was only 36% in 'class 2'. These results highlight the great potential for improving the productivity of the Sicilo-Sarde breed especially in 'class 1' where the animal production system, herd management and the farms' socio-economic characteristics are better.

Keywords: Tunisia, dairy sheep, technical innovation, organizational innovation

Introduction

Raising dairy sheep in Tunisia dates from the colonial period when settlers brought in two Italian breeds, the Sardinian and the Comisana, from southern Italy. The crossbreeding of these two breeds gave birth to the only dairy sheep breed in Tunisia: the Sicilo-Sarde (Khaldi and Farid, 1981). The raising of this breed is concentrated in the hills of northern Tunisia, especially in Beja and Bizerte and a cheese-making tradition has developed in the area. Milk production is limited, but the entire yield is used for industrial or artisanal cheese manufacture. Djemali *et al.* (1995) estimated the average milk production to be 72 kg/ewe/lactation for an average of 124 milking days and a suckling period of 114 days. Over several years, total milk production has stagnated at around 1,700 T/year (Kayouli, 2006). Low productivity is due mainly to the extensive management of the herds (Snoussi., 2003). Faced with this stagnation, efforts have been made to improve the breed's performance and indeed to save it from extinction, as the total number of animals has decreased by almost 90% (Mohamed *et al.*, 2007). Scientific research in this area has demonstrated the important role of early weaning in increasing milk yield. Khaldi (1987) showed that early weaning of lambs at 42 days rather than at 90 days, significantly increased ($P<0.001$) the milk yield in both single suckling ewes (82 vs. 62 l liters) and twin suckling ewes (118 vs. the 66 liters). The average total milk yield (residual+milked) may reach 176 liters when lambs are weaned at 90 days (Mohamed *et al.*, 2008). Adequate feeding management including supplementary feeding with concentrate at appropriate periods could also increase the milk yield (Khaldi., 1981; Rouissi., 1989; Al Jassim *et al.*, 1999; Bocquier *et al.*, 2004). In 2002, an association of the Sicilo-Sarde Breeders (AESS) was created in the region of Beja to enhance livestock products and to ensure the sustainability and organization of the entire dairy sheep sector. The main objectives of the association, as stated in its 'master plan', for the period 2005-2010, are to increase the size of the total herds belonging the members of the association from

10,000 to 30,000 female units and to improve the milk yield/ewe/year from 90 liters to 150 liters (Sâadoun. 2007). Based on the fact that these actions represent innovations for the farmers and that research in this domain in Tunisia has been limited to purely zootechnical aspects, technical and socio-economic analyses of the impact of these innovations on dairy sheep farming were needed. This study thus aimed to:

1. identify the rate of adoption of both technical and organizational innovation;
2. analyze the impact of these innovations on the performance of the Sicilo-Sarde breed and on the sector as a whole;
3. determine why the innovations are not always adopted.

Methods

On the assumption that low dairy performance is related to inappropriate herd management, in 2006/2007 an exhaustive survey was conducted of 105 farms evenly distributed between the regions of Bizerte and Beja. The investigation addressed aspects related to dairy sheep activity, its history and its place in the production system, herd management, milk production and its destination (sales, processing and marketing), and membership of the AESS.

The farms were classified on the basis of the average milk yield. Each class was then characterized by the rate of adoption of technical and organizational innovations. Available technical innovations provided by research and used for the analyses were:

- Supplementary feeding during the critical periods of mating (400 g/ewe/day) and before lambing (400 g/ewe/day).
- Early weaning of the lambs ($\leq$90 days old).

The adoption of organizational innovations consisted in the farmers becoming members of the AESS.

The adoption rate (AD) of the innovations was calculated as follows:

$$AR = \frac{\text{(the number of farms that adopted the innovation)}}{\text{(the total number of farms)}} \times 100$$

The impact of technical innovations was measured by comparing the productivity of the dairy sheep farms in the survey with results obtained at an experimental station in northern Tunisia in optimal production conditions. The implementation rate of the research results (IRRR) was calculated as follows:

$$IRRR = \frac{\text{(average milk yield/ewe/lactation) obtained by the farms}}{\text{(average milk yield/ewe/lactation) obtained at the experimental station)}} \times 100$$

The difference between the two classes was explained by technical and socio-economic variables related to individual farms and their environment.

Results and discussion

Characterization of dairy sheep breeding in the farms surveyed

Place of dairy sheep breeding

A total number of 19,126 ewes, 84% of which were purebred, was identified in the farms surveyed. The breeding of purebred dairy sheep is a main agricultural activity in the public farms (57% of which are located at the region of Beja) with a rate of 70%. In the public sector, 6,126 ewes were recorded (Table 1). The private sector comprised 95 farms (of which 68% are located in Beja) with 13,000 ewes. Raising purebred dairy sheep is the main agricultural activity of 53% of the farms of which 78% are located in Beja. However, the breeding of crossbred dairy sheep (Sicilo-Sarde breedxBarbarine local breed) for meat and milk production purposes is widespread and remains the main activity of 47% of the private farms of which 87% are located in Bizerte.

Milk performances

The average milk yield (AMY) of all the farms surveyed was about 78 liters/ewe/lactation for an average of 165 milking days (Table 2). It reached 83 liters in the region of Beja for a milking period of 174 days.

The dairy sheep farms were classified according to their average milk yield (AMY). Two classes were identified (Table 3):
- 'Class 1': AMY>70 liters/ewe/lactation (49% of the total number of farms)
- 'Class 2': AMY≤70 liters/ewe/lactation (51% of the total number of farms).

In class 1, 60% of the farms were located in Beja and 40% in Bizerte. The average milk yield/ewe/lactation was 91 liters for an average milking period of 180 days (Table 3). In the same class, the performance in public sector was slightly better than that of the private sector. This is due to the longer milking period. In class 2 33% of the farms were located in Beja and 67% in Bizerte. In this class, the milking period was shorter (150 days on average) than in class 1, which explains the lower milk yields in class 2 (64 liters/ewe/lactation). Sight differences were observed between regions within the same class.

Table 1. Characterization of the surveyed farms.

	Overall	Public sector			Private sector		
		Total	Beja	Bizerte	Total	Beja	Bizerte
Number of farms	105	10	5	5	95	45	50
Number of ewes	19,126	6,126	2,497	3,629	13,000	8,840	4,160
Main agricultural activity (%)							
Breeding of purebred dairy sheep	58	70	57	43	53	78	22
Breeding of crossbred sheep (Sicilo-Sarde/ Barbarine)	42	0	0	0	47	13	87

Table 2. Milk performances of the surveyed farms.

	Average milk yield (AMY) (liters/ewe/lactation)			Milking period (days)		
	Total	Beja	Bizerte	Total	Beja	Bizerte
Overall mean	78	83	74	165	174	157
Standard deviation	17	18	16	44	42	44

Table 3. Classification of farms according to milk yield (%).

	Class 1						Class 2					
	AMY (liters/ewe/lactation)			Average milking period (days)			AMY (liters/ewe/lactation)			Average milking period (days)		
	Total	Beja	Bizerte	Total	Beja	Bizerte	Total	Beja	Bizerte	Total	Beja	Bizerte
Overall mean	91	93	89	180	185	172	64	64	65	150	154	148
Standard deviation	14	13	16	39	38	40	5	6	4	43	44	44
		Private sector	Public sector		Private sector	Public sector		Private sector	Public sector		Private sector	Public sector
Overall mean		91	94		175	201		64	-		150	-
Standard deviation		15	12		42	12		5	-		43	-

Adoption rates of technical and organizational innovations

According to the hypothesis that the production variability between class 1 and class 2 is related to the implementation of innovations, the calculated adoption rate (Table 4) showed that in class 1, the techniques of flushing and steaming supplementary feed at critical periods have been adopted by respectively 47% and 55% concerning the two food supplementation techniques. The quantity of feed provided to the animals is still insufficient (300 g/ewe/day). The rates of adoption in class 2 were lower: respectively 25% and 44% concerning the two food supplementation techniques. The early weaning of lambs is practiced by nearly half the farms in class 1 as opposed to only 19% in class 2.

Generally speaking, the adoption rates of technical innovations were higher in the region of Beja. In class 2, the steaming technique has been adopted by the same percentage of farmers (44%) in Bizerte and Beja. In class 2, the animals are generally crossbred with the main aim of producing vigorous lambs at birth and of increasing meat yield.

In class 1, the adoption rate of technical innovations by public farms was 100%, but only 17% of private farms practiced early weaning. Food supplementation was practiced by almost half the private farms in class 1. These results show that the low performances of the Sicilo-Sarde breed are linked to the low adoption rates of technical innovations by private farms, especially early weaning of lambs. Although the adoption rates in the public sector were higher, milk performances were still low.

Concerning organizational innovations, more than half the farmers in class 1 (55%) are members of the AESS, 46% of whom belong to the private sector. This rate reached 90% for public farms. In class 2, only 31% of the breeders belong to the AESS (Table 5). This could be explained by the fact that farmers are not yet accustomed to this new concept of an 'association'.

Impact of the adoption of technical innovations

Comparison with research results

Research results (RR) obtained with early weaning (before three months) and food supplementation during critical periods (see above) produced an average milk yield of 176 liters/ewe/lactation (normal + residual milk). The comparison of the performances of the surveyed farms with research results,

Table 4. Adoption rates of technical innovations per class (%).

Food supplementation technique	Class 1: AMY >70 liters/ewe/lactation			Class 2: AMY ≤70 liters/ewe/lactation		
	Total	Beja	Bizerte	Total	Beja	Bizerte
Flushing	47	48	44	25	38	19
Steaming	55	48	66	44	44	44
Early weaning (≤90 days)	49	69	44	19	31	13
		Private sector	Public sector		Private sector	Public sector
Flushing		47	100		25	*
Steaming		55	100		44	*
Early weaning (<90 days)		17	100		19	*

Table 5. Adoption rates of organizational innovations (membership of the AESS) as a function of yield (%).

	Class 1: AMY >70 liters/ewe/lactation	Class 2: AMY ≤70 liters/ewe/lactation
Total	55	31
Beja	66	63
Bizerte	39	16
Private sector	46	31
Public sector	90	*

revealed significant differences: 85 liters in class 1 and 112 liters in class 2. The implementation of the results of research was 52% in class 1, but only 36.4% in class 2 (Table 6). Depending on the region, it is clear from this analysis that there is no great variability, but farmers in Beja still performed slightly better than those in Bizerte.

Even if the performance was better in class 1, it was still below the potential of the breed which can reach 181 kg (milked+residual milk) in the case of semi-weaning at 45 days of age and steaming using a 400 g/ewe/day ration of concentrate (Mohamed *et al.*, 2008).

Explanatory factors

The higher implementation rate of research results in class 1 (AMY >70 liters/ewe/lactation) can be explained by several factors related to the farmer, (production system, herd management), and to the farm (the socio-economic environment which was far better than in class 2) (Table 7). Indeed, in class 2, the average herd is relatively small, with an average of 99 ewes as opposed to 270 ewes in class 1. In class 2, the Total Cultivated Area (TCA) was on average 19 ha as opposed to 234 ha in class 1. The pasture and forage areas reserved for dairy sheep farming were respectively 13 and 3 ha. In class 1, the number of hectares was nearly 3 and 5-fold higher. In class 2, herds had more crossbred animals (65%) than purebred. The culling rate was very low (6%) compared to class 1 (12%). In class 2, animal housing (77%) was made of sheets of metal and twigs and only partially covered, thus exposing the animals to cold in winter and high temperatures in summer. In class 2, family farms in which the shepherd is a member of the family are the most common types of farms (Mohamed *et al.*, 2007). Only 27% of the farms in class 2 use specialized manpower (60%). Furthermore, the farmers' level of education was low (83% were illiterate). Problems of marketing their milk were reported by a third of the farmers in this class.

Table 6. Comparison of the performances of surveyed farms and research results.

	Difference in AMY between RR and farm performances (liters)		Implementation rate of research results (%)	
	Class 1	Class 2	Class 1	Class 2
Beja	83	112	53	36.4
Bizerte	87	111	50.5	37
Private sector	85	112	52	36.4
Public sector	82	*	53.5	*

Table 7. Characterization of the farming systems as a function of yield.

	Class 1: AMY>70 liters/ewe/lactation		Class 2: AMY≤70 liters/ewe/lactation	
	Average	S.deviation	Average	S.deviation
Criteria related to the farmer				
Average age	46		49	
Illiteracy rate (%)	47		83	
Criteria related to the production system				
Herd size (ewes)	270	319	99	98
Total Cultivated Area (ha)	234	660	19	29
Pasture (ha)	36	59	13	16
Forage area (ha)	16	24	3	4
Criteria related to the herd management				
Crossbreeding of the Sicilo-Sarde breed with other breeds	38		65	
Animal housing in good condition (%)	62		23	
Average culling rate (%)	12		6	
Use of specialized manpower (%)	60		27	
Criteria related to the farm socio-economic environment				
Problems with marketing milk	12		30	

Organizational innovation

In recent years, several attempts have been made in Tunisia to rehabilitate the dairy sheep sector. The principal one was the creation of the AESS in the region of Beja. A 'master plan' was established for the association for the period 2005-2010. The main objectives of the plan are to increase the combined herd size of the members of the association from 10,000 to 30,000 ewes and to increase the milk yield/ewe/year from 90 liters to 150 liters (Sâadoun, 2007). Collaboration between the AESS and research institutes plays an important role in ensuring the sustainability of the dairy sheep industry. Several actions have been implemented to motivate farmers, including membership of the association, control of yield and the improvement of pastures, managed by the OEP (Office of Livestock and Pastures), and the organization of training and information days in collaboration with the Animal Production Services, Higher Education and Scientific Research Institutes to prepare and distribute data sheets for better herd management.

Negotiations with government services on the price of milk led to an increase in the sales price of milk from 1.100 DT to 1.440 DT. An increase of about 31% has been reported (Mohamed, 2008). This has helped reduce the monopoly that had been imposed by the only industrial cheese processing company in Tunisia. Whereas farmers used to process their milk individually or sell it to the industrial cheese processing company, they can now sell it to an artisanal processing unit with an average capacity of 300 liters/day that was set up in the region of Beja in 2007. The creation of this unit ensures higher milk prices for members of the AESS. The ricotta and the artisanal cheese produced in this unit are sold at 7.500 DT and 3.800 DT instead of 3.700 DT and 2.800 DT, i.e. an increase of 103% and 36% respectively. With the collaboration of the Research Institutes, the Animal Production Services of the OEP and the AESS, semen of Italian rams (Sardinian breed) was imported for artificial insemination to reduce inbreeding in the ovine dairy herds. This has enabled seven farmers, who are members to the AESS, to inseminate 1,600 Sicilo-Sarde ewes. This insemination led to higher fertility rates (59%) and higher prolificacy rates (156%) (Sâadoun, 2007) compared to those reported in the literature (90% and 144% for the fertility rate and the prolificacy rate respectively (Khaldi and Farid, 1981; FAO, 1995).

The increase of 5.6% in average milk yield which reached 94 liters/ewe/lactation instead of 89 liters/ewe/lactation among the AESS members in class 1, is already an achievement (Table 8). In class 2, the adoption of organizational innovation has still not resulted in an improvement in milk performance because animals are generally raised for mixed purposes (milk and meat). In this case, farmers mainly join the AESS to receive financial aid, grants, etc.

Conclusion

The low performance of the dairy ovine breed 'Sicilo-Sarde' in Tunisia underlined the need to identify its main causes. This article presented the technical and organizational innovations at the heart of this issue and analyzed the impact of these innovations on milk production. The implementation of technical and organizational innovations was shown to have a positive impact on the improvement of Sicilo-Sarde productivity, but the real potential of this breed has not yet been reached. Data from an exhaustive survey of 105 farms (19,126 ewes) was organized in two performance classes based on the average milk yield (AMY): class 1 AMY >70 liters/ewe/lactation (49% of farms) and class 2: AMY ≤70 liters/ewe/lactation (51% of farms).

The results of this study showed that:
- The low performance of the Sicilo-Sarde breed was related to the adoption rate of technical innovations, which remains low especially in the private sector.
- Although the adoption rates of technical and organizational innovations in the public sector were high, performance remains low.
- Membership of farmers in the AESS remains limited even in class 1.
- Despite the higher adoption rates in class 1, which enjoys better production conditions (production system, farm management and socio-economic environment), the average milk yield remains below the real potential of the Sicilo-Sarde breed as demonstrated by the research results.
- Organizational innovation appears to have slightly improved the performance of the herds.
- Opportunities exist for improving the productivity of the Sicilo-Sarde breed, particularly in class 1 by educating farmers about the importance of practicing severe culling, providing adequate, balanced and good quality animal feed and early weaning of lambs. For class 2, the situation is more difficult because the animals raised by small-scale farmers are generally crossbred. However, they are also a vital source of income for these farmers.
- There is an urgent need for a genetic improvement program based on milk quantity and milking facility, which would allow the maximum quantity of residual milk kept back by ewes during milking to be recovered.
- Saving the Sicilo-Sarde breed from extinction requires greater awareness among farmers of the issues involved. This could be achieved especially by breeders' associations, which should be supported through grants, facilities for the technical training of farmers, the collection of milk and

Table 8. AMY (liters/ewe/lactation) as a function of the adoption of organizational innovations.

	Class 1		Class 2	
	Average (liters/ewe/lactation)	S. deviation	Average (liters/ewe/lactation)	S. deviation
Members of the AESS	94	14	62	4
Other farms	89	14	65	5

the creation of other industrial processing units. A ban on slaughtering females of this breed and the encouragement of its breeding through donations (small flocks, for example) are also essential.

- Improving the income of small and medium farmers still requires incentives such as grants or loans in order to increase the size of their herds.

References

Al Jassim, R.A.M., D.I. Aziz, K. Zorah and J.L. Black, 1999. Effect of Concentrate Feeding on Milk Yield and Body-Weight Change of Awassi Ewes and the Growth of their Lambs Effect of Concentrate Feeding on Milk Yield and Body-Weight Change of Awassi Ewes and the Growth of their Lambs. Ani. Sci. 69(2): 441-446.

Bocquier, F., Ph. Guillouet and F. Barillet, 1995. Alimentation hivernale des brebis laitières: intérêt de la mise en lots. INRA Productions Animales. 8: 19-28.

Djemali, M, I. Ben M'sallem and R. Bouraoui, 1995. Effets du mois, mode et âge d'agnelage sur la production laitière des brebis Sicilo-Sarde en Tunisie. CIHEAM, Cahier Opt. Medit. 6. L'élevage ovin en zones arides et semi-arides.

Khaldi, G. and M.F.A. Farid, 1981. Encyclopédie des ressources animales dans le monde arabe. Cas de la république tunisienne (en langue arabe). Organisation Arabe de l'Education et de la Culture: 214 p.

Khaldi, G., 1987. Influence de l'âge au sevrage et du mode de naissance des agneaux sur la production laitière des brebis de race Sicilo-Sarde pendant les phases d allaitement et de traite. Ann. INRAT: 60 (14): 16 p.

Kayouli, Ch., 2006. La race Sicilo-Sarde: comment sauvegarder une race généreuse d'être dégénérée dans son berceau Béjaois. Document de vulgarisation. Institut National Agronomique de Tunisie.

Khaldi, G., 1981. Encyclopédies des ressources animales dans les pays arabes - République. Tunisienne (en langue arabe). Organisation Arabe de l'Education et de la Culture.pp: 13-19.

Lemdellel, M., 2007. Injection des gènes Sardes chez la race Sicilo-Sarde par insémination intra-utérine. Mémoire de projet de fin d'études de l'Institut National Agronomique de Tunisie: 47 p.

Mohamed, A., R. Khaldi, Z. Rached and G. Khaldi., 2007. L'élevage ovin laitier dans les systèmes de production du Nord de la Tunisie. Communication Orale au 6ème Séminaire International: Changes in Sheep and goat Farming systems at the beginning of the 21th Century. Ponte de Lima, Portugal, 15-17 Novembre 2007(communication orale).

Mohamed, A., S. Khaldi, B. Rekik and G. Khaldi., 2008. Normal and residual milk yields in Sicilo-Sarde ewes: effects of the litter size and the weaning age of lambs. Research Journal of Animal Sciences. Medwell Journals 2 (5): 144-148.

Mohamed, A., 2008. L'élevage ovin laitier en Tunisie: analyse de la situation actuelle, contraintes, moyens et perspectives de développement. Thèse de Doctorat de l'Institut National Agronomique de Tunis, 159 p.

Rouissi, H., 1989. Infl uence d'une complémentation en concentré sur les résultats d'engraissement des agneaux de Noire de Thibar: Bulletin d'information. ESA de Mateur.

Sâadoun, L., 2007. Simplification du contrôle laitier chez la brebis Sicilo-Sarde et injection de nouveaux gènes par insémination artificielle intra-utérine. Mémoire de Mastère de l'Institut National Agronomique de Tunisie: 79 p.

Snoussi, S., 2003. Situation of sheep farming in Tunisia and the role of Research. Reflection on the improvement of a systemic approach. Cahiers d'études et de recherches francophones/ Agricultures. Volume 12, 6, 419-28, Novembre-Décembre 2003. Available from Internet: http://www.john-libbey-eurotext.fr/en/revues/agro_biotech/agr/e-docs/00/04/02/94/article.md?type=text.html.

Tradition and innovation in the Mediterranean pastoralism: recognition of its multiple roles for the sustainable development of rural areas

C. Ligda[1], E. Tchakerian[2], E. Zotos[3] and A. Georgoudis[4]
[1]Centre for Agricultural Research of Northern Greece, NAGREF, Thessaloniki, Greece
[2]Institut de l'Elevage, Montpelier, France
[3]Genetic Improvement Centre of Animals, Ministry of Rural Development and Food, Ioannina, Greece
[4]Department of Animal Production, School of Agriculture, Aristotle University of Thessaloniki, Greece

Abstract

This paper presents the results of the Interreg III project PASTOMED on pastoralism in Mediterranean regions. In this project, partners from eight Mediterranean regions worked together to define the conditions and characteristics of pastoral systems, study the impacts of policies and measures applied to pastoral systems, and present successful innovative actions that could be extrapolated to other regions. In this frame, representative models of pastorals systems in the participated regions were analyzed. Given the different levels of information in the different regions, a common set of data was created, including the structure of the farm and certain technical and economic data. Common elements in the production systems between the regions were identified, but also differences that were mainly the level of exploitation of agro-environmental measures. The future prospects of pastoralism were studied by analyzing the impacts of the CAP and of agro environmental measures in different sectors including the quality and recognition of the products, generation renewal, social recognition of the profession and improvement in the farmers' working and living conditions.

Keywords: pastoral systems, sustainable development

Introduction

During the period from 2005 until 2007, the Pastomed project on the future of Mediterranean pastoralism was implemented in the framework of Interreg III. In this project, partners from eight Mediterranean regions worked together to define the conditions and characteristics of pastoral systems, study the impacts of policies and measures applied to pastoral systems at the European and national levels, and present successful innovative actions that could be extrapolated to other regions. Common elements in the production systems between the regions were identified, but also differences that were mainly the level of exploitation of agro-environmental measures. The future prospects of pastoralism were studied by analyzing the impacts of the CAP and of agro-environmental measures in different sectors including the quality and recognition of the products, generation renewal, social recognition of the profession and improvement in farmers' working and living conditions.

The results of this work showed that even though pastoral systems are a traditional way of farming in the Mediterranean region, they contain elements that may prove to be very valuable in today's conditions, especially given the need to protect the environment and biodiversity and respond to consumer demands concerning the safety of the products. The necessary evolution of pastoral systems implies the overall modification of their components so that the farming activities will provide sufficient income for the farmer in acceptable work and living conditions. In this context, the differentiation of the products from mass production, the valorization of their specificities and the clear definition of the relationship of the different activities connected with pasture lands (livestock production, forest, hunting, tourism) are necessary steps. All actions should have as a clear target the recognition of the profession in connection with the position of pastoralism in the local communities, in the economic, social and cultural sector.

This paper presents results mainly concerning the characteristics of pastoralism in the different regions that participated in the project, along with a brief description of the current situation and the future prospects of pastoralism in the region of *Epirus*, the Greek partner in this project.

Material and methods

One of the main components of the project aimed at the in-depth understanding of the current situation of pastoral systems in the regions concerned by the porject. Information was collected on the natural, social, economic and agricultural context of each region including pastoral farming practices, the sectors, marketing of the products, and the organization of the profession. Data were also collected on the exploitation of the CAP measures in the different regions.

A questionnaire was designed including technical and economic questions for administration on representative farms to compare the situation in the different regions.

The following information as collected in face-to-face interviews:
1. Number of workers and work time over the year, with a distinction between family and paid labor.
2. The total surface area of the farm, with a distinction between the different land uses.
3. The surface areas used for grazing and their status, the practice of transhumance, the type of animals raised.
4. The existence of other activities on the farm.
5. A description of the main buildings.
6. The production, volumes and type of marketing.
7. Some zootechnical data on the management of the animals and how they are fed.
8. Economic results including the volumes of production, the price per unit, the total amount, the incentives received and farm expenses.

Given the different levels of information between the regions, a common set of data was created, including the main characteristics of the structure of the farm and certain technical and economic data.

Results and discussion

The results indicated that there is a wide range of production systems across the regions. But common elements exist that provide a basis for a common understanding and planning of strategies for the evolution and sustainable development of these regions. Among these, it should be mentioned that the pastoral systems exploit marginal areas, thus contributing to the environment and promoting sustainability, but on the other hand these systems are characterized by difficult working conditions and in most cases by the lack of the necessary infrastructures thus contributing to the isolation of people working in this domain and the generally unappealing nature of the profession.

The significance of mountain and marginal areas was observed in all regions, among which mountain zones cover the largest area. Apart from the two French regions, more than 70% of the regions concerned are located in marginal zones (over 90% in *Epirus* and 100% in *Alentejo*) (Figure 1).

In the regions that participated in the project, there was a difference in the weight of each species, with *Entre Douro e Minho* and *Languedoc Roussillon* having the highest proportion of cattle. Dairy sheep farming is more important in Sardinia and Epirus, while meat sheep farming is important in PACA. Goats represent a marginal activity in most of the regions, except Andalusia and Epirus, where goats make up over 20% of total livestock (Figure 2). However, a decreasing trend was observed (19%) in Epirus and (-6%) in Andalusia. Figure 3 shows changes in the number of livestock between 1990 and early 2000. The largest decrease in population was observed in Abruzzo. Alentejo and Languedoc Roussillon differed from the other regions due to a significant increase in livestock units, of 24%

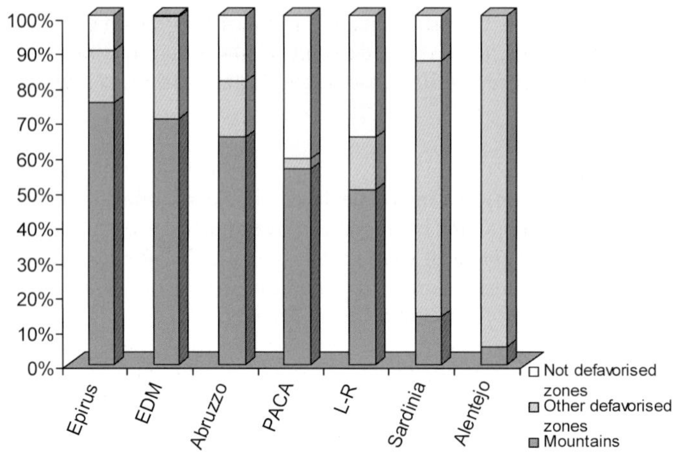

Figure 1. Defavorised zones in the regions.

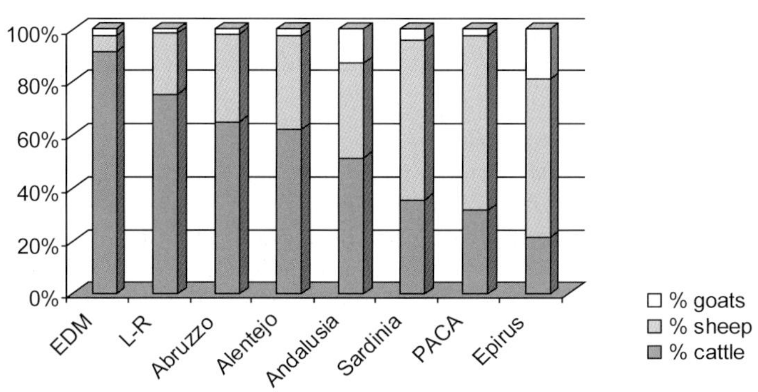

Figure 2. Species distribution in the different regions.

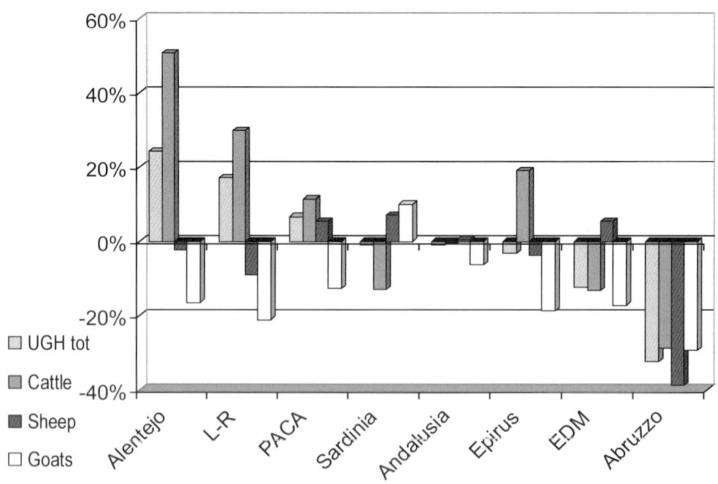

Figure 3. Changes in the number of livestock in the different regions.

and 17%, respectively. In both cases, the cattle sector was mainly responsible for this increase, as this sector has taken advantage of the relevant incentives and also of the fact that the cattle require fewer hours of work. The same trend was also observed in Epirus and PACA, where increases of respectively 21% and 11% were observed (Tchakerian, 2007).

The age of the farmers is an important factor in the future propoects of the profession. To estimate the degree of ageing of the farmers, age classes were defined and the proportion of farmers over 55 to farmers younger than 35 was calculated. In Entre Douro e Minho, Abruzzo, Epirus and Sardinia this proportion was higher than 10, while in Languedoc Roussillon and PACA it was about 5 (Figure 4).

In all the participating regions, despite the different levels of development and the level of exploitation of agro-environmental measures, it is accepted that some of the difficulties pastoral systems are facing are due to social and structural impediments. Figure 5 shows the proportion of incentives received under the 1st and 2nd pillars to total revenues in the sheep raising sector. The impact of the different incentives varied with the region and with the production system. The impact was lowest (12%) in Entre Douro e Minho in semi-intensive dairy sheep farms, higher in the transhumance systems in Epirus, Abruzzo and Andalusia (25%, 27% and 32, respectively), with similar levels in the dairy sheep farms in Sardinia (23%) and highest (51% and 55%) in PACA and Languedoc Roussillon respectively. The two last regions differed from the others, as they have exploited the premiums of the 2nd pillar to a greater extent, an experience that it would be very useful for the other regions to follow. Alentejo stood out among the regions, as total income also includes the income from crop production, and pemiums, those of the 1st pillar already reaching 41%.

The current situation in Epirus

There follows a brief description of the different production sectors in Epirus. Generally speaking, *Epirus* is characterized by the high proportion of its animal production sector in comparison with the rest of the country. A total of 33% of the farms in the region are specialized in the raising of ruminants and 9% combine crop production and ruminants.

The cattle sector is located in mountainous and semi-mountainous regions and is almost completely oriented towards meat production. Calves are slaughtered between 6 and 12 months (up to 100 kg) and between 12 and 18 months (150 kg). When the farmers raise local breeds (Brachykeratiki cattle), they benefit from agro-environmental measures and receive incentives for endangered breeds. With total decoupling, the increasing trend in the number fo livestock, which was to a large extent due to

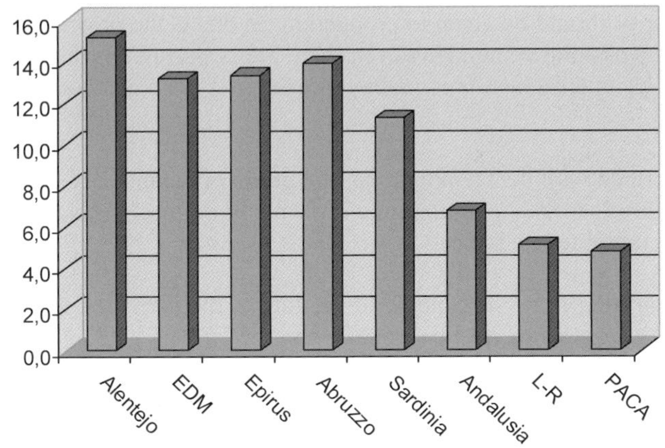

Figure 4. The index of ageing in the different regions.

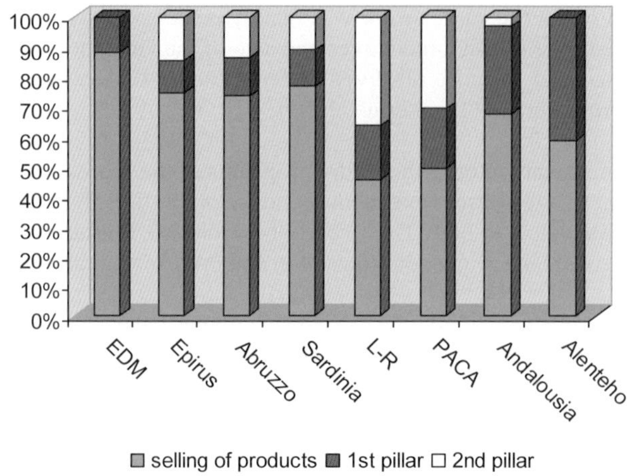

Figure 5. Percentage of premiums received to total income.

the incentives for suckling cows, is expected to be reversed. Although the products are differentiated and are appreciated by the consumers, a system of certification has not yet been developed.

In semi-mountainous and mountainous regions, the sheep production systems are based on grazing, in certain cases with transhumance, mainly in the mountainous zones. Usually the flocks are large (300-800 head in the semi-mountainous zone) and larger (more than 500 head in the mountainous zone), with medium milk production. The delivery of milk to the dairy ends in June, when the animals move to the mountain pastures. The remaining milk is used for family consumption or locally. The lambs are sold as milk-fed lambs between 30 and 60 days at 11 to 14 kg live weight. The price varies over the year, with higher prices at Easter and the end of the year. Lambings are concentrated in the autumn, which corresponds with the return from the summer pastures in the transhumance systems, and to change of pasture in regions with better quality pastures near the farms with possible complementary feeding. The majority of the systems manage to maintain significant number of livestock in relation to the low quality of pastures and to cover the 75% of their needs by grazing. As mentioned earlier, the goat sector is decreasing, as goats are linked with low productivity, ageing farmers and difficult social conditions (Georgoudis *et al.*, 2006; Ligda, 2007)

All partners agreed that priority should be given to production, as this is the principal function of all livestock production units, pastoral or not. However, there are opportunities that remain to be exploited, including agro-environmental measures, mainly measures for the protection of endangered breeds.

A SWOT analysis was performed with the collaboration of a group of technicians and breeders in Epirus, to identify the major threats to and opportunities offered by the pastoral systems in the region. The results of the survey are presented in Table 1.

In conclusion, for pastoral farming to be sustainable in the future, the activities should have the ability to provide the market with labelled quality products and the farmers with sufficient income in an environment that offers the necessary social infrastructures. Specific measures are proposed, aiming at the integrated development of mountain zones. These measures include the establishment of a separate production chain, the provision of incentives to compensate for the drop in the income (due to the lower production level), and the creation of infrastructures to facilitate the farmer and his family.

Table 1. SWOT analysis in Epirus.

Strengths	Weaknesses
• Quality products (milk and meat) • Conditions for the production of certified products • Rich biodiversity • Local breeds adapted to the environment	• Lack of sales structures • The image of the product is not linked with the producers • Lack of professional organisations • Weak application of certification procedures • Difficult living conditions
Opportunities	**Threats**
• Improvement of living conditions and infrastructures • Extension services to the farmers • Innovations and modernization of farms • Exploitation of CAP measures • The increasing trend towards certified quality products • Diversification of agriculture	• The continuation of the movement of farming towards the plain • Lack of strategies of professional organisations • Need to adapt to the new requirements of CAP

This implies investments in:

- added value, mainly related to research on the products, the market, the production chains, etc.;
- productivity, to identify the balance between the performance per work unit, management etc.

Furthermore, true collaboration is required between the different actors to reach agreement on the objectives of the production sector and objectives concerning the environment. In this context, efforts will continue to be made concerning innovations at different levels and in different fields, as all those involved agree that pastoral systems; with their production practices, their products and their positive image (although this is not always recognised), are a response to today's questions concerning sustainability. The necessary technologies and know how are available; the main issue is the contribution of active farmers who, with the support of agricultural technicians, will have to take the lead if we are to achieve the target of the sustainable development.

References

Georgoudis, A., C. Ligda and J. AlTarayrah, 2006. Genetic characterisation of local genetic resources and its use for sustainable management. EAAP Publication, No. 119, 137-144.

Ligda C., 2007. Epire – Description de son contexte régional, 2007. Séminaire final de PASTOMED, Octobre 2007, Marseille, France.

Tchakerian E. 2007. Pastoralisme méditerranéen, état de lieu et perspectives. Séminaire final de PASTOMED, Octobre 2007, Marseille, France.

PDO project for Corsican pork as an innovation for the whole marketing chain

A. Lambert-Derkimba[1,2], E. Verrier[2] and F. Casabianca[1]
[1]INRA, Laboratoire de Recherche sur le Développement de l'Elevage, Quartier Grossetti, 20250 Corte, France; derkimba@corte.inra.fr; fca@corte.inra.fr
[2]INRA/AgroParisTech, UMR Génétique et Diversité animale, 16 rue Claude Bernard, 75005 Paris, France; verrier@agroparistech.fr

Abstract

A PDO project for Corsican pork began at the end of the 1990s. The objective of this paper is to show how this project could play a dynamic role by creating different types of innovations. We analysed the different steps in the construction of the project in the non-structured pork production sector in Corsica. We interviewed the main actors of the project and attended technical and political meetings organized by the PDO association. The materials (discourse, grey literature) we analyzed enabled us to identify three innovations introduced by the project: (1) In a project based on the valorisation of the 'terroir' and of the territory, a change in the status of the local resources, in particular of the local breed; (2) in a project designed to fit European regulations, restructuring of the marketing chain with the creation of a new institution (the PDO association) and a new strategy for the sector; (3) in a project based on technical rules, producers have to respect the new production system including the exclusion of some producers who are unable to comply with the rules. Even though the PDO project has not yet ended, we wondered if it could be a real driving force for the pork sector in Corsica.

Keywords: local breed, marketing chain, origin product, collective action, livestock system, pig, Corsica

Introduction

A PDO project concerning Corsican pork meat began at the end of the 1990s. The project targeted a non-structured sector in which pig husbandry has long been – and still is – a family activity. The objective of this paper is to show how the PDO project could play a dynamic role by creating different types of innovations.

We first describe the steps in the construction of the project. We then explain the resulting innovations at the level of the territory, the sector and individual households: the change in the status of local resources, changes in the structure of the marketing chain, and the new conception of the production system as a whole.

Such a huge change in one local sector gives an overview of what could be called a 'radical innovation' covering a large range of changes and modifying the shape of the regional production system.

Context

In Corsica, pigs are traditionally reared extensively, with animals using pasturelands for most of the year before a fattening period in the autumn when the pigs are pastured in the forest where they eat chestnuts and acorns. During winter, pigs are slaughtered and the meat is processed into local products that are culturally rooted in Corsican gastronomy. This represents an 'eco-cultural' system (Commandeur and Casabianca, 2007) which contrasts with much conventional pig production in Europe today but which shares some characteristics with local Spanish and Italian systems (Casabianca and Fallola, 1994).

In this type of eco-cultural system, the producers are themselves responsible for all the operations that are normally distributed throughout the supply chain: farrowing, fattening, slaughtering, processing, ripening and selling the final product (Casabianca, 2004). Although this means the individual farmer has a great deal of work, he or she benefits from the added value of each activity. This presupposes that the farmer is able to find customers who will come to the farm to buy authentic products, and who are ready to pay a high price for a rare and unique specialty product. In very difficult areas like the Corsican mountains, raising pigs is the main activity and provides a reasonable income for farming families.

This robust system is very stable due to the independence from other producers, from the needs for inputs and from the final market. On the other hand, it leads to individualism that could be an obstacle to the collective organization of the sector.

After years of effort involving many breeders and several generations of animals, the Corsican pig breed was recently officially recognized by the French Commission for Genetic Improvement (CNAG). This recognition is both the final step of the previous stage and the first step of a new stage: managing the genetic resource. To this end, managers of the local breed are in the process of stabilizing technical choices concerning breed genetics (Casabianca *et al.*, 2000). The creation of a regional organisation is the first collective initiative in this sector, which, up to now, has been characterised by individual initiatives. Hopefully, it will provide an example for future collective actions. Creating a regional organization in charge of this type of management is the first professional skill to have been used collectively in a sector that has few points in common.

More specifically, the regional association was created to handle official applications for a PDO for Corsican processed meat. In this new project, a code of practices was required in order to draw up the rules governing production.

For the application, the low-growth local breed was obviously chosen to ensure a strong link with the 'terroir'. Originally, the application was to include not only the use of pure-bred animals but also cross-bred. It was also planned to accept two different diets for finishing: chestnuts and/or acorns vs. barley. To this end, a set of four possibilities (two genotypes and two finishing feeds) was proposed, with a distinctive label identifying the top level product of the PDO for pure-bred animals fed with chestnuts and/or acorns during the finishing period. This complex feature, which represents real breeding choices and feeding practices, was inspired by regulations that have already been accepted in Spain for the use of the Iberic breed for the production of dry cured hams originated from Extremadura, Salamanca and Andalucia. In these four PDOs, two genotypes (pure-bred and cross-bred ¾ Iberic × ¼ Duroc) and three finishing feeds (concentrate, acorns, and a succession of acorns and concentrate) are mentioned in the specifications, i.e. six types of products are accepted within the same PDO.

But during the application procedure, the French institute in charge of PDO applications (*Institut National de la Qualité et de l'Origine* INAO) refused to accept such a wide range of products under the same PDO label. In France, up to now there has been no application covering such a wide range of products and no pork products have yet obtained a PDO label. The Corsican applicants were thus obliged to choose only the pure breed as they wished to keep the two finishing feed options. In this context, new stakes appeared concerning the genetic resources, because of the fundamental role played by the breed in the PDO project. In the rest of this paper, we show how the PDO project led to several innovations concerning the local resources as well as the marketing chain and the production system.

Methods

We conducted several interviews with the main actors of the two projects:
- breeders (belonging to the PDO project, or not);
- technicians involved in the management of the breed, or belonging to the PDO association;
- animators.

We collected and compared the different points of view of the actors about technical choices for the PDO project based on the content of the Code of practices. At the same time, we analyzed changes in the regulations for breed stock management in the breeding project.

The second tool used for the study was participant observation at meetings. We observed many meetings attended by the breed managers and the PDO applicants in the two past years. We also observed board meetings at which producers discussed political choices and made strategic decisions. During technical committee meetings, we heard the arguments used by technicians and animators when discussing technical choices. We were thus able to observe in real time the whole set of arguments used by producers, technicians and animators during the negotiation process and before the final agreements were reached.

Third, we reviewed written documents such as regional development projects, technical historical papers and association reports.

Results

By analysing all the documents and data, we were able to identify three main kinds of innovation introduced by the PDO project at the level of the territory (local resources, the marketing chain, and the production system).

Change in the status of local resources

As the project targets the valorisation of the *terroir* and the territory, the resulting change in the status of the local resources, (chestnut and oak trees for finishing) is a major innovation. Chestnut trees previously represented added value for milled chestnut flour (already recognized as a PDO product), and we observed genuine management of chestnut trees by the farmers concerned. Can chestnut trees used for fattening pigs also become a valuable resource? This could lead to true management of the trees because of their importance for the project. But as it has also been proposed that barley be permitted for the production of PDO products, it is in fact difficult to envisage improvement in the management of chestnut trees for the production of PDO pork products.

In the case of the local breed, the change in status is much clearer. We observed a dialectic relation between the breed and the PDO:
- The PDO label will enable the development of the breed by ensuring some financial support for breed management, and will increase the market for pure-bred animals.
- The organisation of the breed will also have an impact on the PDO project by increasing the market and prices for products from pure-bred animals, giving new legitimacy to PDO applicants due to the symbolic importance of the local breed as a collective heritage.

The PDO project (based on pure-bred animals) should lead to an increase in the demand for pure-bred pigs. All producers who wish to join the PDO project will have to stock their farms with pure-bred animals. As a consequence, the value of the pure breed should increase (Langlois, 1998). The breed is already well known to all producers but until now only as an informal resource. All pig producers in Corsica – as well as all inhabitants of Corsica – have quite a clear vision of what the Corsican pig breed should be. However, the breed has been frequently crossed with breeds with much higher performances. So the image of the local breed is a little fuzzy and the work of the managers during the recognition project was to reconstruct a base of pure-bred animals by looking for such animals all over Corsica.

However, extensive systems require pigs that are able to deal with the constraints of such a hard life. For some breeders, a Corsican sow is above all a sow that looks after her piglets and protects them from predators. Others prefer an intelligent sow that knows her territory well and is capable of teaching her piglets to use the right tracks, waterholes and sleeping areas. For still others, a good sow ensures strong piglets (even if only a few) of good size after weaning. For these reasons, the use of more specialized breeds did not result in improvements in terms of the required hardiness of the animals. And the local breed was consequently kept in production, especially sows.

In fact, this breed has always had a strong identity because of the traditional pork production system based on subsistence management by individual families. Even when it became a production system for some farmers, the breed and the pork products remained linked to the Corsican identity.

But with the need for protection for Corsican pork which was recognized by a large proportion of farmers, an association was created for the management of the pure breed (ARGRPC): the members declared themselves to be in a legitimate position to create management rules for the breed. When it was created in 1992, the association was limited to southern Corsica (where this type of production is more common) but in 1998 it became a regional association. From an informal status, the breed achieved a formal status (Lambert-Derkimba, 2007). The breed has to be clearly identified: a pure-bred animal is not an animal that only looks as if it belongs to the breed. Pure-bred animals have to be identified following a specific procedure and cannot be kept on the same farm as cross-bred animals. Pure-bred pigs have to be registered in the herd book for the local pig breed (LIGERAL) and pass the standard test. Moreover, for mating, careful attention has to be paid to the sow's and the boar's families: with such a small breed, producers need to avoid consanguinity.

With this project, the status of the local breed has undergone a real change. But in fact, the choices made by the ARGRPC have made the change applicable to all those who use the breed, whose value has gone up thanks to the management rules (LAMBERT-DERKIMBA *et al.*, 2007). In addition, the breed managers drew up rules under which only they themselves have the right to breed reproducers.

The rules are very simple:

- breeders (from ARGRPC) have the exclusive right to breed reproducers;
- breeders (from ARGRPC) have the exclusive right to sell reproducers;
- producers cannot self-renew their own reproducers.

These rules have several consequences: each year all the farmers in the PDO project have to compensate for the turnover of their herd, since only pure-bred pigs are accepted in the project and only breeders can produce them (Table 1). The scarcity of the breed means these animals have a really high value. This guarantees an income for breeders, who are certain to earn money for their job of selection each year. But this also means that there is a limit to the number of animals available (reduction in the potential market) both for reproduction within the herds and for the production of PDO products. This high value thus also represents a risk for the development of the project and, in the long term, for the breed itself. The dialectic between the breed and the PDO could become a vicious circle.

Another impact of the PDO project on the breed project has been the choice of the breed name: Nustrale (meaning 'ours'). This name was chosen to avoid any confusion linked to applications for the Protected Designation of Origin (PDO) for products that will use the protected name 'Corsica'.

Table 1. Consequences of the PDO project for the genetic structure of animals on the farm.

	Without PDO	With PDO
Composition of the herd	Pure and cross-bred pigs	Only pure-bred pigs
Identification of animals	Animal with IPG tag	Herd book and standard test
Structure of the farm	No specific constraint	Constraints for linkage

Such a decision by the breed managers shows that the PDO project and the breed project are interlinked: to preserve the PDO project, the breed managers agreed to change a fundamental aspect of the breed, its name.

Changes in the structure of the whole chain

To comply with European regulations several changes have been made in the structure of the marketing chain.

At the beginning of the project, the roles appeared to be reasonably clear: small-scale farmers were the most authentic feature and it was consequently important that it should not be difficult for them to comply with all the rules of the future PDO. All small-scale farmers were directly concerned by the protection offered by the name 'Corsica' and were thus assumed to be in favour of the PDO project.

Conversely, the industrial processors, who invested a lot of money in factories and used imported raw material, had no place in the PDO project. They produced imitations of the genuine product and it was thus assumed they would all be opposed to application for a PDO label.

In practice, after some months of preparing the first draft of the application, it became clear that a large proportion of farmers, even those located in the mountains who used extensive systems, would not spontaneously support the PDO project. The majority did not have an official activity, their pigs were not identified, a lot of products were sold without a label, with no invoice and no tax paid. Moreover, the word 'rules' was unacceptable to some breeders, their animals were not under real management and even though the rules were based on traditions, those concerning the future PDO code of practices were quite far from existing practices.

On the other hand, some processors who had made a name for themselves because they were located in a mountainous area were in fact interested in the PDO project; not necessarily for all their products, but because it would enable them to sell a small part of their products with a PDO label. They joined the collective organization and participated actively in the meetings of applicants.

As a result, the former categories were no longer applicable and new categories appeared during the application process for the PDO label. For the first time, small-scale producers discussed with processors and artisans. They discovered that some agreement is possible and the resulting alliances are now in the process of restructuring the supply chain. Some breeders are considering selling some of their animals for slaughter to processors involved in the PDO project and the better prices paid for such specialized local raw material are creating new relationships. The association was obliged to respect INAO requirements including several sections that are representative of the sector i.e. producers, processors, artisans, and retailers, and this standard structure is now conceivable for the PDO board.

In the same dynamic way, the creation of a new institution (the PDO association) has opened the way for a new strategy within the sector, which is now recognised by local development institutions:

human resources have become available and financial resources are being mobilized. A feeling of success has suddenly emerged in a sector that has always complained of being left out and forgotten.

Designing rules through the involvement of actors

In France, a PDO project has to be build by a collective represented by an *Organisme de Défense et de Gestion* (Organisation for Defence and Management). The association has to use traditional knowledge and practices to build a project with two objectives: (1) to demonstrate the legitimacy of the PDO project, (2) to choose a set of rules to define the PDO product. The first point leads producers to search their collective memory and learn the history of their practices. The second obliges producers to consider their own practices and technical culture collectively. They have to use reflexivity for this exercise and have to reach a compromise among themselves. They need to find a new vision of the production system, i.e. to use traditional approaches to build a project for the future. They have to choose a code of practices for the project representing what they would call: 'local, true and constant' practices. This does not mean doing exactly what people did 40 years ago, but reinventing 40-year old practices in a production system that appears to be logical to the majority of farmers and is strongly linked to the territory. As the same time, in order to propose a viable project, the specificity of the context has to be made clear.

As an example, we can take genetic type and the feeding of pigs for finishing. The first choice of the association was to allow cross-bred animals and to specify the finishing of animals with chestnuts and acorns (and barley only when necessary). Cross-bred pigs have been raised on the farms for more than 30 years. The original intention was to continue to accept cross-bred pigs in a system in which animals known to improve carcass characteristics could be used, and where characteristics of behaviour and exploitation of the territories would be provided by the sow (intended to live for several years on the farm and to transmit her knowledge to her piglets). The type of feed was part of the traditional system, i.e. animals found the food they needed in the chestnut and oak woods. At that time, feed supplementation was merely a way of managing risk in the case of crisis.

In agreement with INAO policy, the association finally decided on a transition period of eight years with the long-term aim of having herds of 100% pure Corsican breed, with the pigs fattened in chestnut and oak woods with the addition of barley. One might have expected authorization for cross-bred pigs and only the use of chestnut and acorns for finishing, knowing that crossing dates back nearly 50 years, as does exploiting natural resources for food (Lambert-Derkimba *et al.*, 2006). This scenario would have been more consistent with the spirit of PDO projects which claim to be based on local practices. But one can also say that the choice of pure-bred animals and of supplementary feeding with barley is an obvious adaptive strategy (it is easier to manage a pure breed than it is to fatten pigs only using natural resources (De Casabianca and Santucci, 1989)). All this collective work is the main innovation for producers who are not used to working in collective organisations. Moreover, few have already practiced reflexivity with respect to their own practices.

The project necessarily implied the exclusion of producers who were unable to deal with these rules: so how to choose the appropriate rules for inclusion? In our example, the possibility of excluding producers on the feeding criterion was rejected. Instead, the choice was made to exclude producers on the pure breed criterion. For some rules, farmers apparently do not always realize the potential difficulties involved in respecting the PDO conditions. For example, the obligation to only use pure-bred pigs means that it is not possible to keep cross-bred pigs on the farm. Today, even for pure breed stockbreeders, there can only be pure-bred pigs on the farm (registered in the herd book and with special monitoring) and no cross-bred animals (which were formerly kept because they were easy to raise, their price was low and because of the tradition of breeding). That means that such farms will have to convert: even if the regional policies support some producers (subsidies to improve the structure of the farm, to buy reproducers of pure local breed) these subsidies will not continue forever and not every producer will receive them. In fact, a lot of producers could have difficulty complying with all the rules required by the code of practice.

Some insight into the question of innovation

The PDO project should clearly be considered as an innovation process. It is based on the modern interpretation of traditional practices: a pig that develops slowly feeding on natural resources and a very long production cycle. These traits result in very low productivity and high costs. The good reputation of the cured products provides added value for the breeder. But until now there has been no real income from pig production itself.

The perspective opened by the PDO label gives breeders the opportunity to balance their income between raising pigs and processing pork. This change is crucial to ensuring some specialization at each step of the chain, and to scale up the volume, to increase the number of producers and the size of the market. In doing so, breeders will not be under the control of the processors (as is the case in some other situations). This should enable the fair distribution of added value among all the local actors. Even if such a concrete change is not yet apparent, all the necessary conditions are now in place.

However, the technical rules for PDO application are expected to cause some conflict in the local sector. As already mentioned, there is a real risk of eliminating a large number of producers. The length of the application process for the PDO label could be viewed as a transition period which will be useful for learning and to enable a progressive change in the structure of production. Obviously, it is impossible to reach unanimity and some exclusion will have to be accepted while the PDO system is being set up. But such exclusion should be reasonably acceptable thanks to the transparency of the process and to the support provided for producers who are slow to adapt. From this point of view, the PDO project is mobilizing all available means and reflections (even those opposed to the project) in the region, and should thus be qualified as a 'radical innovation'.

Conclusion

Our analyses of the three types of innovations underlined one fundamental aspect: the PDO should be seen as a creation. It is creating dynamics concerning the management of the resources, the structure of the sector, and the way producers consider their own work. Technical changes created by the PDO are tightly linked with organizational changes, because the production rules that have to be respected in order to belong to the PDO project emerged from the collective organization. For all that, institutional and financial support appears to be needed. Without such support, it would have been impossible to achieve recognition of the local breed and ensure its future development. And, as a consequence, the PDO project would have been more difficult to build.

Although the PDO project is not yet achieved, could it be a real driver for the pork sector in Corsica? The full consequences of the project for resources management are still difficult to envisage but there is no doubt that they could be really positive but also involve certain risks. Above all, livestock activities have now a strategic path to follow that seems to be coherent with the development of Mediterranean mountainous areas: exploiting territory and local know-how to produce a high added value product.

A project of this type provides an opportunity for the local technical culture to no longer be considered as archaic but instead as a vital base for creating and capturing territorial rent. Until now, pig producers, processers and policy makers did not have such a clear vision of the future. The PDO project is thus a vehicle of change: organisational, technical, and political change. This is not a project of the past but a project turned towards the future that can be expected to contribute to sustainable development.

References

Casabianca, F., 2004. Elevage Sylvo-Pastoral Du Porc En Corse. In: 'Porchi E Cignali; Saveurs Et Mystères Des Suidés' (Albiana And M. D. L. Corse, Eds.), pp. 73-92, Corte.

Casabianca, F. and A. Fallola, 1994. Avenir De L'elevage Porcin Mediterraneen Extensif Et Coordination Des Filières De Production - Orientation Des Activités D'élevage Par La Qualité Des Produits Transformés. In: 'The Study Of Livestock Farming Systems In A Research And Development Network' (A. Gibon And J. C. Flamant, Eds.), Pp. 61-73. Wageningen Press Eaap Publications.

Casabianca, F., A. Poggioli, J.D. Rossi and O. Maestrini, 2000. L'amorce D'une Gestion Collective De La Race Porcine Corse. Construire Un Standart Et Elaborer Les Contrôles Des Reproducteurs. In 'Options Méditerranéennes: Tradition And Innovation In Mediterranean Pig Production', Vol. 41, Pp. 23-34.

Commandeur, M.A.M. and F. Casabianca, 2007. Diversity In Pig Farmers' Logic In Corsica. In: '6th International Symposium On The Mediterranean Pig', Pp. 414-419. Nanni Costa L., Zambonelli P., Russo V., Capo D'orlando, Italy.

De Casabianca, F. and P. Santucci, 1989. Une Expérience De Recherche-Développement En Zone Marginale: La Castagniccia. Greghje E Rughjoni 16, 151-165.

Lambert-Derkimba, A., 2007. L'inscription Des Races Locales Dans Les Conditions De Production Des Produits Animaux Sous Aoc: Enjeux Et Conséquences Pour La Gestion Collective Des Races Mobilisées., Agroparistech, Paris.

Lambert-Derkimba, A., O. Maestrini, M. Poggi, E. Verrier and F. Casabianca, 2007. Corsican Pig Breed And Pdo Project For Processed Meat: Main Challenges On Collective Management. In: '6th International Symposium On The Mediterranean Pig', Pp. 403-406. Nanni Costa L., Zambonelli P., Russo V., Capo D'orlando, Italy.

Lambert-Derkimba, A., E. Verrier and F. Casabianca, 2006. Aop Charcuterie De Corse: Un Projet De Valorisation Qui Interroge La Gestion De La Race Porcine Locale. In 'Colloque International Sur Les Systèmes Agroalimentaires Localisés, Alter', Baeza, Espagne.

Langlois, M., 1998. Rareté, Valeur Et Utilité, Approche Economique. In 'Pression Sur Les Ressources Et Raretés', Vol. 6, Pp. 69-81. Orstom, Montpellier.

Innovation systems and processes in the field of agricultural direct marketing: a cross-national analysis between France, Italy, the Netherlands and Switzerland[11]

C. Couzy, A.-C. Dockès and A. Guillaumin
Institut de l'Elevage, 149 rue de Bercy, 75595 Paris cedex 12, France

Abstract

Studying collective innovation in agriculture and in the rural area, identifying the main facilitating and hampering factors to their development are the main objectives of the In-Sight project (EU funded). This paper presents and compares innovations in France, Italy, the Netherlands and Switzerland on the field of Direct Marketing in agriculture. These new initiatives involve consumers and new profiles of farmers. They emerge from a few pioneers and develop by diversification, horizontal growth or up-scaling, but always remain at a novelty or a niche scale. The innovation dynamic, the actor systems and the evolution of the consumption models in different European countries, constitute the main key explanatory factors of our findings.

Keywords: direct marketing, European countries

Introduction

Agricultural direct marketing is a traditional way of marketing, developed on the basis of relations of proximity amongst families and neighbours. As such it progressively became a separate market channel in which especially vegetable and wine producers where involved. For these farmers, direct marketing appeared as a necessary way of getting enough income. Even if it often concerned many producers, the agricultural direct marketing remained a small-scale, traditional and limited channel for a long period. In the Netherlands this marketing method even disappeared almost completely during the 20th century, when food processing was moved almost completely from farms to factories.

Recently however, a new demand appeared for farm produces, favoured by the urbanization, the urban way of life and the needs of alternatives to the industrial food products. Meanwhile some farmers were seeking new opportunities to improve the added value of their products, particularly in small farms; this new demand supported new dynamics for the agricultural direct marketing, related to:

- Re-development of existing direct marketing frameworks: weekly farmers street markets and various types of on-farm selling (self-picking, self-service, tasting, etc.).
- New types of direct marketing (or diffusion of pioneer initiatives): community-supported agriculture, baskets of products for tourists, shops of producers, etc.

These new initiatives involve consumers (often from alternative circles) and new profiles of farmers (rural entrepreneurs, creators ...). They emerge from a few pioneers and develop by diversification, horizontal growth or up-scaling, but always remain at a novelty or a niche scale. They can be considered as innovations.

Studying collective innovation in agriculture and in the rural area, identifying the main facilitating and hampering factors to their development are the main objectives of the In-Sight project (EU Funded). The paper at hand presents the main findings about innovation in direct marketing in agriculture. The focal issues are formulated as:

[11] This paper is based on collaborative work of all colleagues involved in the EU-funded IN-SIGHT project. The original work can be found in project reports at www.insightproject.net.

- How do researches on innovation processes in the agricultural direct marketing sector contribute to our understanding of rural innovation processes and systems?
- What kinds of policy recommendations can be formulated on the basis of the research on innovations in the sector?

In the present paper we firstly present our methodology. Secondly we describe the main innovations in direct marketing in agriculture, their characteristics and dynamics. Thirdly we present the socio-technical systems of the innovations. Finally we describe the main facilitating or hampering factors for innovation and propose some recommendations for the information and knowledge systems as for the policy makers, according to the stage of the innovation process. The innovation dynamic constitutes one of the main key explanatory factors of our findings.

Methodology

Four countries: France, Italy, the Netherlands and Switzerland were involved in this research work. A systemic case study approach was adopted, focussed on innovation systems and networks. The work was organised in three steps:
1. elaboration of collective guidelines for all countries;
2. country studies, each including an in depth review of the field and a specific innovation case study;
3. synthesis among country reports to identify the main findings of the work package.

The main methodological tools we used were the following:
- short literature, grey literature, websites and media review;
- interviews of experts and stakeholders involved in the studied fields through semi-structured or structured interviews and questionnaires;
- workshops with relevant actors, stakeholders and experts.

The following innovative topics were specifically studied:
- solidarity groups (in Italy) or community supported agriculture (in Switzerland): citizens or consumers are directly at the initiative of the innovation and aim at promoting new consumption practices;
- direct sales organised by producers, through collective shops (in France) or a small scale cooperative (the Netherlands).

General overview of direct marketing and innovation in the field

We understand the direct marketing in agriculture as the different ways for producers to sell their products directly to consumers. The relation can rely on new technologies of information like the Internet. The direct relation between producers and consumers may be completed with semi-direct relations in which the producer sells his products to a collective system of marketing (involving several producers) or to restaurants or shops or even supermarkets, but with a personal relation between the producer and the buyer.

In the different countries, the number and percentage of farms practicing direct marketing is very different according to the national statistics (see Table 1): from 600,000 farms and nearly 40% of the farms in Italy to 4,500 farms (5.5%) in the Netherlands. France and Switzerland are in a medium position, with respectively 15 and 20% of farms practicing direct marketing.

Table 1. Direct marketing according in the different countries.

	No. farms	No. of direct marketing farms	% on farms that commercialize
France	660,000	102,000	15%
Italy	1,600,000	600,000	37.8%
The Netherlands	82,000	4,500	5.5%
Switzerland	63,000	13,000	20.8%

The dynamics of recent innovations in direct marketing

For farmers who are at the initiative of a direct marketing innovation (collective shops, Internet sales …), it is at first a mean to improve their income with the direct selling and to get free of the agribusiness firms. Some farmers also stress a more idealistic motivation of promoting local production and marketing, as a way of preserving the village identity. This idea fits also some consumers' demand to develop direct contacts with producers.

As developed by Brunori *et al.* (2007), when innovation cycles are repeated, interactions between people, tools and natural resources become more and more structured. Four levels of organization of the socio-technical network can be identified (see Figure 1):

- Novelties are localised 'breaks of the routines'. They are limited by external constraints like laws, actors, norms.
- Niches are the result of an aggregation of different smaller systems. Niches are the places where new paradigms emerge as an effect of learning processes.
- Regimes are paradigms turned into practices, the incorporation of given paradigms into concrete socio-technical systems. Networks are structured and coordinated by rules.
- Landscapes can be changed as an effect of supranational policies or scaling up of radical changes, but more often changes in socio-technical landscapes are important drivers for radical innovation.

In the field of direct marketing, the innovation cases we studied have an history. At first, it emerged as a novelty. Then, several types of evolutions can occur:

- By diversification or horizontal growth (including new farms, new products in a collective shop, new products in a solidarity group and thereby broadening one initiative's scope of activities) as shown on the Figure 1 in the round arrow. Diversification seems to be favoured by the actors of solidarity groups projects, with a high involvement of volunteers and a will to remain in an alternative situation;
- By up scaling either with the involvement of new actors and the support of the agricultural knowledge system or the local development system (many collective shops in a region …), or by the organisation of a network of small initiatives (network of shops, …) as shown on Figure 1.

Supporting innovation in the field of direct marketing

A few individuals are often at the emergence of an innovative project in the field of direct marketing (one of the first solidarity groups or collective shop in an area), they rely on their own skills and social capital. They find some individual ways to cope with the regulatory framework which is rather hampering. They benefit from very little support. But more structured coaching could be helpful.

In case of horizontal growth the innovation spreads by direct contacts between actors sharing the same values and projects, and sometime with involvement and support from local policy makers.

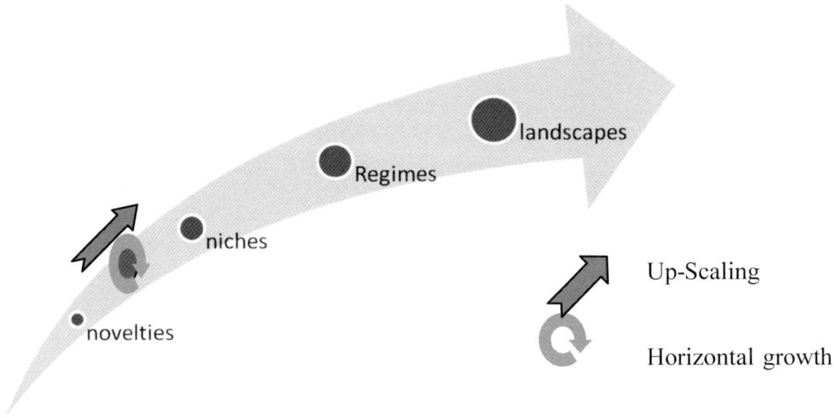

Figure 1. The dynamics of innovation.

The embeddedness mainly concerns alternative circles. Active networking can be helpful, as open subsidy systems.

For up-scaling, involvement of the agriculture knowledge system and opening between agricultural and non agricultural knowledge and support systems seem essential to impulse or support new projects and networks between projects. A regulatory framework is also essential. This regulatory framework may evolve to take into account the specificities of the innovation project, and either prove supportive or constitute a real hampering factor. Communication and promotion methods, training of actors, project management methods, availability of technical and economic references constitute important development factors.

To summarize, development of innovations can rely on at least 3 diverse methods:
1. Distribution between peers: between farmers or/and consumers, without any outside framework. For instance, visit of a pioneer initiative gives other producers the idea of engaging with this form of marketing.
2. Development supported by local development: local authorities or alternative actors becomes involved in an innovation and looks for actors to set up new projects with support.
3. Development with the help of extension services: extension services, through their diversification services propose accompaniment and follow-up for groups they can either answer to actors demands or initiate new projects.

Synergies between local and agricultural development are essential to support innovation in agricultural direct marketing.

References

Arce, A. and T. Marsden, 1993. The social construction of international food: A new research agenda, Economic Geography, 69 (3), pp. 293-311.

Brunori, G., 2007. Local food and alternative food networks: a communication perspective, Anthropology of Food. URL: http://aof.revues.org/document430.html.

Couzy, C. and A.C. Dockès, 2008. Are farmers businesspeople? Highlighting transformations in the profession of farmers in France. Int. J. Entrepreneurship and Small Business, 2008.

Murdoch, J. and T. Marsden and J. Banks, 2000. Quality, nature, and embeddedness: Some theoretical considerations in the context of the food sector, Economic Geography, Vol. 76, No. 2, pp 107-125.

The role of the Mediterranean networks of cooperative research in animal production in favor of innovation: situation and outlooks for the future

J.-P. Dubeuf[1] *and J. Boyazoglu*[2]
[1]*INRA -SAD, LRDE, 20250 Corte, France; dubeuf@corte.inra.fr*
[2]*06500 Menton, France; jean.boyazoglu@wanadoo.fr*

Abstract

The animal production sector in the Mediterranean, as in other areas has to face new challenges related to environment, productivity and social issues. By considering the hypothesis that more and urgent innovation is needed, to solve the complexity of these problems, the organization, activities and production of some permanent networks of Cooperative Research in Animal Production are described and their possible role in these processes are discussed. It would mean that these networks could involve scientists from several disciplines and other actors cooperating. The development of networking would probably be essential to enlarge a more strategic and open approach of the questions to answer. The present disciplinary organization, the real involvement of a variety of actors, their impact on innovation and some suggestions for their evolution are analyzed to outline the existing difficulties to change the situation.

Keywords: networking, organization

Introduction: the main challenges of animal production around the Mediterranean and the creation of the Mediterranean networks

Animal science has built its legitimacy in the last 50 years upon its results to increase animal performances in controlled systems. Many innovations leading to an increase of productivity were a great success both in meat and milk production. For the animal production sectors, the Mediterranean production systems have often been considered marginal. The dominant models were exogenous and the Mediterranean characteristics, know-how and practices were not involved in them. From a general point of view, the external intensive production systems have been promoted and developed as a priority: feedstuffs fed cattle or dairy cows, intensive pig and poultry are now a majority and have gathered most of the investments. These systems have few territorial links and are high consumers of energy, water and imported food (Benoit, 2007).

On the other hand, it is well known that the Mediterranean production systems, based on the use of rangelands, local breeds and local practices, and a low specialization of animal production and crops have been considered as being not sufficiently productive and for this reason as 'relics of the past'. For more or less 30 years, the 'public services' have only lightly changed their direction with the promotion and certification of local products, actions in favor of pastoralism both in the Northern and Southern banks. In spite of a real success for pigs in Spain, sheep and goat or buffalo cheeses in France, Italy, Spain Cyprus and Greece, the private investment is still very low and the lack of coordination between research, administration, training, professional organizations and industry has often been underlined (Couzy *et al.*, 2000).

In order to take in account these specific situations and their diversity, several scientific networks have been created for more than 30 years thanks to the awareness of some academics, researchers in animal production and pioneers in their specialty. Such networks are initiated from France and Germany at the beginning, but followed by others from Italy, Spain, Greece and also from Eastern Europe. They have been designed very close to extension activity and to the world of breeders and professional organizations. In the light of their academic position, they could have acted in their own name but also with the institutional support of their organization.

We have considered the fact that the Mediterranean region is at the heart of geostrategical, demographic, social and environmental issues not limited to the agricultural sector: organization of trade and exchanges between the North and Southern region, migratory and social problems related to demography and economical imbalance, erosion of soils and desertification of many areas (both in North and South), lack of water and management of water resources, food auto sufficiency (South), disappearance of cultural heritage and know-how. By considering these elements, we will analyze the activity and organization of these networks, how their community has evolved and we discuss in what way they could increase their participation in the innovating process.

A short history of the Mediterranean Networks of Cooperative Research in Animal Production

The dynamism of networking in the Mediterranean Animal Production has been a reality during several decades and several groups have shown to be very active since the 70-80's through many publications, meetings, and thematic exchanges. The initial objective was to improve consultation and dialogue around the mutual understanding of the challenges of researches at economic and geopolitical levels but with individual spontaneous initiatives. We have considered 3 permanent supports for networking specifically involving Research, Development in Animal Production in the Mediterranean: EAAP, CIHEAM and FAO. All other existing networks do not focus on this geographical area. We have checked that there is no other global organization specialized in the Mediterranean region.

The EAAP[12] and their mediterranean group

According to its website (2008), the main aims of EAAP are to promote, for the participating European and Mediterranean countries, 'the advancement of scientific research, sustainable development and systems of production; experimentation, application and extension...by means of active co-operation between its members and other relevant international and national organisations' and to 'encourage the involvement of young scientists and technicians.' This Mediterranean working group has mainly organized thematic symposiums as recently in Badajoz (on quality products in 1996), in Hammamet (on dairy sectors in 2000), in Santarem (2006) and now in Corte (2008). This working group has no formal institutional constitution and is working on the basis of the volunteer involvement of the participants.

The FAO-CIHEAM networks

These networks are officially supported by two institutions, CIHEAM and FAO.

- CIHEAM is organized as a network of 13 countries and 4 national institutions for Mediterranean research and high studies sharing strategic themes, Bari, Chania, Zaragoza (for animal production), Montpellier (for economics, development and public policies).
- FAO is the Agency of the United Nations for Food and Agriculture and has supported the Mediterranean networks through the system of ESCORENA Networks and the regional office for Europe. It is now more oriented towards the Southern Basin and developing countries.

In animal production, FAO and CIHEAM have mainly supported 4 networks on cooperative research and development: the network on sheep and goats (with 3 sub networks, nutrition, production systems and genetic resources) involves 225 researchers from 60 institutions from 35 countries; the

[12] European Association of Animal Production.

network on pasture (sub networks on plain pastures, Mediterranean forage resources and mountain pastures); the network on local rabbit breeds and the network on Mediterranean pig created by 5 institutions from 4 countries after the conclusions of a research project. These networks also work on a volunteer basis.

The more important network on sheep and goats was created in 1979 and reorganised in 1991 with a clear displayed objective for development: to 'define research objectives, to improve the income of farmers, support sustainable production systems, create new products, promote cooperative research projects, give information to all users and organise symposiums and meetings'.

Discussion and conclusion: present organization, present difficulties and possible changes in favor of innovation

These networks were created as a support for training within a geographical and a political environment and with an institutional international (FAO, CIHEAM, EAAP) or national (INRA, GTZ, IZC, IZCS) formalization but they have been a success above all thanks to a real individual involvement. This seems to be the key point to their success but also contributed to their ambiguity. During several decades, they have produced key documents and have added value thanks to the personal enthusiasm of their leaders, all with a global vision and strong local roots (considered as a key for success in networking by Torre, 1996). They have created very diverse and dynamic communities with strong interpersonal and professional relations.

To be accepted by their institutional support, their activity has always been presented for research *and* development with a strong relation to extension services, a wide spread information, and consultancy for decision making, etc. During their meetings it has often been possible to build punctual collective projects but frequently without achieving long term activities and programs. In other words and due to their structural situation there is a gap between the intentions and the reality and in most cases the networks have worked as conventional scientific associations.

- The symposiums and seminars have been the main chance to meet and to interact collectively, being still easy to organize with local and institutional support.
- Most of the services and permanent activities have worked irregularly, (newsletters, more recently websites) by lack of permanent financial or human resources.
- Very little specific financial support is given by the supporting organizations to develop the program. Providing no financial support they have little control because there is neither real evaluation nor institutional steering, and they run the risk of self-legitimacy.

Aware of such a situation, the sheep and goat networks and EAAP have formally supported the creation of CIRVAL, in 1993, as a Resource Centre dedicated to information and realizing their own objectives but with little involvement of the Institutions or the members of these networks to associate their national or professional organizations in a real multilateral financial support (Dubeuf, 2006).

The recent evolution has also changed the profile of the scientists in animal production, making it more specialized, with more academic constraints and less relations to the professional and extension services. If we observe a generally good participation during the symposiums, the permanent involvement of the younger scientists in exchanges and actions is less constant. Although no specific measuring has been done for these networks, such a reality is coherent with the results of Barbier et al. (2008) showing (for an other network) that only 5% of the authors of the International Farming Systems Association (IFSA) symposiums have produced 36% of the communications (many partipants publish once or twice only without continuity) and wondering if the IFSA network is a real community. In the case of the FAO-CIHEAM network on sheep and goats, the number of members has increased from 192 (Morand-Fehr *et al.*, 1996) to 225 in 2008 but with less permanent active participation.

Historically, in spite of a strong awareness of the realities of the Mediterranean region, these networks had a strong mono-disciplinary bio-technical origin (mainly genetics, nutrition, physiology of reproduction); their vision was a direct consequence of the up and down approach of innovation. But, although this pattern is still deeply in the mind of many, today innovation is more considered at a larger extend. The present vision of innovation is more this of a collective construction to face the realities and complexity of the problems and stakes, with simultaneous environmental, social and economical issues. Although, colloquially, the word 'innovation' is often used as synonymous with the output of the process, an innovation is not only technological and 'scientific' but can be also at an organizational or commercial level or all elements of change. Animal production (as other sectors) will have to face important challenges, in the near future and perhaps more dramatically in the Mediterranean regions (and in arid areas). So it seems logical to consider that more and more innovation will be needed urgently at all levels. How could the networks follow these new challenges and patterns? In spite of several punctual initiatives to introduce new questions and debates related to the socio-economical conditions of animal production and their relations to the application of scientific results, presently the activity is still above all academical and focused on bio-technical results.

According to Callon, (1996) innovation also needs the participation of all the actors to be accepted but also a system of exchanges to enable contradictory discussions and debates. We could logically wonder if these networks could change and develop these exchanges as the basis of their activity. Firstly, new scientific knowledge is not equal to innovation and the personal or professional motivation of the scientists participating in these networks in favor of these orientations is not sufficient. Secondly, such activities are presently not always well acknowledged for the scientific evaluation and management of their careers. Besides, through several examples, it seems that very often the participation of the extension services or professional services is announced by the scientists without their real involvement.

Nevertheless, other types of organizations are needed. Between research and development, Le Masson *et al.* (2006) have proposed to insert another function, innovating conception at the company level. The hypothesis is that specific organizations and tools would be needed to transform knowledge (and especially scientific knowledge) in real applied innovations for the companies (farms or organizations) with the support of development and extension agents. Innovation is no more a question for isolated people; it would be a full collective activity that requires integrating technical, scientific and management patterns. That does not mean that the present networks could substitute these new functions for the present ones simply by voluntary declarations. We have to be aware that developing innovative approaches (and so achieve their declared objectives) is not only dependent on the internal forces and goodwill of the leaders of these networks and that developing bridges with social and human sciences is not easy.

There is still a long way to go to make these views fully accepted particularly in the Mediterranean basin. On the other hand, in case of lack of initiative, the permanence of these networks could be threatened at a short term. These networks could more easily realize science in society if more complex organizations involving both economic actors and politics would be created concretely with a clear political and economical support. The concrete planning of such new organizations has been more advanced when there is a strong economic sector as the dairy industry for instance. In these cases, it would push their activity, give more faith, real sense and enthusiasm for innovative practices in these exchanges.

References

Barbier, M., V. Garandel Batifol and M. Bompart, 2008. A textual Analysis of Scientometric mapping of the dynamic of kowledge in and around the IFSA community. communication presented in the '8th IFSA European Symposium; 6 to 10th July, 2008; Clermont Ferrand.

Couzy, Ch., 2000. L'Organisation et le fonctionnement des Services techniques et de Conseil à l'élevage ovin-caprin en Europe du Sud. CIRVAL, 53 pp.

Dubeuf, J.-P., 2006. 10 ans d'activités au CIRVAL; enseignements de l'expérience et propositions pour le futur; 30 p. (rapport interne INRA non publié).

Le Masson, P., B. Weil and V. Hatchuel, 2006. Les processus d'innovation - conception innovante et croissance des entreprises, 470 p. Hermes Sciences éditeurs, Paris.

Morand-Fehr, P., D. Gabina and R. Rubino, 1996. The FAO-CIHEAM network on sheep and goat activities on small ruminants products; in: Les dossiers du CIRVAL, n°1 Information et innovation pour les filières dispersées'; p. 117. Corte.

Part 4.

Innovations dedicated to enhance environmental positive impacts of livestock activities from Mediterranean natural and human resources

Innovative strategies of organic beef producers in Italy

F. Ansaloni
Dipartimento di Scienze Ambientali, Università di Camerino, Via Fidanza 15, 62024 Matelica (MC), Italy; francesco.ansaloni@unicam.it

Abstract

In Italy in recent years, organic animal rearing research projects have centred largely on technical aspects while increasing market demand for organic food products has changed business and farmers marketing strategies. The purpose of this study based on the results of a number of Italian national and regional research projects is to analyse income from beef, sales distribution channels and breeder strategies. Our research shows that breeding farms offer a considerable number of products, especially livestock and value added products (fresh meat cuts, salamis) sold directly to consumers in addition to providing tourist and recreational services. However because organic beef producers have little negotiation power, traditional livestock sales channels are not economically rewarding. The organic beef supply chain stops at the farm stage and therefore organically reared animals are sold at the same price as traditionally reared animals. This situation has motivated some farmers to add meat processing to their normal rearing activities. However not all farmers are able to successfully handle the hardships that this short supply chain requires. As an alternative, the farmers could organise themselves in associations to maximise their negotiation power on the market or manage the short supply chain by outsourcing carcass processing to local abattoirs.

Keywords: organic beef, supply chain, value added product

Introduction

After the adoption of EC Regulation No. 2092 in 1991, the first European law aimed at protecting organic beef production, Italian research projects centred largely on technical aspects. Prior to this EC regulation, organic beef production was simply governed by associations and voluntary certification. In the same period, the major changes which influenced production were a growing consumer demand for organic food products and variations in technical production costs. Furthermore and in order to improve the competitive stance of the beef supply chain, the behaviour of large commercial food businesses and the strategies of individual small organic farms have also undergone changes. Other issues keenly felt in Italy are the need to improve income and market distribution of organic products.

This paper sets out to:
1. illustrate some of the strategies used by farmers;
2. discuss the income of organic farms;
3. analyse organic beef distribution channels.

The themes discussed are of interest not only to farmers but also to public institutions as the latter are mainly concerned with safeguarding organic farms within the territory. This is indirectly true because and thanks to high environmental production sustainability these institutions protect the environment and natural resources to a greater extent. Furthermore, increased environmentally sustainable farming methods favour local, traditional and quality products. The thrust to process farm products (value added products) and use direct sales (short distribution chain), are ways to satisfy consumer demand and ensure that income remains within the territory where it was generated.

Methodology and results

The innovative strategies used by organic beef farmers are presented in a synthesis of the results of three research projects currently underway: the first project, Equizoobio, is an analysis of the income from various organic farm case histories; the second project, Canalibio, examines organic beef distribution channels; the third project, Recupero bovini fine, illustrates the economic benefits gained by in-farm processing of live animals into fresh meat and other processed products.

Analysis of results

Equizoobio is an interdisciplinary research project with 22 partners that is analyzing the technical aspects of organic farming and cross disciplinary issues such as the environment, animal welfare and economic benefits. The analytical method is used to identify and solve problems involved in the creation and study of organic farm case histories in Italy.

Economic analysis was carried out in 17 of the 20 farms available: five in northern Italy, six in the centre and six in the south. These case histories are extremely varied since the farms are located in different types of terrain (mountain, hill and plain), production lines ranged from one specific animal to several types (milk for human consumption, milk to be processed, beef). The quantity and the quality of farm resources also varied significantly (land, animals, farm family labour, outsourced labour, type of farm). Lastly multi-activity farms are widespread. The variable nature of these characteristics makes it imperative that farmers are willing to work with the experiments set up within the projects (animal feed, genetics, crop rotation, environmental sustainability and animal welfare). From an economic point of view, these farms do not always represent success stories or real farm models distributed through the local territory.

The aim of the economic analysis was to determine net farm income: total compensation for farmer-owned production means actually employed in production. Technical and accounting data were collected during a number visits to the farms where personal interviews were conducted and an identical questionnaire was administered for all types of farms. The methods used to analyze income were also identical. The data made it possible to calculate net farm income and also income per individual family member (Ansaloni *et al.*, 2006; Ansaloni *et al.*, 2007). In particular data was obtained to quantify income-generating factors in production (agricultural raw material, processed products, services, bonuses and benefits), the inputs required to calculate total production costs (land, animals, buildings, machinery, labour, type of farm) and elements to facilitate understanding of the technical aspects of breeding which have a bearing on costs, and finally the strong points and objectives of the farm.

As far as production is concerned the farms can be divided into two main types. The first are farms with relatively high labour availability compared to a limited amount of land and/or number of animals. These farms are small family enterprises, medium sized businesses, or socially oriented cooperatives.

In farms with less than 20 hectares of cultivated land per full time work unit our research noted a large number of animal types and consequently the production of a wide variety of agricultural raw materials and in-farm processed goods (cereals, milk and cheese, meat, etc.), along with other attractions (restaurants, tourism, recreational, cultural, sport, retail). This rather extensive product line is due to the increased labour available with respect to the farm resources, and especially the area of land that is cultivated.

The second group are farms with limited labour resources for a large area of available land and the number of animals raised. The majority of these farms are commercial businesses or cooperatives.

On farms with more than 20 hectares of cultivated land per full-time work unit, the product line has been reduced to a limited number of raw materials in order to simplify farm organisation. Furthermore these farms have few and sometimes only one customer.

The reference year used to calculate net income was 2005, but in some cases income was averaged over a two-year period (2005-2006). It is interesting to note that income was extremely varied. In 15 of the 17 case histories, income was in the black (Ansaloni *et al.*, 2007). The net income per farm is the sum of the compensation for use of farm means available on the farm and for labour by family members involved in farm activity (land, capital, labour, farm risk); income is calculated as the difference between farm income for the sale of products and services and the total costs sustained in production calculated at market cost. Income improvement depends mainly on CAP (Common Agricultural Policy) contributions and the sale of processed organic products. Two farms in the study had low incomes with balance sheets in the red. This was due to extraordinary investments to increase herd size and for construction of buildings to start up bed and breakfast businesses. Income depends primarily on animal production and services. In a large part of the cases studied, crop production was primarily used as animal feed for the farmers' own animals. The differences in the results are due to the qualitative and quantitative resources available to the farms.

Analysis of organic beef sales distribution channels

Analysis of the organic beef distribution channels was conducted during the two-year period 2004-2005 in the Italian region of the Marches. At that time, in Italy, the total number of organic farms was 49,859 of which 89.7% were agricultural producers. The number of heads of cattle raised for milk and for meat was 222,516 (SINAB, 2005). The region of the Marches has long been among the areas that use development policy to favour the growth of organic farms. In 2005 in the Marches, the number of organic farms was 2,651 (Pyszny *et al.,* 2007), 314 farms reared all types of animals and 117 farms farms raised cattle (Assam). According to official Italian statistics 'ISTAT 2001' there are 66,283 farms in the Marches; of which 3.99% over the entire region are dedicated to organic farming.

Data concerning the distribution channel were collected in 2007 directly from cattle rearing farms with at least five hectares of arable farm land. This criterion made it possible to exclude small non-market oriented farms. Seventy-five out of a total of 109 farms were sampled (Ansaloni *et al.*, 2008a). Data was compiled in direct interviews with the farmers on farm premises using a semi-structured questionnaire as a guideline. This made it possible to define farm resources, income level, processing activities and product distribution channels.

Results showed that farmers have limited negotiation powers. Forty-four percent of farmers sell calves to other farms where they are fattened, 33.3% of the farms sell baby beef cattle ready for slaughter, 12.0% sell fresh meat directly to the consumer, 5.3% sell fattened lamb and 5.3% sell breeding calves to other farmers.

There are two distribution channels for live beef cattle. In the first channel, five-month old calves weighing between 200 and 250 kg are sold to other farms or intermediaries for fattening. The price for these animals is the same as that for animals reared conventionally. The second distribution channel is for baby beef cattle ready for slaughter weighing 500-600 kg. Again these animals are sold at the same price as conventionally reared animals. In the Marches the organic beef chain stops at the production stage.

Sixty-five percent of the farmers sampled reported difficulties in selling organically reared live cattle. This was primarily due to the unwillingness of consumers to pay higher prices for organic meat and to price instability.

The major innovation was the direct sale of fresh meat. In this case the farmer pays a service charge per live head to the local abattoir. After slaughter the farmer collects the carcass and processes it at in-farm health-certified facilities. The farmer sells the fresh meat at his farm shop or personally delivers ten kg packages of varied cuts directly to customers or in some cases, sells beef quarters to families. In general and because it is less complicated and sales are more regular, the farmer prefers to sell fresh meat to school kitchens, restaurants and local agricultural Bed and Breakfasts. Furthermore in some cases direct sales mean that the farmer-processor can increase the sale price of fresh meat 10-15% over the price of the conventional product.

One of the problems noted is that the short beef supply chain is extremely fragile mainly due to two problems. Firstly the farmer has to assume many roles (processor, packager, distributor and sales promoter, etc.) and frequently cover a long distance between the city market and the farm. In 23% of the cases sampled, farmers encountered difficulties in finding production resources (labour, organic fattening stock and other organic raw materials). The technical efficiency of organic farming was not completely satisfactory due to the need for grazing, health factors, animal welfare and the neonatal death rate. Lastly in some farms and in contradiction with organic farming methods, animals are still kept tied up in barns as this practice is currently permitted as an exception to the law.

Some areas, especially those producing animal feed, are ideal for creating favourable conditions for organic animal rearing. The self sufficiency of the animal feed farm eliminates, or greatly reduces, dependence on third party sources for animal feed. In small mountain or hill country farms there is no substantial difference between organic and conventional farming due to the use of natural fertilizer (manure) rather than market purchased technical means, limited financial resources and modest yields. It becomes clear that organic animal rearing is more economically rewarding when applied over large areas as it reduces production costs.

In-farm live beef cattle processing: fresh meat and other products

To increase their income, some farmers are experimenting with the short supply chain in particular for milk (Coldiretti, 2008), fresh meat and the corresponding processed products (Prober, 2008; Pyo, 2008). The sale of fresh meat (value-added product) allows the farmer engaged in this activity to ask for a higher price than the normal farm price. Processing demands complex farm organization with more labour since processing is added to the existing rearing activities. The short supply chain is therefore not feasible for all farmers.

The short supply chain involves a decision by the farmer to diversify products and attempt to move into a niche market where hopefully sales prices will be significantly higher than ordinary farm market prices. Direct sales by eliminating wholesalers and other intermediaries means abandoning traditional distribution channels. The most common direct sales method is the farm shop, followed by the town market, agricultural Bed and Breakfasts, purchasing groups and the participation in and or supply of processed products to farm fairs and local festivals. The advantages for the farmer are instant encashment and steady cash flow making it possible for the farmer to depend less on bank loans and enjoy greater profit margins.

The third project, an interdisciplinary research project, 'Reuse of cattle at the end of working life' focused on the cost effectiveness of in-farm live cattle processing of fresh meat and other processed products such as fresh meat from end of working-life milking cows, pigs, lamb, chickens and rabbits, mince, hamburger patties, mixed pork and beef sausage meat, mixed pork and beef salami and bresaola along with other farm shop sales (Ansaloni *et al.*, 2008b).

Because of difficulties encountered in collecting accounting and economic data from the farmers, the research was limited to two farm families which raised Marchigiana beef cattle and other breeds organically. The reference year for this project was 2007. The average number of animals slaughtered in one year is 27 baby beef, six end of working life cows, nine light pigs, 19 heavy pigs and 65 lambs. Since 1995, these two farmers have adopted meat processing, innovative sales systems such as 'door to door' and 'family packs'. If the sales price of the processed product is known, it is possible to calculate the cost effectiveness of live animal in-farm processing compared to simply selling live animals. The analysis that needs to be done is to calculate the cost of in-farm processing. (Bernardini, 1960; Cianferoni, 1978; De Benedictis *et al.*, 1979; Giacinti, 1993; Zucchi, 2006).

The first factor that needs to be calculated is the total farm processing costs. This is the difference between processing costs and the net value of the product obtained (e.g. cuts of fresh meat, total number of kg rendered) multiplied by the sales prices. Processing costs are strictly limited to butchering and do not include herd value for cattle to be slaughtered and processed. Some costs directly related to processing were allowed in the calculation of processing costs. These costs are calculated separately from the rearing costs of the farm that produced the animal.

The second factor to be calculated is the processing price per kg of meat. This is calculated by dividing the total value of processing by the total number of kg of fresh meat ready for the consumer. If the processing price is greater than the farm market price of the live animal per head expressed in kg of fresh meat, processing would be cost effective, otherwise it would not. In our particular case the market price of baby beef weighing 760 kg live weight expressed as fresh meat was equal to net farm encashment of €2,731.00 with a yield of 319.20 kg or €7.43 per kg.

The first farm studied was located 607 metres above sea level in the Marches Apennines. The farm is not near any urban centre and is it frequented by price sensitive local customers who are not particularly interested in buying organic products. In contrast, the second farm is located closer to urban centres and to the Marches coast. Customers here are more interested in organic products and willing to pay higher prices for them. To simplify matters, the two farms are referred to as rural for the first farm and urban for the second.

Our research showed that prices for conventional or non-organic products sold in retail butcher shops in the rural area are equal to or less than those sold by the rural farm. When the rural farm was compared to the urban farm, results showed that the rural farm practices lower prices than the urban farm for all products except bresaola.

The economic results were positive. All processed products with the exception of one product per case history, have been profitable. Covering costs is an interesting item. In both cases, labour costs even when supplied by family members and costs for technical processing equipment are the same as market costs. Often in family-run farms the amount of labour available is excessive with respect to the needs of the farmer-processor. In order to avoid not using this labour resource the farmer-processor is willing to accept lower than market compensation and ignore fixed costs (management costs, quotas, interest, use of health certified facilities).

The rural farm recorded a slight loss for fresh baby beef as its sales price was lower than that of the urban farm. Accepting a small underpayment for some factors realigns profits. In addition, fixed costs have a greater influence in the urban case since the weight of the animals and the amount of the meat sold are lower. Furthermore these farms have few customers interested in purchasing live animals or willing to pay the price of organic products. Some of the strengths of fresh meat are quality, rearing method and long aging period. The weaknesses include short shelf life and the reduced number of customers (school kitchens). The rural farm did not process meat from end of working life animals since its customers did not appreciate this type of meat. The urban farm on the other hand made a profit; its customers were mainly local butchers.

There was no demand for the sale of organic live lambs on the market at either farm. The farm price was the same as the conventional product market price. As far as chickens and rabbits are concerned, only the urban farm raised these animals. The profit from chickens was positive while that from rabbits was negative. The latter was the only negative result for this farm. Modest quantities of lamb, chicken and rabbit respond to the need to offer clientele cuts of meat other than beef. In both cases the labour hours for in-farm processing largely supplied by family members were quite high.

The growth of organic farms depends on specific facts other than resources common to conventional farms and CAP contributions. These specific factors are strong demand for food products and the location of the farms in the area.

Product quality as perceived by consumers is one of the main factors that explain strong market demand. Consumers perceive, recognise and prefer organic method certification. Furthermore the consumer recognises the bond with the area (typical products and DOP production regulations) in addition to a trustworthy supplier. The market areas adjacent to urban centres or located close to tourist areas, panoramic sites, historical areas, archaeological sites and natural reserves (natural parks and protected areas) help draw more customers. Consumers equate the territory with quality. They also identify with a number of environmental and ethical values. Direct sales of local food products satisfy demand for low mileage points of sale and help reduce the consumer's carbon footprint. In the short food chain or on the stalls in the local farmers' market, and in agreements with GAS (collective purchasing groups), the sales price is variable and depends on product diversification and provenance.

Conclusions

Results showed that the organic rearing method and in-farm processing are economically sustainable. In a large proportion of the case studies in Italy, however, the organic beef supply chain stops at the production stage.

In Italy in future, if supply chains are not developed for the sale of organic meat products with consumers willing to pay a higher price than for conventional products and taking into consideration the reduction in direct support through CAP payments, we can only expect a return to conventional farming methods by some farmers presently engaged in organic farming.

In-farm processing can improve income and offer opportunities to develop the short supply chain. However, the complexity of this production method along with direct sales cannot be applied to all farmers. Furthermore a good understanding of both meat processing techniques and selling techniques is necessary. One alternative to individual in-farm processing would be to set up inter-provincial associations of farmers capable of insuring a constant supply of live animals. Carcass processing and preparation of processed products could be outsourced to local abattoirs which in turn would be paid for their services. Frequently the small number of animals reared on the farms is insufficient to cover the fixed costs of an in-farm processing plant. Products that need aging like bresaola and salami could be returned to the farm for the aging process. Fresh meat could be stored before sale in freezer warehouses belonging to associations. The problem with this solution is overcoming the individual farmer's diffidence and the need to promote direct sales to urban consumers. Prober (2008) provides an interesting example in the province of Bologna where beef and other food products are sold online and orders from customers are taken from the web site (http://www.mangiocarnebio.it/). In order to reward positive effects from organic rearing and to support demand for organic meat, public institutions could help set up farmers associations by offering consultations, providing tax advantages and encouraging promotional campaigns to inform consumers about product quality and the complex nature of organic farming.

Acknowledgements

We would like to thank the farmers who participated in these projects and also the region of the Marches for its financial support for the three research projects: Equizoobio (PI III Fase, Sviluppo Rurale, Zootecnia Biologica, L. 499/99, 2006-2008), Canalibio (ASSAM, PI III Fase, Sviluppo Rurale, Zootecnia Biologica, L. 499/99, 2006-2007) and Recupero bovini fine carriera (BRS, L.R. 37/99, DGR 1234/05, Regione Marche 2007-2009).

References

Ansaloni, F., M. Chiorri, A. Menghi, F. Pyszny, G. Schifani and R. Zanoli, 2006. Analisi economica di imprese zootecniche biologiche in italia: primi risultati del progetto di ricerca interdisciplinare E.Qu.I.Zoo.Bio Convegno SIDEA, Assisi, Poster.

Ansaloni, F., M. Chiorri, F. Galioto, G. Guccione, A. Menghi, F. Pyszny and G. Schifani, 2007. Strategie imprenditoriali delle aziende zootecniche biologiche 3° Workshop GRAB-IT, Sostenibilità e qualità delle produzioni agricole biologiche, Roma, in corso di stampa.

Ansaloni, F., F. Pyszny and U. Testa, 2008a. Market Relationship Organic Beef Cattle Breeders in the Region of The Marches (Italy), 16th IFOAM Organic World Congress, Modena, Italy.

Ansaloni, F. and F. Pyszny, 2008b. Il reddito delle aziende da carne aumenta con la filiera corta, L'Informatore Agrario 33: 25-28.

Assam, 2005. Data Bank, Agency for the services of the agricultural and the food industry sectors of the Marches Region Marche, Ancona, Italy.

Bernardini, C., 1960. Il valore di trasformazione in Esercitazioni di economia agraria, Istituto di Economia e Politica Agraria, Officine Grafiche Calderini, pagg. 190-191, Bologna.

Cianferoni, R., 1978. Guida allo studio dell'Economia e politica agraria, Cedam, Padova. Coldiretti 2008 Elenco della localizzazione dei distributori automatici di latte fresco, http://www.coldiretti.it/.

De Benedictis, M. and V. Cosentino, 1979. I metodi di pianificazione parziale in Economia dell'azienda agraria, pagg. 601-605, Il Mulino.

Giacinti, R., 1993. L'amministrazione economica dell'azienda agraria, pagg. 155-156.

Guthrie, J., A. Guthrie, R. Lawson R. and A. Cameron, 2006. Farmers' markets: the small business counter-revolution in food production and retailing British Food Journal Vol. 108 7:560-573.

Hutchinson, K., 2008. Opportunities for Organic Farmers in Local Food 16th FOAM Organic World Congress, Modena.

Istat, 2001. Censimento generale dell'agricoltura 2000, Presentazione dei dati definitivi Marche, http://censagr.istat.it/dati.htm.

Morel, B., R. Le Guen, A. Lambert, N. Schieb-Bienfait, G. De Fontguyon and R. Euzen, 2003. Diagnostic et perspective de developpement de la filiere viande bovine biologique des pays de la Loire, Rapport n.2003/08, Programme INRA_DAPD Arpent de la Loire.

Povellato, A., 2006. La zootecnia biologica bovina e suina in Italia: tecniche e mercato, (editor), ESI, INEA.

Prober, 2008. Vendita diretta on line di carne fresca e prodotti trasformati http://www.prober.it/mangiocarnebio/.

Pyo, 2008. Pick Your Own, Raccolta diretta dei consumatori presso le aziende agricole, http://www.pickyourown.org/.

Pyszny, F., F. Ansaloni, F. Cecconi, U. Testa and P. Bonanni, 2007. I bio allevamenti marchigiani continuano a crescere AZBio, 12: 49-53.

Sans, P., G. De Fontguyon, B. Sylvander, A. Le Floc'h, M. Auersalmi and O. Schmid, 2004. Is it easy for producers to market organic beef meat? The case of Biobourgogne Viande (France), SAFO, Florence (Italy), 5-7 September 2003, pp. 51-65.

Sinab, 2006. L'agricoltura biologica in cifre al 31/12/2005, http://www.sinab.it.

Veysset, P., S. Ingrand and M. Limon, 2008. Direct marketing of beef in organic suckler cattle farms: economic results and impact on breeding system management, 16th FOAM Organic World Congress, Modena.

Wynen, E., 2006. Organic beef production and marketing in Australia, Journal of Organic Systems, 1.

Zucchi, G., 2006. Zooeconomia, Economia del sistema delle produzioni animali, pp. 258-259, Ed. Avenue Media.

Use of homeopathy in parasite control in a flock of Zerasca sheep

M.N. Benvenuti[1], F. Pisseri[2], J. Goracci[1], L. Giuliotti[1], F. Macchioni[4], P. Verità[1] and G. Guidi[3]
[1]Dipartimento Produzioni Animali, Viale delle Piagge 2, 56124 Pisa, Italy
[2]Medico veterinario omeopata - Scuola CIMI-Koinè, Via Ugo Bassi 20, 00153, Roma, Italy
[3]Dipartimento Clinica Veterinaria, Via Livornese 1097, 56122 S. Piero a Grado, Pisa, Italy
[4]Dipartimento Patologia Animale, Profilassi e Igiene degli Alimenti, Viale delle Piagge 2, 56124, Pisa, Italy

Abstract

The control of gastrointestinal parasites in livestock relies almost exclusively on multiple and regular dosing with anthelmintics. This leads to problems linked with residues in food products and in the environment and has encouraged researchers to search for alternatives or new helminth control methods (Ketzis *et al.*, 2006). Among these medthods, homeopathy could play a strategic role in solving health problems; however, at present, there is a lack of scientific results with validated techniques used on a large scale. The aim of this study was thus to investigate the feasibility of homeopathy in the control of the parasite burden in sheep. The study lasted 12 months and involved 30 Zerasca ewes divided into two groups. Sabadilla MK was administered twice. A faecal egg count (FEC) was performed of gastrointestinal strongyles and blood parameters were characterized. Statistical analysis was by ANOVA and relations between FEC and hematocrit (HCT) were estimated by Pearson's correlations. FEC was influenced by the sampling date, confirming seasonal fluctuation. On the whole, homeopathy did not influence the parasite burden, but a significant decrease ($P \leq 0.05$) in egg output was observed at the second and at the last sampling following the administration of the remedy. Data concerning blood parameters did not show appreciable differences between the two groups. FEC and HCT showed an inverse correlation even if mean values were within the normal range. The study revealed the advantages of monitoring parasites and the interest of homeopathy in the control of gastrointestinal nematodes.

Keywords: sheep, homeopathy, gastrointestinal strongyles, blood parameters

Introduction

Small-ruminant husbandry is becoming increasingly important since it plays a fundamental role in preventing the decline of rural land and populations (Ronchi and Nardone, 2003). However, ovine livestock systems must ensure an acceptable level of sustainability with regard to animal health, environmental impact, and the quality of products, and profits.

The control of parasitic diseases is one of the most important goals in animal health and welfare as far as environment safeguards are concerned. Unfortunately, parasitic infections are too often treated with chemical products, resulting in anthelmintic resistance, ecotoxicological effects and deleterious effects on the development of natural immunity against helminths (Ketzis *et al.*, 2006). The main cause of drug resistance is overuse, misuse or incorrect application of the medication (Waller, 2006). These facts, linked with the problems of drug residues in food derivates, underline the need for thorough parasite monitoring and for integrated parasite control. In this context, homeopathy could play a strategic role in overcoming health problems; however, at present there is a lack of scientific results indicating validated techniques that could be used on a large scale.

All these considerations have become increasingly important for local sheep farms, where the environment and the quality of production are closely linked. The availability of sustainable control of gastrointestinal parasites along with limited use of conventional therapy could prove to be of great value to farmers and beneficial to the environment (Ambrosi, 1995; Urquhart, 1999).

The aim of this study was to investigate the feasibility of homeopathy in the control of the parasite burden in a flock of Zerasca sheep.

Material and methods

The study was carried out on Zerasca sheep, an Italian local sheep breed, whose farming system enables the sustainability of land resources in a mountainous area in northern Tuscany. The farm studied was located at an altitude of 800 m a.s.l. with a climate midway between Mediterranean and continental conditions. Two groups of 15 animals were randomly selected in the farm; one control group (C) with no treatment, the other (H) was treated with a homeopathic remedy.

A homeopathic examination of both the farm and the animals was performed at the beginning of the study to identify the appropriate remedy: all relevant information on the behaviour of the flock, management practices, animal-human interactions and pathological farm history were collected in order to fill in a homeopathic repertory with the appropriate remedies applying the 'similar principle' of Hahnemann, (1920). Homeopathic repertory consists in a database with the results of treatments that are used as homeopathic evidence. The obtained record includes the use of homeopathic remedies as a function of the patient's symptoms according to the method of complexity (Mac Repertory 5.5, 2002). From the repertory the vet chooses the remedy that shows the most similarity with the symptoms (Mangialavori and Marotta, 2004).

The profiles of the animals was synthetically characterized using the following features: calm behaviour in the presence of the farmer and outsiders, unease towards wolves, tight bond with humans, symptoms related to gastrointestinal parasites, heat delay, chronic cough, skin abscesses in summer; occasional mastitis events, thinness.

Sabadilla MK was prescribed in three administrations. As a consequence of problems linked to the farm organisation, the remedy was in fact only administered twice, one after the first sampling and the second before final sampling. Farm health inspections were repeated at regular intervals. The study lasted 12 months during which faecal and blood samples were collected twice monthly. Faeces were taken directly from the rectal ampoule and individually examined for the faecal egg count (FEC) of gastrointestinal nematodes using a modified McMaster technique (Permin and Hansen, 1998). Blood samples were drawn from the jugular vein and blood parameters were estimated by cell counter (HeCo SEAC) to assess possible damage due to parasites. Data on the growth traits of the offspring were recorded throughout the year to assess animal productive performances.

Statistical analysis was performed by ANOVA. The factors included in the model were date of sampling, treatment, date nested in the treatment and physiological phase. Data referring to FECs were logarithmically transformed [$y = \log(EPG + 25)$] to normalize error (Baker *et al.*, 1997). Correlations between FEC and PCV were estimated by Pearson's correlations.Statistical analyses were performed with JMP software (JMP, 2002). Results and discussion

Mean FEC was 360 (±355), which is very similar to that reported in a study previously conducted on Zerasca sheep or on other breeds (Benvenuti *et al.*, 2005), even if in the present study no chemical treatment was performed.

Data showed significant differences for date of sampling, treatment nested in each sampling and physiological phase. No significance was evidenced for the 'treatment' factor. The season of sampling significantly influenced the parasite burden (Table 1), in agreement with our previous observations on Zerasca sheep (Benvenuti *et al.*, 2006) and Massese (Benvenuti *et al.*, 2004). The highest parasite level was observed in spring (March) which is the period with the highest risk of infestation (known as the 'spring rise') (Kahn *et al.*, 2003; Ambrosi, 1995; Brunsdon, 1970).

Table 1. Parasite dynamics over the year.

	Mean	S.E.
April '06	539^a	99.69
July '06	219^b	83.16
September '06	200^b	103.47
November '06	160^b	118.86
January '07	230^b	85.89
March '07	656^a	77.72

^{a,b} Data with different superscripts are significantly different (P<0.05).

Although there was no statistical evidence of the efficacy of the treatment in terms of a general reduction in the FEC, there was noticeable decrease in EPG output after the first administration of the remedy (Table 2). Focusing on the trend in the parasite burden, we observed that the FEC reached values above the threshold of 600 EPG only in the control group, with possible zootechnical and animal health risks. At the last sampling which corresponded to the second homeopathic treatment, the 'H' group had a lower FEC.

The mean FEC of Zerasca ewes in the peri-parturient phase was significantly higher (429 S.D. 455) than in non peri-parturient ewes (285 S.D. 326).

These observations regarding FEC during the peri-parturition period are in agreement with those of other authors who reported a significant rise in faecal egg output around lambing time and throughout the early lactation period (Benvenuti *et al.*, 2004a; Ng'ang'a *et al.*, 2004; Ambrosi, 1995). During this period, immunity declined in adult ewes not only as a result of climatic conditions, but also of lambing and stress-related problems (Abbott *et al.*, 2004).

Data on blood parameters did not reveal any differences between the two experimental groups as shown in Table 3. All values were within the normal range of values and there were no major fluctuations.

FEC and PCV were inversely correlated (r= -0.262; $P \leq 0.001$) and this may underline the parasite effect in altering haematocrit values; however PCV levels were within the normal range in 95% of the ovine population tested, and revealed no damage caused by the parasite.

Table 2. Parasite dynamics in the experimental groups.

Date of sampling	FEC			
	Controls		'H'	
	Mean	S.E.	Mean	S.E.
April '06	463	147.84	605	182.12
June '06	306	112.52	198	66.89
September '06	270	161.14	240	83.03
November '06	165	70.28	340	131.15
January '07	204	37.45	320	88.16
March '07	716	185.08	516	140.18

Table 3. Blood parameter values.

Parameters	Controls		'H'		Range of values
	Mean	S.E.	Mean	S.E.	
WBC (K/µL)	9.44	0.34	9.18	0.34	5.2-13.9
RBC (M/µL)	8.43	0.14	8.61	0.13	6.9-10.9
HGB (g/dL)	9.19	0.20	9.26	0.14	7.4-11.8
HCT (%)	28.2	0.46	28.06	0.43	22.0-34.2
MCV (fL)	33.51	0.26	32.78	0.30	28.4-37.0
MCH (pg)	10.78	0.25	10.81	0.11	9.4-12.4
MCHC (%)	33.27	0.18	32.87	0.17	30.4-36.3
RDW (%)	16.01	0.06	16.01	0.08	15.1-17.2
PLT (K/µL)	146.2	9.75	159.0	9.99	107.6-324.8

The weight of lambs (Verità *et al.*, 2006) at slaughter (25.5 kg) was satisfactory and was reached at age of 71.5 days (Table 4).

Conclusion

There was no evidence that homeopathy reduced the parasite burden in the long term, but it was able to keep FEC under the threshold of health and zootechnical damage. The fundamental principles of homeopathy are its ability to balance the relationship between the animal and parasites and the environment, whereas anthelmintic efficacy is based on the maximum number of parasites eliminated from the host. It is also important to consider that in our experimental period, no animal was treated with the chemical drugs that are normally administered twice or more a year, proving that an effective monitoring schedule can avoid any unnecessary anthelmintic treatment when there is an acceptable balance between the host, the parasites and the environment.

The good performances reached by the animals and the normal level of haematic parameters revealed the breed's resistance characteristics in terms of tolerance and resilience.Homeopathy needs new evaluation approaches because its aim is not the eradication of parasites but rather to maintain the level of parasites below the threshold of economic and health problems. Homeopathy tries to enhance animal defences and minimize the damage encurred by the hosts. The homeopathic approach does not treat diseases directly but makes the individual itself able to fight adverse conditions (Braghieri *et al.*, 2006). As a consequence, this method of study should take into consideration all information about metabolic, immunological and stress related parameters.

Table 4. Performance traits of all offspring.

Parameter	Male		Female	
	Mean	S.D.	Mean	S.D.
Weight at slaughter (kg)	26	3.52	25	4.23
Age at slaughter (d)	67	10.06	76	12.54

The use of homeopathy in animal health programmes could help achieve the following goals:

- no chemical residues in animal derivates;
- no environmental impact;
- avoidance of problems concerning drug resistance;
- increasing animal welfare;
- reducing the cost of treatments;
- correspondence to organic farming requirements.

This study represent a contribution to the scientific application and evaluation of homeopathy in agriculture which is lacking today. The interest shown by the farmers in the Zeri area in the development of this type of study points to the feasibility of participatory research programmes.

References

Ambrosi, M., 1995. Parassitologia zootecnica. Ed agricole, Bologna, Italia.

Baker, R.L., 1997. Resistance genetique des petits ruminants aux helminthes en Afrique. INRA Prod. Anim. 10: 99-110.

Benvenuti, N., L. Giuliotti, J. Goracci and P. Verità, 2005. Study of gastrointestinal parasite dynamics in Zerasca sheep aimed at reducing anthelmintic treatment. Int. Symp. on comparative advantages for typical animal products from the Mediterranean areas, Vale de Santarem – Portogallo, vol. 119: 283-287.

Benvenuti, N., L. Giuliotti, J. Goracci and D. Cianci, 2004a. Le strongilosi gastrointestinali e la produzione del latte in ovini di razza Massese allevati con metodologie biologiche. Ob. e Doc. Vet. 9: 23-28.

Benvenuti, M.N., F. Pisseri, D. Cianci and S. Perrucci, 2004b. The dynamics, prevalence and impact of strongyle (nematode) infections in a farm of massese sheep and evaluation of the efficacy of a homeopatic treatment. Atti XVI Congr. SIPAOC, Siena (Italy), 29/9-2/10: 131-132.

Brughieri A., C. Pacelli, M. Verdone, A. Girolami and F. Napolitano, 2007. Effect of grazing and homeopathy on milk production and immunity of Merino derived ewes. Small Ruminant Res. 69: 95-102.

Brunsdon, R.V., 1970. The spring rise phenomenon: seasonal changes in the worm burdens of breeding ewes and the availability of pasture infection. New Zealand Vet. J. 18: 47-54.

Hahnemann, S., 1920. Organon dell'arte del guarire. VI Edizione, Ed. Red.

JMP, 2002. J.M.P. User's Guide ver. 5.0, S.A.S. Institute Inc., Ed. Cary (NC), USA.

Kahn, L. P., M.R. Knox, G.D Gray, J.M. Lee and S.W. Walkened-Brown, 2003. Enhancing immunity to nematode parasites in single bearing Merino ewes through nutrition and genetic selection. Vet. Parasitology. 112, 3: 211-225.

Ketzis, J.K., J. Vercruysse, B.E. Stromberg, S. Athanasiadou and J.G.M. Houdijk, 2006. Evaluation of efficacy expectations for novel and non-chemical helminth control strategies in ruminants. Vet. Par. 139: 321-335.

MacRepertory 5.5, 2002. Kent Homeopathic Associates Inc., San Rafael, CA, USA.

Mangialavori, M. and G. Marotta., 2004. Praxis. Un nuovo metodo in Medicina Omeopatica: la ricerca della coerenza nella manifestazione dei fenomeni clinici. 1. Matrix Editrice.

Ng'ang'a, C.J., W.K. Munyua, N. Maingi and P.W. Kanyari, 2004. Occurrence of peri-parturient rise in trichostrongylid nematode egg output in Dorper ewes in a semi-arid area of Kajiado District of Kenya. Acta Trop. 2004 Nov-Dec. 92(3): 213-218.

Permin A. and J. Hansen, 1998. Epidemiology, diagnosis and control of poultry parasites. FAO Animal Health Manual.

Ronchi B. and A. Nardone, 2003. Contribution of organic farming to increase sustainability of Mediterranean small ruminants livestock systems. Liv. Prod. Sci. 80: 17-31.

Urquhart, G.M., 1999. Parassitologia Veterinaria. Ed. U.T.E.T., Torino (Italy), 370 pp.

Waller, P.J., 2006. From discovery to development: current industry perspectives for the development of novel methods of helminth control in livestock. Vet Paras. 139: 1-14.

Verita, P., M.N. Benvenuti, J. Goracci and L. Giuliotti, 2006. Zerasca. In 'Risorse genetiche animali autoctone della Toscana'. Ed. A.R.S.I.A., Firenze, 175 -182.

Innovating on welfare: an added value?

A. Velarde, A. Dalmau and X. Manteca
IRTA, Finca Camps i Armet s/n, Monells, 17121 Girona, Spain

Abstract

The concern for animal welfare is based on the belief that animals can suffer, and has become a major element in consumer perception of food quality. In Europe, legislation has been the commonest way of protecting the welfare of farm animals, ensuring that animals do not endure avoidable pain, distress or suffering and obliging the owner or any others who deal with animals to respect minimum welfare requirements. However, growing consumer demand for more ethical food production has meant that the welfare of farm animals is emerging as an area of potential added value for producers, retailers and other actors in the food chain. The Welfare Quality® project (www.welfarequality.net) aims to develop scientifically based tools to measure animal welfare, to convert these measures into accessible and understandable information, and to improve animal welfare through species-specific strategies. The project is based on the assessment of 12 welfare principles grouped under four main criteria (good feeding, good housing, good health and appropriate behaviour) according to how they are experienced by animals. The data obtained from the measures are combined in an overall assessment. At the end of the process, farms can be categorised according to the assessed level of welfare of their animals. The welfare assessment system could help stakeholders to certify farms and specify their markets, e.g. enhanced farms for a general quality label or excellent farms for a niche market. It can also provide farmers with a broad picture of the welfare status of their animals as well as identifying aspects requiring their attention.

Keywords: animal welfare, assessment, livestock

Definition of animal welfare

Definitions of animal welfare can be grouped into three main approaches: a 'feeling-based' approach, a 'functioning-based approach' and a third set of approaches in which welfare is measured by assessing whether the animal can live according to its inherent 'nature' (Duncan and Fraser, 1997).

According to the 'feeling-based' approach, animal welfare involves the subjective feelings of animals, so that welfare will be reduced by negative subjective states such as pain and fear, and will be improved by positive states. The task for science is therefore to study the subjective experiences of animals (Duncan and Fraser, 1997). However, the main problem with this approach is that subjective experiences cannot be measured directly. This raises the long-standing question of whether trying to study them falls within the realm of science. A thorough discussion of the arguments involved in this debate is beyond the scope of this paper and the reader is referred to Dawkins (1990) and references therein for further information.

Many scientists agree that although an animal's experiences of suffering are the defining traits of its welfare, such experiences are difficult to assess, whereas measures based on biological functioning and on the animal's ability to cope with the environment provide relevant information. Indeed, one of the most frequently cited definitions of animal welfare is that given by Broom (1986) according to which the welfare of an individual is its state as regards its attempts to cope with its environment. This definition refers to both how much has to be done in order to cope with the environment and the extent to which coping attempts are succeeding. Impaired life expectancy and reduced ability to grow or reproduce are examples of indications of failure to cope. Attempts to cope include emergence physiological responses and a variety of behavioural changes. All these measures can be integrated in an assessment of welfare that is objective and independent of moral considerations (Broom and Johnson, 1993).

It is important to emphasize that, although the different approaches to animal welfare do not always lead to the same conclusions, they very often do. In particular, research indicates that the feeling-based and functioning-based interpretations often correspond (Duncan and Fraser, 1997). Indeed, pleasant and unpleasant feelings – including suffering – are part of the experience of an animal when it attempts to cope with its environment (Broom and Johnson, 1993). According to Dawkins (1990), for example, suffering occurs when the animal is unable to perform those behaviours that would allow it to cope with its environment.

Concern for animal welfare

The concern for animal welfare is based upon the belief that animals can suffer, and has become an increasingly important issue due to the fact that citizens in many parts of the world demand that animals are reared, transported, and slaughtered in a humane way (Appleby and Hughes, 1997). Nowadays, consumer perception of food quality is not only determined by its overall nature and safety but also by the welfare status of the animals from which it was produced. For example, a very recent survey in Europe showed that more than 63% of the 29,152 people interviewed in the 25 member states of the European Union and four candidate countries showed some willingness to change their usual place of shopping in order to be able to purchase more 'animal welfare friendly products' (Eurobarometer, 2007). This concern over the welfare of animals is by no means restricted to Europe. For example, animal welfare was identified as a priority in the 2001-2005 Strategic Plan of the World Organisation for Animal Health (OIE), an international organization with more than 160 member countries throughout the World (OIE, 2006). Also, the International Financing Corporation (IFC, World Bank Group) has recognised that 'animal welfare is gaining increased recognition as an important element of commercial livestock operations around the world'. According to the IFC, high standards of animal welfare are important to enhance business efficiency and profitability, to meet consumer expectations and to satisfy domestic and international markets (IFC, 2006).

Retailers and producers are increasingly recognising animal welfare as a fundamental aspect of product image and quality (Kjaernes *et al.*, 2007). Producers consider looking after their animals and ensuring their wellbeing an essential aspect of their job. They insist on the need to satisfy the physiological needs of the animals, and their health but also general conditions (comfort, the atmosphere in the farm buildings) and psychological aspects (absence of stress, good relations between humans and the animals). Depending upon the type of production, they additionally insist on freedom of movement, access to fields and outside grazing areas. They also wish for more recognition of their professional knowledge of and engagement in animal welfare and often feel blamed and stigmatized. Animal welfare issues have been considered by European retailers as a commercial opportunity and are 'sold' in a variety of forms. A dominant strategy in some countries is to include the animal welfare standard as part of the branding strategy.

European legislation

In Europe, legislation has been the commonest way of protecting the welfare of farm animals, ensuring that animals do not endure avoidable pain, distress or suffering, and obliging the owner or any others dealing with animals to respect minimum welfare requirements. Legislation at community level is primarily founded on the EC Treaty's Protocol on Protection and Welfare of Animals that recognises animals as 'sentient beings' and requires that full regard is paid to their welfare requirements in formulating and implementing the community's policies.

In 2006, the European Commission adopted an action plan on the Protection and Welfare of Animals that, through five main areas of action, sets out the path to take in order to ensure continuing improvement in animal welfare in the coming years:

- upgrading minimum standards for animal welfare;
- promoting research and alternative approaches to animal testing;
- introducing standardised animal welfare indicators;
- better informing animal handlers and the general public on animal welfare issues;
- supporting international initiatives for the protection of animals.

That means that, in the context of the EU food chain policy, the further development of the existing legislation will continue to be a cornerstone of the policy and any initiative will be based on the best available scientific advice, taking into account public expectations, socio-economic consequences and trade concerns. However, one of the key messages of the Action Plan is that this approach should be complemented by initiatives to improve consumer information and awareness. With their purchasing behaviour, informed consumers can support the development of sustainable farming practices. The Action Plan suggests introducing improved marketing, labelling and communication strategies in order to meet this goal.

Welfare assessment

Retailers and producers are increasingly recognising animal welfare as a fundamental aspect of product image and quality which creates a need for reliable systems for monitoring animal welfare status and providing guarantees on appropriate production conditions. During the last decade, numerous trade groups (producers, processors, retailers and restaurant chains) have developed certification systems with their suppliers. Many of these inspection schemes include a section on animal welfare. However, there is no common standard for assessing animal welfare and for providing consumers with the relevant information. Nowadays, consumers are unclear about what different assurance schemes and labels actually say about the quality of the animals' lives. There is a clear need for the harmonised, comprehensive and reliable welfare assessment and product information systems.

The Welfare Quality® project (www.welfarequality.net) integrates animal science with societal concerns to identify standards and measures for improved animal welfare. The project aims to develop scientifically based tools to measure animal welfare, to convert these measures into accessible and understandable information, and to improve animal welfare through species-specific strategies. The standards for on-farm welfare assessment and information systems are based upon consumer demands, the marketing requirements of retailers and stringent scientific validation, with the objective of linking informed animal product consumption to animal husbandry practices on the farm. The project therefore adopts a 'fork to farm' rather than the traditional 'farm to fork' approach'. In the project, considerable effort is focussed on analysing and addressing the perceptions and concerns of consumers, retailers and producers about animal welfare.

Forty-four institutes and universities (representing 13 European countries and four Latin American countries) with specialist expertise are developing a standard approach to measure animal welfare in cattle, pigs and poultry on the farm and at slaughter: the Welfare Quality® Assessment Tool. Welfare assessment is a multidimensional concept that requires the combination of several measures or indicators (Broom and Johnson, 1993), and should ideally reflect the different definitions of animal welfare.

In 1993, the United Kingdom Farm Animal Welfare Council proposed the so-called five freedoms (FAWC, 1993), which have become a widely used framework to define and assess welfare. The five freedoms read as follows:

1. Freedom from thirst, hunger and malnutrition – by ready access to fresh water and a diet to maintain full health and vigour.
2. Freedom from discomfort – by providing a suitable environment including shelter and a comfortable resting area.
3. Freedom from pain, injury and disease – by prevention or rapid diagnosis and treatment.
4. Freedom to express normal behaviour – by providing sufficient space, proper facilities and company of the animals' own kind.
5. Freedom from fear and distress – by ensuring conditions that avoid mental suffering.

The five freedoms combine elements from the three approaches to welfare explained above and are a very useful framework to identify the main welfare problems in a given production system and also as an starting point to select the main welfare indicators.

Welfare Quality® is based on the assessment of 12 welfare principles grouped into four main criteria, according to how they are experienced by animals (Botreau *et al.*, 2007; Table 1). These principles are assessed throughout the animal's entire lifespan, the rearing period, which runs from birth to weaning, the beginning of production (e.g. milk, eggs); the production period, during which a dairy animal will produce milk, a hen will lay eggs and animals farmed for meat will be fattened; and the end of the life of the animal, where it will be transported and slaughtered.

Previous monitoring systems and legislation focussed mainly on 'what' or 'how much' of different resources are given to animals (i.e. type of housing, allocation of resources, stocking density etc). However, resource-based and management-based measures are a poor guarantee for good animal welfare in a particular situation. Because the subjective experiences of an animal are important in most definitions of animal welfare, the assessment system places its emphasis on the experiences of the animal. The protocol includes mainly measures which are as close as possible to the experiences of the animal, i.e. animal-based parameters (as opposed to only management- or resource-based

Table 1. Welfare principles and criteria identified in Welfare Quality®.

Welfare criteria	Welfare principles	Meaning
Good feeding	1. Absence of prolonged hunger	Animals should not suffer from prolonged hunger
	2. Absence of prolonged thirst	Animals should not suffer from prolonged thirst
Good housing	3. Comfort around resting	Animals should be comfortable, especially within their lying areas
	4. Thermal comfort	Animals should be in good thermal environment
	5. Ease of movement	Animals should be able to moving around freely
Good health	6. Absence of injuries	Animals should not be physically injured
	7. Absence of disease	Animals should be free of disease
	8. Absence of pain induced by management procedures	Animals should not suffer from pain induced by inappropriate management
Appropriate behaviour	9. Expression of social behaviours	Animals should be allowed to express natural, non-harmful, social behaviours
	10. Expression of other behaviours	Animals should have the possibility of expressing other intuitively desirable natural behaviours, such as exploration and play
	11. Good human-animal relationship	Good human-animal relationships are beneficial to the welfare of animals
	12. Absence of general fear	Animals should not experience negative emotions such as fear, distress, frustration or apathy

measures). Furthermore, animal-based measures enable comparison of the welfare of animals kept in different systems, are sensitive to fluctuations in the welfare status of animals on these farms, and reflect the welfare status of the herd as a whole, and remain transparent to stakeholders. However, management- and resource-based measures are used to complement the animal-based ones or to substitute for them in cases where no valid animal-based measures are available, and also to help diagnose the causes of poor welfare and advise farmers on ways to improve the welfare of their animals (Botreau *et al.*, 2007a). All measures that are included in the monitoring scheme have been previously evaluated with regards to their validity, repeatability and feasibility. Validity means that it really does say something about the welfare of the animal. Repeatability means inter- or intra-observer repeatability and how robust the measure is to external factors such as time of day or weather conditions. Feasibility means the possibility to carry out the measure under practical conditions.

The assessment protocol also enables the identification of welfare problems and the development of practical improvement strategies that will provide producers with the opportunity to enhance the welfare status of their stock. The project has initiated appropriate studies in important areas like handling stress, injurious behaviours, lameness, temperament etc.

Integration of the measures into a scheme

The implementation of the collective knowledge generated by this project will support the development of welfare-friendly husbandry systems and genotypes. This, in turn, will contribute significantly to the diversification and societal sustainability of farm animal production in Europe. However, to achieve this aim, the standardised welfare measures should be converted into accessible and understandable information that allows clear marketing and profiling of the products. For this reason, the data obtained from the measures have been combined into an overall assessment through a hierarchical evaluation model, that interpret the data firstly in terms of the 12 welfare criteria identified, and then the four welfare principles (Botreau *et al.*, 2007a). The methodology for doing this relies on a series of expert evaluations of the relative welfare concerns caused by different welfare problems. At the end of the process, farms can be categorised according to the assessed level of welfare of their animals. Four such categories have been proposed:

- Excellent: the welfare of the animals is assessed as very good. An animal unit fitting this category may primarily supply a niche market, via a label assuring consumers of very high quality products, (this label could be dedicated to animal welfare).
- Enhanced: the welfare of animals is good (but not excellent). Good practices are applied and are sufficient to ensure a good level of welfare within a more general quality label (which could include other product characteristics like food quality, food safety, environmental impact, etc.).
- Acceptable (or basic): the welfare of animals is acceptable (i.e. above minimal requirements defined for a compulsory label), but insufficient for the animal unit to enter a certification scheme inclusive of animal welfare.
- Not classified: the welfare of animals is low and considered unacceptable.

A welfare assessment system incorporating these four categories could be used in various ways. For example, it could help stakeholders to certify farms and specify their markets, e.g. enhanced farms for a general quality label or excellent farms for a niche market. It can also provide farmers with a broad picture of the welfare status of their animals as well as identifying aspects requiring their attention.

Regulating animal welfare by way of the market offers some farmers the opportunity for distinction in the market, and therefore of obtaining added value for their products. Not only can improved animal welfare conditions contribute to the generation of higher commodity prices, but lower welfare conditions are proving costly. That cost comes through harmful effects on the animals' health, productivity and product quality, and when producers are unable to access higher value markets or respond to consumer demand. Therefore, many food producers and suppliers welcome the growth of welfare conditions as a component of product and brand differentiation. The establishment of an EU label for animal welfare is becoming an option to be investigated in the near future which could promote products elaborated under high welfare standards thus facilitating the choice of the consumers between products obtained with basic welfare standards (the minimum standards laid down in EU legislation) or with higher standards (contained in voluntary codes of practice or Member States' legislation going beyond EU minimum rules).

References

Botreau, R., P. Perny, J. Capdeville and I. Veissier, 2007. Construction of product information from animal welfare assessment. Proceedings of the Second Welfare Quality_R Stakeholder Conference, pp. 33-36.

Broom, D.M., 1986. Indicators of poor welfare. British Veterinary Journal 142: 524-526.

Broom, D.M. and K.G., Johnson, 1993. Stress and animal welfare. London: Chapman and Hall.

Duncan, I.J.H. and D. Fraser, 1997. Understanding animal welfare. In: M C Appleby and B O Hughes (eds.) Animal Welfare. Wallingford: CAB International.

Eurobarometer, 2007. Attitudes of EU citizens towards animal welfare. Brussels: European Commission.

Farm Animal Welfare Council, 1993. Second report on priorities for research and development in farm animal welfare. Tolworth: MAFF.

International Financing Corporation, 2006. Good Practice Note. Animal Welfare in Livestock Operations.

Kjaernes, U. and B. Bock, 2007. Societal concerns on farm animal welfare. Proceedings of the Second Welfare Quality_R Stakeholder Conference, pp. 13-186.

OIE, 2006. Terrestrial Code, 15th edition. Paris: OIE.

The use of grazing pastures in goat production: development of an approach to combine optimized use of the forage resource and the control of related risks

M. Napoléone[1], H. Hoste[2] and Y. Lefrileux[3]
[1]INRA, SAD Ecodeveloppement, Domaine St Paul, Agroparc, 84914 Avignon cedex 9, France
[2]INRA, UMR 1125, INRA ENVT, IHAP, 23 chemin des capelles, 31076 Toulouse cedex, France
[3]PEP Rhône Alpes, Le Pradel, 07170 Mirabel, France

Abstract

This article describes an approach to provide support for changes in herd management to increase the proportion of grazed forage in the diet of dairy goats. This research was conducted in collaboration with researchers, agricultural advisors and farmers. The use of grazed forage introduces many sources of uncertainty for the farmers (fluctuating resource, risks of drought or parasite infestations). It is thus difficult to propose a robust technical model that is reliable in a variety of circumstances. The aim of the approach described here is to make the progressive adaptation of herd management feasible. The approach is based on the formalisation of the farmer's practices in time and in space, and makes it possible to analyse combinations of the different grazed resources in relation to indoor feeding, production management and the epidemiology of gastrointestinal nematodes, used as a model of parasite infections. These representations help diagnose the system, identify the possibilities of adaptation and discuss corrective actions. In supporting and guiding such innovative processes, researchers and agricultural advisors have to modify the angle from which they observe the situation. The tool was initially developed from data acquired in Mediterranean conditions. However, it could be of interest for other production systems and/or in other regions, in particular for farmers who currently use intensive systems and who have lost the know-how required for managing a grazing herd, but who are concerned by a return to less intensive management systems in order to fulfil the criteria of sustainable development.

Keywords: goat production, herd management, grazed resources

Introduction

In goat farming, the development of grazing is a major issue, whether from the viewpoint of returning to less intensive practices, reducing farming costs, organising work, or making best use of links with the 'terroir'. However, the use of grazing is accompanied by risks related to the evolution of the grazed resource or increased exposure to risks of parasite, among others.

Developing grassland obliges livestock farmers to manage their herd in situations of uncertainty, requiring them to move from a logic where the objective is to control the principal parameters of herd management (calculating the ration to adjust feed inputs to the needs of the animals or managing health hazards such as parasite infestation), to a posture in which they come to terms with risks by seeking balances. Until recently, grazing had become marginal in the major dairy goat areas in France, but back in the 1970s it was one of the driving forces for neo-rural people setting up pastoral goat-farming systems in areas with woods and rough grazing in Mediterranean regions. By using grazing land, Mediterranean goat farmers,developed alternative management methods to the technical models that dominated goat farming at that time. To compensate for their relative lack of control over each resource, the farmers combined a range of spontaneous and cultivated resources, whether daily, or at the level of the whole grazing period.

In this context, for about 15 years, work has been underway in close cooperation between researchers, agrocultural advisors and farmers, with the aim of developing approaches to strengthen farmers' abilities to manage herds using a range of different grazing land. This work relies on the study of farmers' practices, and is based on the identification of keys for interpreting and analysing the situation.

Taking the above work as our basis and using a concrete example, here we present an approach for planning the use of grazing land, estimating the risks associated with the management of milk production and the risks of parasite infection. The approach is based on the co-construction of graphs depicting herd management to facilitate analysis of the existing situation, and to help in the diagnosis and search for corrective scenarios.

In the first section, we present the general principle of the approach. In the second section, we stress the need for technical diagnoses to optimize the use of resources using two different angles of analysis: the management of dairy production and the consideration of health hazards, taking as a model infestation by digestive tract strongyles. In the third section, we discuss the role of the intermediate object and the indicators, to help elaborate a shared diagnosis process.

Methods

Characterising the technical management of a grazing herd

In the South of France, a great deal of research and development has focused on understanding how pastoral farms function so as to be able to work out situation diagnoses. They study the way in which the process for developing productions is organised over time. To understand the organisation set up by the farmers, and to make this analysis easier, some authors have recommended the adoption of calendars to represent practices (Landais and Balent, 1993). Thus, Bellon *et al.* (1999) developed grazing calendars to study the use of the different fields of the farm. These representations, carried out at the scale of the farming production year, make it possible to characterise the farmer's management entities and the way in which he combines and prioritises his practices. These authors then interpret the coherence underlying this organisation. So these representations make it possible to connect a practice carried out at a given moment (such and such a field used in rotation) to the overall organisation over the farming year (the role of a given field in the grazing calendar). Several studies have focused on grazing strategies (Bellon *et al.*, 1999), feed strategies (Hubert *et al.*, 1993), or production strategies (Napoléone, 1999).

The livestock diagnosis: a question of viewpoint and angle of analysis

Characterising herd management over time makes it possible to identify the strengths and weaknesses, and ways to improve the situation. However, the same situation can give rise to different diagnoses and lead to quite different courses of action depending on whether it is evaluated from a technical or a practical point of view (Darré, 1999).

Technical diagnoses and their angles of observation

A diagnosis can be made from a technical point of view, from references and indicators measuring a difference between a standard or a model and what is actually carried out, or estimating a risk, taking into account what is known about the subject. Observing a situation, technicians[13] note a number of parameters on which to base their opinion and make their evaluation. Of course, the parameters will vary with the question asked. For example, a grazing calendar can be analysed to optimise plant production. In this case, the focus will be on the number of days the herd spent grazing the fields concerned to estimate production, which can then be compared with regional references. The same calendar can be analysed from the point of view of managing the milk production of the grazing

[13] Technician: someone with technical knowledge such as an extension agent or a scientist.

herd. In this case, the focus will be on how the unfolding of the calendar is linked to production. On the other hand, if the question is how to control parasite infestation, the parameters will seek to estimate the risks associated with the use of the grazing land. A technical diagnosis can be formulated at several levels. It is directly related to how the question is formulated at the outset, which itself determines the technical indicators chosen for the analysis.

The diagnosis of the one carrying out the practices expresses implicit and varied factors

The farmer's diagnosis is made from the point of view of the person carrying out the practices. He evaluates the situation, taking into account his production target, his constraints, the way he organises his work, his way of perceiving risks... In other words, it is a set of implicit or explicit factors that contribute to his assessment of the situation.

Making a shared diagnosis to plan the use of grazing land by managing the associated risks

A diagnosis of the existing situation that is shared by a technician and a farmer is difficult since it involves constructing a judgement from two points of view of the same situation (Darré *et al.*, 2004). To get round this difficulty, the grassroots approach we present is based on four steps undertaken in collaboration with the farmer:

- *Represent the practices on graphs:* in collaboration with the farmer, a chart is created to be used as a basis for dialogue between the technician and the farmer.
- Analyse the chart to understand – and formalize – *the method of organisation and the underlying logic*
- *Establish a diagnosis,* identify the key points and the problems, estimate risks, etc. from both points of view
- *Imagine ways to act, to test a new method of organisation to reduce the risks* (e.g reorganisation of grazing) *that is compatible with the operation of the farm.*

This diagnostic approach, carried out in year N, is based on a comprehensive interview lasting approximately two hours. Initially the farmer describes how his practices unfold. A calendar is used to represent this process visually. In the second part of the interview, the calendar helps analyse the situation and making a combined diagnosis. In year N+1, the same approach can be used again to make a new assessment, to learn from the corrections undertaken, or to readjust the system according to the objectives.

Here we describe the aid to diagnosis on the use of grazing land using a concrete example, 'M' (Box 1), based on participatory research concerning the technical management of a grazing herd (Institut de l'Elevage, 2007) and associated risks (PEP, 2007). We first describe the steps that characterise the use of grazing land and herd management. We then tackle the question of the diagnosis, developing two successive points of view: (1) analysis of the production calendar from the viewpoint of herd production management; (2) analysis of the calendar from the viewpoint of gastro-intestinal infestation by strongles (SGI). We show the interest of managing the two assessments together to identify prospects for organisation taking these two aspects into account.

Box 1. Use of grazing land.
'M': a goat farmer who produces organic cheese in the *Drôme Provençale*, France.
Herd: 70 Alpine goats.
Production: 720 l/year/goat and 50,000 l/year for the herd.
Territory: 18 ha of fields that can be mowed, 35 ha of woods and heath.
Cheese processing on the farm and direct sale of Picodon AOC.

Understanding herd management and the use of grazing land

Representing the grazing calendar and herd management

The first thing to be understood is the way in which the farmer structures and divides up his territory. Thus in the example in Figure 1, the farmer creates units of use, to which he gives precise names. These units of use are defined by the way the farmer uses them. They are not necessarily based on agronomic properties or cadastral plots. Their periods of use can be marked on a grazing calendar.

Based on the calendar, grazing can be linked to management practices (reproduction, feed, pathology, grazing etc.), or to the evolution of herd production. In this way, a factual representation is obtained of how herd management unfolds (Figure 2).

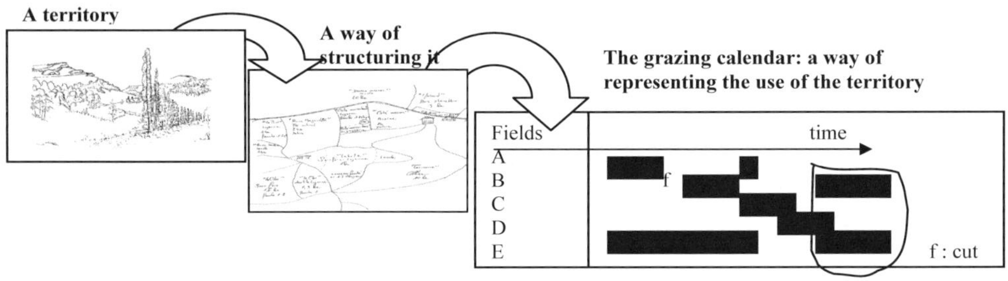

Figure 1. From the description of the territory to the grazing calendar.

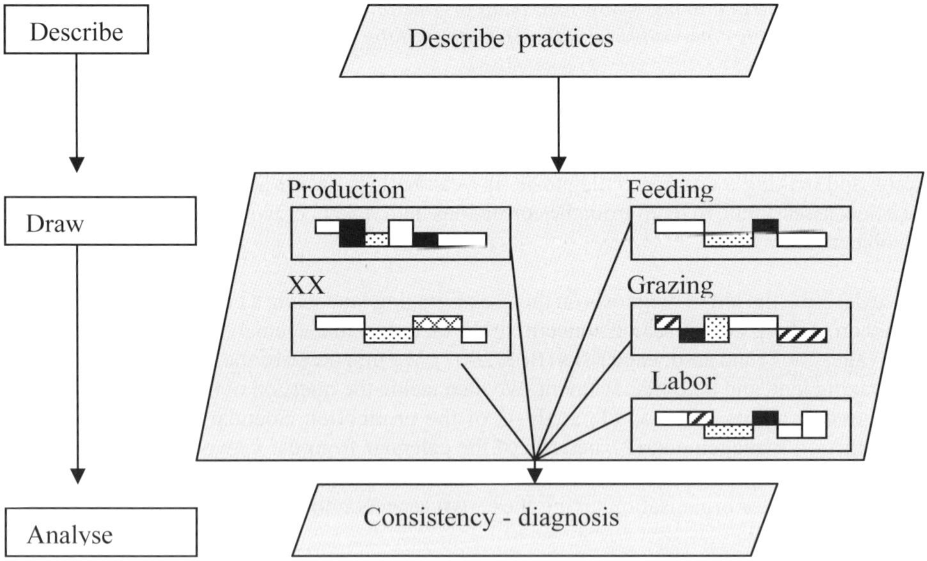

Figure 2. A factual representation of herd management.

Characterising the organisation of grazing in relation to herd management

The calendar makes it possible to analyse the combinations and successions of practices and then to formalise herd management. So, to take our example again (Figure 3), Box 2 formalises the progress of herd management and grassland management set up by 'M'.

Diagnosing the use of grazing

From the viewpoint of herd production

During grazing, herd production can vary to a greater or lesser extent. An abrupt change in the quality or quantity of grazing results in a marked change in production. The analysis of the temporal agreements between management sequences and production management makes it possible to diagnose the links between practices and production and plan courses of action with the farmer. In the example used, (Figure 2) the diagnosis carried out in year N highlighted the fact that difficulties in maintaining production in spring are directly related to the management of grazing land (exhaustion of the resource of sainfoin used since the herd was put out to grass, continuous grazing of an large area, no more grazing on forage at the end of May). Three courses of action are studied: re-examine the balance between cutting/grazing and assign an additional field for spring grazing, subdivide and organise rotations in the natural grasslands and heathlands, grow a vetch/oats mixture specifically for the transition period at the end of spring.

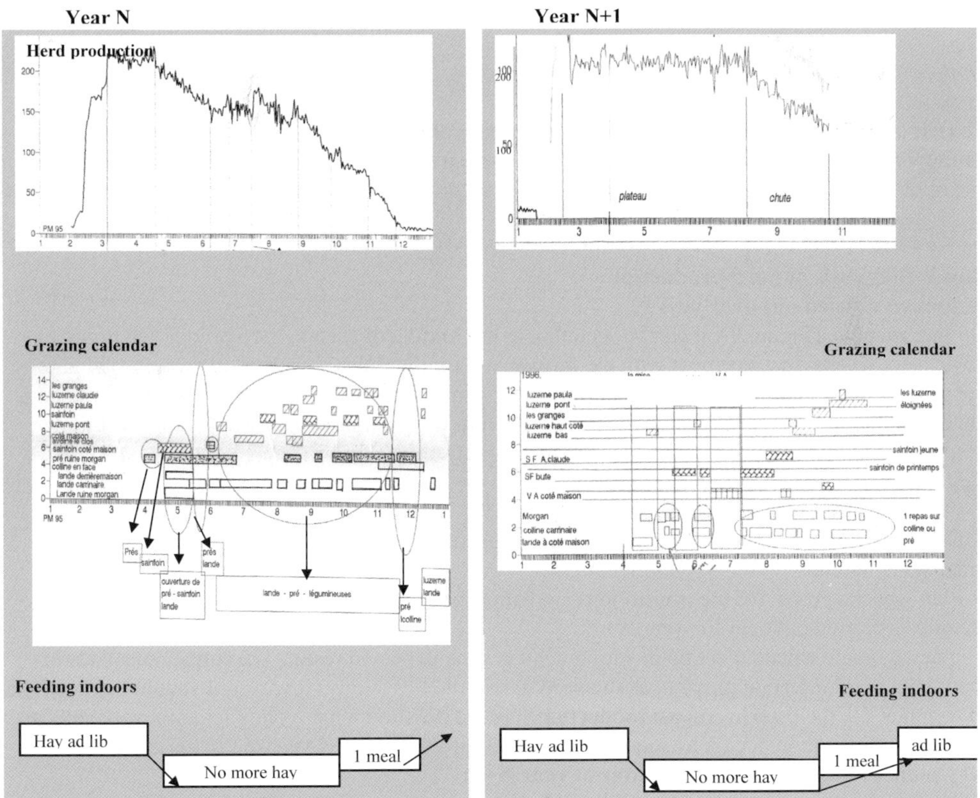

Figure 3. An example: calendar of herd management.

Box 2. Progress of herd management and grassland management.
The organisation of herd management, year N
The herd is managed in a single batch, with births grouped by season. The herd grazes from early April to December. The grazing land provides the whole diet from mid April to the end of September. During the lactation period 'M' supplements the diet with 700 g of grain (1/3 barley, 2/3 concentrate). He keeps a field of sainfoin for spring grazing. The other fields of pulses, mown in the first cut, are only grazed in the summer and autumn.
Organisation of grazing: *every season* 'M' *associates pulses, natural grasslands and rough grazing.*
The flock is put out to grass on natural grassland in early April. This transition period lasts for 10 days until the end of hay.
In spring (mid April to mid May): Continuous grazing on four fields of grass and heath, with the supplementary use of the field of sainfoin.
Late spring (mid May to the end of May): the sainfoin is finished. The herd grazes on the grassland they have used since they were put out to grass and on ungrazed heathland.
Transition from spring to summer. The oats/vetch mixture is used to supplement spring and summer grazing on heathland and natural grassland.
Summer-autumn: One meal on pulses (used in rotation) and one meal on grassland and heathland.

The new organisation in year N+1 solves the problem of production management in full spring. However, the analysis of its progress highlights a new problem: the risk incurred if the only resources to be grazed in summer are fields of sainfoin planted in the current year (see Box 3).

Optimising grazing with respect to parasite infection

The use of grazing land exposes the animals to the risk of parasites, the most common being digestive strongyles. In conventional livestock farming, the control of this parasite is based on the regular use of chemical worming treatments. However resistance to these molecules is increasing in worm

Box 3. Diagnosis of herd production.
Diagnosis carried out over year N
On the calendar, (Figure 3), it can be seen that in the middle of spring, herd production drops by 25% in six weeks. In agreement with the farmer, we assumed that this marked drop is related to the progressive lack of sainfoin and continuous grazing on too large an area. At the end of spring, the effect on production of feeding green oats for four days is immediately apparent. In summer, production stabilizes as soon as the second leguminous crop (alfalfa) can be grazed.
Key points to plan the calendar for year N+1
1. Organise rotations on the spring fields.
2. Do not under-estimate the fields of legumes needed for spring grazing. Reconsider cutting vs. grazing.
3. Plan 'special crops' for the transition period after the end of spring (approximately 1 month).
Diagnosis carried out over the year N+1
The new grazing calendar set up in spring achieves the expected result. The farmer succeeds in managing regular feed at grazing, as the production kinetics show. However, it should be noted that in summer: (1) the drop in August production starts at the time when young leguminous plants are used; (2) stabilization occurs as soon as the animals are moved to 3-year old fields of alfalfa.
Key points to include in the calendar of year N+2:
Is it necessary to retain the grazing of 1 meal on a good area, on *young* leguminous plants in summer? In this case, would it be necessary to supplement this meal by another area (grassland, alfalfa in year 2, sorghum…)?

populations (Jackson and Coop, 2000). In organic farming, there are strict quotas concerning the use of synthetic molecules. The management of parasites therefore has to reconcile two contradictory aspects: *greater exposure of the animals to risks of parasites and a reduction in the possibility of treatment*. Whether in conventional farming or organic farming, it is important to develop innovative diagnostic monitoring tools at the level of the animals or of the management systems to avoid massive recourse to chemical substances. We sought to develop a tool for the diagnosis of parasitic risk factors that can be used by farmers and technicians; it is based on the analysis of herd management practices in relation to the kinetics of herd milk production.

The infestation of pasture by nematodes is related to their use by the animals

The infestation of a pasture depends on the number of larvae present which, in turn, is linked to the number of eggs deposited by the animals that graze there. The nematode cycle lasts for about three weeks but the larvae can continue to infest the land for several months. The infestation of a field will thus depend on the stocking density (number of animals/ha), the length of use by the animals, and the condition of the animals that graze there. The more the animals use a field, the more they contaminate it. And when an animal has just used a contaminated field and moves onto another field, it will contaminate the new field. However, certain agricultural techniques (ploughing, mowing), or significant and durable weather phenomena (frost or long drought) can help reduce the rate of infesting larvae. Consequently, it is particularly important to identify periods in a grazing calendar when the fields are left to rest, while keeping in mind aspects related to avalaibility of ressources and their valuation to reach expected production rates. in terms of infestation, in temperate conditions, management of grassland with short rotations using a number of small fields has the same consequences as simultaneous and uninterrupted use of all these fields. In this case, with respect to the risk of parasites, these fields can be considered as a single block.

Estimating risks from practices

Armed with this knowledge, a group of agricultural advisers and researchers designed a diagnostic approach to the risk of parasites inspired by HACCP approaches and a previous approach (PEP, 2007; Hoste *et al.*, 2005). From the analysis of farming practices and the grazing calendar, it is possible to estimate the state of infestation of fields and of the animals, in order to calculate the risk of parasite infestation. To return to the example of 'M's' farm, the grazing calendar for year N+1, which was satisfactory as regards herd production management, may face a risk of parasite infestation. In 'M's' context, it is useful to question:

- The regular use of the 'Morgan' field after putting the animals out to grass for the winter. The periods when this field is not used are too short to have a positive impact on the rate of larvae present.
- The return in early summer to the 'Bute' field of sainfoin that was used in spring.

Taking the above knowledge into account, we sugguest that the way of managing these two fields could lead to a risk of infestation in summer. To limit this risk, two methods are possible: (1) subdividing the fields of natural grassland called 'Morgan' and managing them as fodder areas, with rotations rather than continuous use and (2) leaving a longer interval between the two uses of the 'Bute' field of sainfoin.

Simulating new organisations taking into account both production needs and risks of parasites

With these two angles of analysis (production and health risk), the farmer can refine his organisational principles to design the grazing calendar and new organisational rules. Thus, in our example:

- associating leguminous crops (1 meal after midday) and heath land (1 meal in the morning), throughout the grazing season;
- not under-dimensioning the fields of legumes reserved for the spring;
- managing the fields in rotation, and leaving sufficient time before returning to the same field;
- specifically allocating resources for the transition period in late spring;
- carefully watching the use of young fields that are sensitive to drought in summer.

The farmer can also define ways of monitoring the progress of the farming year by identifying the critical periods, the objects that have to be watched, and corrective measures that can be applied. For example:

- Carefully monitor the spring-summer transition period: *If* I see a persistent drop in production and the resource in the fields becomes insufficient in spring, *then* I can mobilize a particular 'buffer' resource while waiting for second cut regrowths (e.g. valley floor reserved for this purpose).
- Monitor the condition of the animals if the spring is very rainy and if it is necessary to return to the spring fields: *if* I see that the animals have soft droppings, or their coat is dull *then* I can assume that there is parasite infestation and take the necessary measures.
- Monitor the use of young fields of sainfoin in summer, (because they will be sensitive to drought).

Discussion

Representing the action to help in the interpretation and construction of a shared diagnosis

The diagnostic approach we have just presented is an iterative approach, from action to analysis. The chart of the progress of herd management and the use of grazing land plays the role of an intermediate object (Vinck, 1999), that facilitates dialogue between the technician and the farmer but also helps the farmer to stand back and judge the situation for himself.

The chart makes it possible to link the short and the long term and therefore connect a collection of elements which might seem irrelevant when considered individually. It also helps clarify tactical decisions, and analyse the effect of these decisions on herd management. So it involves a true process of abstraction, from the calendar, which makes it possible to pass from a factual representation to the formalisation of an organisational method, and dialogue with the farmer on the underlying logic.

As recalled at the beginning of this paper, several interpretations can be made of management progresses leading to different diagnoses. The representation makes it possible for everyone to clarify the elements they use to construct their interpretation. Thus the dialogue between the farmer and the technician can begin on these respective keys to understanding. In the end, the diagnosis resulting from this exchange of views on ways of viewing things will be neither completely that of the technician, nor completely that of the farmer. The way of formulating – in writing – this combined interpretation, and the question(s) to treat that result from it, is a crucial point that will influence the courses of action considered.

In the 'search-for-solutions' phase, the chart makes it easier to imagine new organisation scenarios. Here we define 'scenario' as a unit formed by the description of a future situation and by the advance of events – which must be coherent – making it possible to move forward from the original situation to a future situation (Godet, 1991). Several scenarios can be imagined, represented, and discussed, before finally choosing only one believed to be both desirable for different reasons (technical, economic, labour) and feasible in the eyes of the farmer (taking into account farm operating costs, or interactions with other activities).

A long term learning process

In the long term, this approach can be renewed here and there. From these successive analyses, placing the action in relation with the interpretation (Figure 4), the farmer and the technician construct new knowledge they use to design new forms of organisation aimed at improving the points considered to be problematic. They also create means of monitoring and analysing the situation, based on observation of the herd and of the resources, in order to readjust management if necessary. In this way a succession of trail and error is obtained leading to a process of organisational learning.

These reference points and knowledge are hybrid know-how resulting both from observations and analyses of how the action proceeds and from technical models (e.g. the physiology of lactation, vegetative cycles, or parasite cycles). This hybrid knowledge is then formulated in the form of rules that can be analysed in relation with the action.... *If I see that...(and that...)...then...* The formalisation of this know-how, and its organisation, for its subsequent mobilisation remains an open question, whether the aim concerns farm management (for the practitioner) or support and guidance (for the adviser).

This aid to diagnosis will not inevitably move in the same direction depending on the life cycle of the farm concerned. For farmers who are just setting up their business or farmers who wish to make major changes to their herd management, this approach is a way of acquiring knowledge and know-how that is internal to the farm. For example, it may be useful for goat farmers engaged in a process of moving to less intensive management practices. For an experienced farmer who has tested the robustness of his herd management over a long period, the exercise of explaining his organisation and taking time to analyse it with someone from outside the farm, allows him to call into question possible organisational routines, which might imperceptibly lead to risky situations. But this approach can also be a way of anticipating and monitoring a period that this experienced farmer is able to manage carefully.

New roles for the adviser and the researcher

In this role, the adviser guides the reflexion, and helps in the expression the project. His role is not the that of an expert, from whom specific information would be expected to solve a particular problem. This approach can only be used if the farmer and the technician are indeed engaged in a logic of exploration. The role of guide modifies the classic format of 'agricultural advisory service' and the linear dissemination of research knowledge towards technical institutes and farmers. By helping the expression of a particular project, these approaches have shown their effectiveness in the context of research and development. Their extension will depend on the organisation of advisory structures,

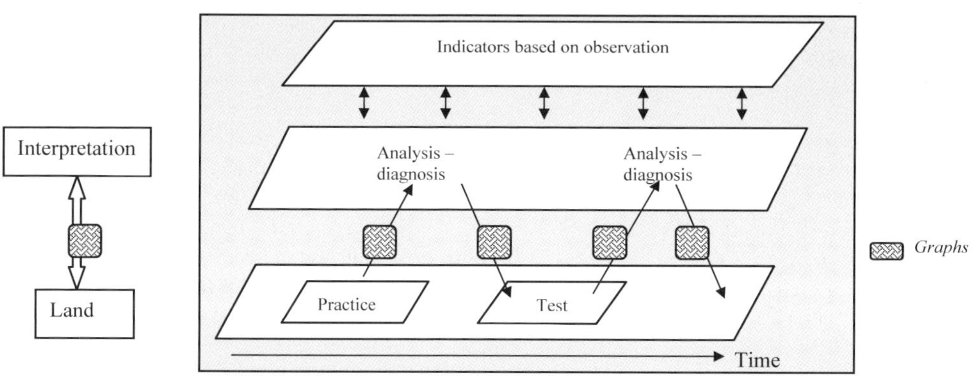

Figure 4. An iterative approach between action and analysis.

and the missions intended for advisers. Finally, the context of Mediterranean regions, in which farmers aim to find a balance between a range of different resources, makes it necessary to search for alternative ways to reinforce the capacities of management of the grazing herd. However, we put forward the hypothesis that it is a challenge for farmers in intensive systems, who have often lost the know-how to manage grazing land and who, taking sustainable agriculture into consideration, are confronted by the necessity of moving to less intensive practices.

References

Bellon S., N. Girard, G. Guérin, 1999 Caractériser les saisons-pratiques pour comprendre l'organisation d'une campagne de pâturage Fourrage 158, pages 115-132.

Darré J.P., 1999. La production de connaissances pour l'action. Arguments contre le racisme de l'intelligence Editions de la Maison des sciences de l'homme, Paris, I.N.R.A., 242 p.

Darré J.P., A. Mathieu and J. Lasseur, 2004. Le sens des pratiques – Conceptions d'agriculteurs et modèles d'agronomes. Ouvrage coordonné par Darré J.P., Mathieu, A., Lasseur, J. Science update INRA ed. 320 p.

Godet, M., 1991. Méthode des scénarios. In: De l'anticipation à l'action - Editions Dunod p 39-55.

Hoste, H., M. Napoleone and Y. Lefrileux, 2005. Construction d une problématique et d'un partenariat sur la gestion du parasitisme lie au pâturage en élevage caprin en agriculture biologique - Symposium National PSDR Lyon 9-11 Mars 2005, 15 p.

Hubert, B., N. Girard, J. Lasseur and S. Bellon, 1993. Les systèmes d'élevage ovins préalpins – Derrière les pratiques, des conceptions modélisables. In Pratiques d'élevages extensifs - Etude et recherche N°27 Ed. Landais et Balent, INRA pp. 351- 385.

Institut de l'Elevage, 2007. Guide pour la conduite du pâturage caprin – Synthèse réalisée dans le cadre du réseau national des techniciens caprins travaillant sur le thème du pâturage Editeur IE collection: « résultats », 146 pages + annexes.

Jackson, F. and R.L. Coop, 2000. The development of anthelmintic resistance in sheep nematodes. Parasitology 120, S95-S107.

Landais, E. and G. Balent, 1993, Pratiques d'élevages extensifs - Identifier – Modéliser – Evaluer, Ouvrage collectif Etude et recherche N°27 Ed. INRA. 387 p.

Napoléone, M., 1999. a: Accompagner les éleveurs dans la gestion du processus de production, in Option méditerranéennes, N°27, série B, CIHEAM-CEE/DG VI pp. 245- 258.

PEP, 2007. Guide pour la maîtrise des risques parasitaires in Guide pour la conduite du pâturage caprin Editeur IE collection: « résultats », pp. 71-104.

Vinck, D., 1999. Ingénieur au quotidien – Ethnologie de l'activité de conception et d'innovation Ouvrage collectif Ed. P.U.G. 232 p.

Silvopastoral innovation in North Mediterranean livestock farming systems: current knowledge and prospects for technical support

C. Aubron[1], *D. Gautier*[2], *G. Guérin*[3] *and F. Launay*[3]
[1]*Montpellier SupAgro, UMR ERRC, 2 place Viala, 34060 Montpellier cedex 1, France*
[2]*Institut de l'Elevage, Maison Régionale de l'Elevage, Route de la Durance, 04100 Manosque, France*
[3]*Institut de l'Elevage, 2 place Viala, 34060 Montpellier cedex 1, France*

Abstract

For several decades, North Mediterranean regions have experienced an expansion in their forest areas. Classic forestry practices are difficult to apply to these zones. Combined with the difficulties faced by the livestock sector in these regions, this context calls for a fresh examination of silvopastoralism. In woodlands, fodder production staggered over the year and its herbaceous component is preserved longer due to the shadows of trees. The selective cutting of trees generates wood products which may be sold through new marketing channels. When carefully thought over, grazing and silviculture operations can be mutually beneficial. At the farm or forest parcel level, supporting the development of innovative silvopastoral projects involves planning a succession of silvopastoral operations in the medium term and evaluating the project with a set of indicators. A technical support tool including these features and their spatial dimensions is currently under construction.

Keywords: silvopastoralism, North Mediterranean region, livestock farming system, silvospastoral project

Introduction

For several decades, forests – principally Downy Oak (*Quercus Pubescens*) and Scots Pine (*Pinus Silvestris*) – have been expanding in northern Mediterranean regions. These forest stands, considered to be relatively unprofitable from a silviculture perspective, often are overlooked by the traditional wood processing industry. This unfavorable context, combined with the difficult economic and climate situations faced by the livestock sector in these regions, calls for a new look at silvopastoralism and associated technical support mechanisms.

Silvopastoral coherence established at the plot level

In forests, fodder production (of herbaceous plants and woody plant leaves) staggered over the course of the year and its standing herbaceous component persists later than in open areas due to the shade of the tree canopy (umbrella effect, see Figure 1). As long as it does not hinder animal movement, palatable underbrush is part of a diversified forage resource that stimulates ingestion (Agreil *et al.*, 2004). Animals thus may find, for example, a forage resource in wooded areas during summer, an appreciable point in a time of climate uncertainty and the increasing cost of feed.

Forest area grazing systems are characterized by one or two pasture seasons a year, often timed to lag in relation to peak grass growth to take advantage of the capacity of the vegetation to remain standing later into the year. Moreover, wooded areas offer considerable flexibility: when needed, pre-cutting or complete pasture in the spring, grazing of second growth and acorns at the end of autumn after a complete harvest in summer, etc.

The selective cutting of trees generates wood products and encourages the development of future trees. Besides conventional outlets such as trituration and firewood, wood production in Mediterranean forests can now generate revenue through new channels, often locally based, such as wood energy

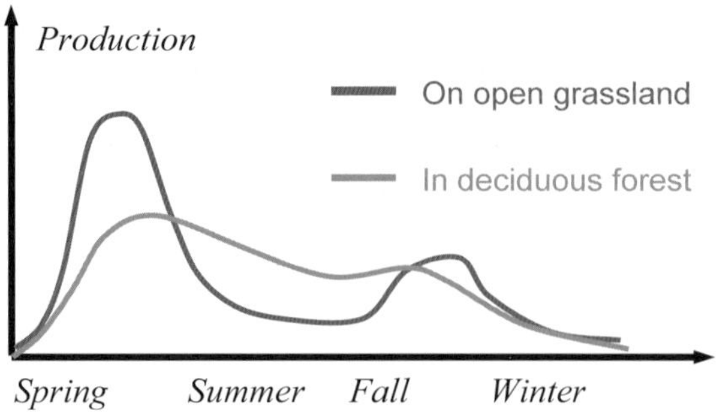

Figure 1. Production lag between grassland and forest: umbrella effect.

(wood pellets) or a construction owing to mobile sawmills. These may be combined to augment the impact on a forest stand and consolidate economic performance.

Planned out together on the same wooded area, grazing and silviculture operations are mutually beneficial. The umbrella effect of trees can benefit pastures; however, it eventually can compromise forage resources due to the closing of the canopy. Silviculture operations increase the quantity of light reaching undergrowth and thereby favors the development of forage resources (grass growth, lower hanging foliage, increase in percent fruit set). They also facilitate animal circulation. Reciprocally, well managed grazing keeps underbrush under control and facilitates later silviculture operations. Thus conceived, silvopastoralism in Mediterranean forests allows two marginal activities – grazing and silviculture – to reinforce each other with considerable technical coherence.

The silvopastoral project at the livestock farm level

While the reasoning developed above is applicable to the plot level, it is insufficient in designing the integration of silvopastoral practices on a livestock farm or in the management of a forest stand. The concept of a 'silvopastoral project' put in place by a livestock farmer and/or a forest owner – each of these stakeholders possibly being a collective identity – corresponds to a combination of silvopastoral operations in space and time. Such a project integrates different constraints and aims at one objective. The space of a silvopastoral project is the forest area managed by the project holder. The timing of a silvopastoral project is the medium term, halfway between the 'short-term of livestock farming' and the 'long-term of silviculture', in other words, between 10 and 15 years (Guérin and Picard, 2005).

Monitoring different silvopastoral sites under the ACTA[14] and CASDAR[15] programmes reveals a range of diverse situations that give rise to varied silvopastoral projects, even if the silvopastoral practices implemented at the plot level are comparable (Guérin and Macron, 2005). For those livestock farmers who already allowed animals to graze their woods before the project began, silviculture operations are often aimed at increasing forage production by hindering progressive crown closure. This essentially pastoral objective sometimes is coupled with a silviculture objective of producing fuel and construction wood for the farm. Moreover, some livestock farmers wish to earn revenues through silviculture operations that maintain forage resources. This is made possible, for example, through the sale of planks produced with mobile sawmills, which is an interesting product when commissioned. Another set of situations is represented by forest owners who wish to earn revenue

[14] Silvopasture of Scots Pine and Downy Oak (*Sylvopastoralisme du Pin sylvestre et du Chêne pubescent*), funded by ACTA, 2002-2004.

[15] In search of balanced local development based on silvopasture: drawing value from forest stocks and setting up innovative livestock systems, funded by CASDAR, 2007-2009.

from their woods and seek to establish contracts with one or more livestock farmers to ensure the grazing of thinned areas and thus the maintenance of underbrush.

Beyond their grazing and silviculture objectives, projects differ in terms of their technical and economic organisation: the timing of silviculture operations during the project period and thus the evolution of the forage resource and distribution of wood production, the organisation of labor (family or contract), cash flows, etc. This raises the question of how such different dimensions may be integrated in support and evaluation activities for silvopastoral projects.

Towards a design assistance tool for silvopastoral projects

In response to this question, the 'Silvopastoralism' programme now underway is working to develop various tools (technical and economic references, silvopastoral typology of tree stands, etc.). A computerized tool is being built to assist in the design of silvopastoral projects. This technical support tool is meant to be used by the technician/project holder pair (project holders being either livestock farmers or forest owners) to formulate and evaluate silvopastoral project scenarios on a forest area (Simon, 2006) and furnish spatial representations.

The forest area is divided into vegetation stands that are homogenous from a silvopastoral perspective, in parks for grazing, and in silviculture units submitted to the same silviculture operations during the project period. Users of the tool then choose to apply a sequence of interventions on these spatial objects over the 10 year project period, drawing from technical fact sheets concerning three types of action: pasture use, silviculture operations, and associated management (fences, circulation routes, water points). This chronologic sequence defines a project scenario. The scenario is evaluated based on calculations made using the technical data sheets. These data sheets pertain, for example, to the change over time in the number of grazing days for a use season in the absence of – or following – a given silviculture operation. The evaluation examines: (1) the seasonal evolution of forage resources over the project period (see Figure 2); (2) wood production during the project period; (3) labor provided by the project holder and its distribution; (4) cash flows during the project period and total profitability over 10 years; (5) the status of the forest stand at the end of 10 years.

This set of indicators should allow users of the tool to judge the pertinence of the project in the light of their own objectives and constraints (limited labor force available for a certain year, urgent need of pasture days in summer, steady silviculture revenue requirement, etc.).

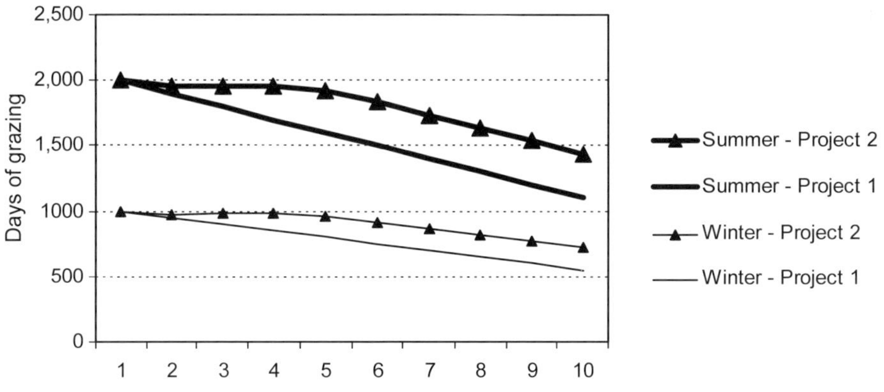

Figure 2. Example of a seasonal grazing profile on a forest park without silviculture operations (project 1) and with silviculture operations (project 2).

Conclusion

Thus conceived, silvopastoralism certainly is an innovation. While some of the practices involved are not new, the innovative quality lies in their careful integration within a context of expanding tree cover, a lack of guaranteed competitiveness in livestock activities, and the reinforcement of climate risks.

The technical innovation concerns the constitution of new wood products sold through new marketing channels, but also in the selective harvesting modes of disperse silviculture and forage resource. The management of 'what slips through' selective silviculture harvests by grazing (and vice versa) is a key facture in the sustainability of these practices and constitutes a field of technical innovation.

At the scale of the livestock system, the organizational innovation lies in being part of a project in which silvopastoralism is combined coherently with other farm operations in terms of cash flows, labor, and the feeding of herds. The tool under construction proposes to evaluate this organizational innovation in order to facilitate its diffusion.

References

Agreil, C., M. Meuret and M. Vincent, 2004. GRENOUILLE: une méthode pour gérer les ressources alimentaires pour des ovins sur milieux embroussaillés. Fourr. 180: 467-481.

Guérin, G. and M.-C. Macron, 2005. Sylvopastoralisme. Les clés de la réussite. Institut de l'Elevage, Techniques Pastorales, Paris, France, 77 pp.

Guérin, G. and O. Picard, 2005. L'enjeu technique du sylvopastoralisme: des échelles d'espace et de temps communes à l'élevage et à la sylviculture. In: Animal production and natural resources utilisation in the Mediterranean mountain areas, A. Georgoudis, A. Rosati and C. Mosconi (editors), EAAP 115, Wageningen, the Netherlands, 59-66.

Simon, C., 2006. Litterature review of scenario methods. Add-TRANS, 80 pp.

The role of secondary compounds in foraging on Mediterranean rangelands

J. Rogosic
University of Zadar, Department of Ecology, Agronomy and Aquaculture, Mihovila Pavlinovica bb, 23000 Zadar, Croatia

Abstract

Goats are primary consumers of the Mediterranean shrubby vegetation, and thereby shape the diversity, structure and dynamics of these extensive ecosystems. Low nutritional quality and high levels of secondary compounds can result in low consumption of Mediterranean shrubs by herbivory. Goats were allowed to choose between Mediterranean shrubs containing tannins and other Mediterranean shrubs containing tannins and saponins. Twelve goats offered tannin-containing shrubs consumed less biomass than animals offered shrubs containing tannins and saponin. Simultaneous consumption of shrubs that contain tannins and saponin may promote chemical interactions that inhibit toxin absorption from the intestinal tract. Besides complementary tannin and saponin interactions, biodiversity of Mediterranean shrubs also plays a positive role in increased biomass intake by goats.

Keywords: Mediterranean shrubs, goats, tannin and saponin interaction, biodiversity

Introduction

Goats foraging on Mediterranean shrubs that contain secondary compounds (toxins) may consume a variety of foods to avoid toxicosis (Freeland and Janzen, 1974). Different toxins have different physiological effects on herbivores (Cheeke, 1998), and herbivores are equipped with various physiological mechanisms to counter toxins (McArthur *et al.*, 1991). A varied diet should enhance goats ability to meet its nutritional needs when foraging on Mediterranean shrubs with secondary compounds provided (1) the toxins have different physiological effects, (2) the toxins are detoxified by independent metabolic pathways, and (3) the toxins do not interact with each other to become more toxic.

A varied diet with plant species containing different kinds of nutrients and toxins may increase food intake and animal production by enabling individuals to select plants that are biochemically complementary (Rogosic *et al.*, 2003).

Tannins and saponins are among the more common classes of phytochemicals used as defences by plants and can be expected to influence herbivore-feeding behaviour. Freeland *et al.* (1985) suggested that herbivores may be able to overcome the toxic effects of tannins and saponins by simultaneously consuming both classes of chemicals.

Goats were provided with Mediterranean shrubs containing tannin and/or saponin in order to determine whether ingestion of condensed tannins in conjunction with saponin resulted in reduced toxic effects, manifested as increased shrub intake. The objective of this study was to determine if goats consuming shrubs with different toxins (tannin+tannin or tannin+saponin) would ingest more shrubs foliage than goats consuming a shrubs with a single toxin (tannin). We used five Mediterranean shrubs that contain tannins and/or saponins in different combinations. Also we studied how different combinations of the Mediterranean shrubs, which contained different amounts of nutrient, toxins and flavor influence foliage intake and show how plant biochemical/biological diversity influences shrub intake by goats.

Material and methods

The four most important shrubs of the Mediterranean maquis plant community *Orno-Quercetum ilicis* (Rogosic, 2000) have been chosen in these experiments. These shrubs included *Quercus ilex* L. (Fagaceae), *Arbutus unedo* L. (Ericaceae), *Pistacia lentiscus* L. (Anacardiaceae) and *Hedera helix* L. (Araliaceae).

Tannin analysis showed that *Pistacia lentiscus* had the highest tannin index (1.50), followed by *Arbutus unedo* (1.26) and *Quercus ilex* (0.96). Beside these shrubs are rich in tannins, they content other components, so for example genus *Quercus* contain a cyclic diterpene alcohol, and quinic acid, species *Arbutus unedo* contains arbutoside and ethyl gallate (which is ethyl 3,4,5-trihydroxy benzoate), while *Pistacia leutiscus* was rich in terpenes including alpha pinene, beta pinene, camphene, trans-caryophyllene, cubebene (or similarsesquiterpene), cadinene, and fatty acids (Rogosic and Clausen, unpublished).

The shrub *H. helix* is reported to contain a mixture of pentacylic terpenoids (Burrows and Tyrl, 2001). These terpenoids are classed into one of three groups: genins, monodesmosides (α-hederin and β-hederin), and bidesmosides (hederacosides C and B). This complex mixture is often referred to saponins. Clinical signs of toxicity include vomition, excess salivation, and diarrhea, indicating both liver and digestive tract irritation (Harshberger, 1920). *H. helix* has also been reported to cause abortions in pregnant sheep and goats. Saponins are generally bitter tasting. Saponins that have a triterpenoid aglycone may not be bitter, but may have a licorice taste as glucuronic acid replaces sugar in triterpenoids.

Experimental animals were Alpine meat goats (average body weight 24.17 kg) with a background of grazing on Mediterranean shrublands on the Croatian coast. Animals were placed into individual pens and maintained on alfalfa pellets and barley on the level of 50% energy diet. Prior to the experiment all animals (n=12 goats) were offered all 5 shrubs from 9:00 to 13:00 for 5 days. Shrub intake was monitored, and the study was balanced by dividing the animals equally into treatment groups based on total shrub intake. We did not change the groups during the experiments but between shrubs feeding (each trial) baseline was reestablished using alfalfa pellets and barley at a maintance level for three days.

Each morning from 08:00 to 08:30 sheep and goats received 100 g barley. Then, from 09:00 to 15:00 hours all animals were fed with 200 g each shrubs biomass in separate food boxes daily for 6 days. We controlled food boxes each half of hour and in each empty boxes we additional offered shrub biomass. Food refusal was collected and shrubs consumption was calculated. At the 15:00 hours animals offered amounts of alfalfa pellets, including the former offered barley to the level of 50% energy diet. Throughout the experiments, animals had free access to trace mineral blocks and fresh water. We conducted eight trials at an experimental station 25 km far from Split in the Central part of the Adriatic seaside.

Trial 1: *Tannins and tannins.* The objective of Trial 1 was to determine if goats consuming three shrubs with the same toxin (tannin), but in different toxin concentration and different flavor would consume more biomass than goats offered one shrub containing a single toxin and flavor. We offered three shrubs *Q. ilex, A. unedo* and *P. lentiscus* to the first group and only *Q. ilex*, which has the lowest concentration of tannins to the second group.

Trial 2: *Tannins/tannins and saponins.* The objective was to determine if goats offered 3 shrubs with the same toxin, but in different amount of nutrients, different tannins concentration and different flavor would consume more biomass than goats offered only one shrub (*Hedera helix)* containing a single toxin (saponins) and flavor.

Trial 3: *Tannins/tannins and tannins+saponins.* The objective of the Trial 3 was to determine if goats offered 3 shrubs with the same toxin but in different flavor and tannins concentration would eat more than goats offered two shrubs containing different toxins-tannin+saponin and different flavor. We offered three shrubs *Q. ilex, A. unedo* and *P. lentiscus* to the first group and *Q. ilex+H. helix* (tannins+saponins) to the second group.

Trial 4: *Tannins (3 shrubs) and tannins (3 shrubs)+saponin (1 shrubs).* The objective was to determine biochemical/biological interaction and influence of the saponin in *Hedera helix* on the biomass intake of the three other shrubs with different tannins concentration and different flavors. So, we offered three shrubs *Q. ilex, A. unedo* and *P. lentiscus* to the first group and the same combination of shrubs+*H. helix* (3 tannins-rich shrubs+saponins) to the second group.

Results and discussion

Goats select diets from a different plant species that vary in nutrient, toxins and flavor. We offered the three shrubs to the first group and only one shrub *Q. ilex* to the second group (Trial 1). There are significant diferences between the first and the second group in regard to shrubs biomass intake. According to this graph we can conclude that goats prefer to eat a greater variety of shrubs that differ in toxin concentration and flavor than one shrub, notwithstanding the very different concentration of toxins and palatability between shrubs. As we can see on the second graph (Trial 2) the first group of goats which were offered three shrubs that contain tannins ate more biomass than the second group of goats which was offered only *H. helix* containing saponins.

Combination of shrubs that contain tannins with *H. helix* that contain saponins may enhance shrub intake because tannin and saponins chelate in the intestinal tract, thereby reducing the aversive effects of both compounds (Freeland *et al.*, 1985). The second group of goats ate more biomass of the shrubs *Q. ilex* and *H. helix* (tannins+saponins) than the first group of goats that was offered a combination of the three shrubs *A. unendo, Q. ilex and P. lentiscus*, shrubs which are loaded with tannins. Tannins and saponnins are one of the major groups of phytochemicals, which are widespread in Mediterranean shrubs. Goats provided with a choice between shrubs containing tannins and saponins consumed combinations of the shrubs that apparently avoided symptoms of toxicity associated with consumption of either toxin alone. This study supports Provenza`s hypothesis (2003) that combining the toxins in diet selection might eliminate toxicity. The nullification of any toxic effects is obviously dependent on the relative proportions of the toxins ingested. Tannins, saponins and other phytochemicals are likely to chelate within the intestinal tract. Tannins and saponins chelate with one another in 'in vitro' systems (Freeland *et al.*, 1985). Intestinal bonding of tannin and saponin may attenuate toxic affects by reducing absorption of the toxin. Herbivores have various mechanisms that prevent absorption or reduce the effect of ingested toxins. For example, induced microsomal enzymes can reduce (not eliminate) the effect of toxins that are absorbed. Polygastric (i.e. ruminant) animals may sometimes have an advantage over monogastric animals in regard to reductions in toxicity because of ruminal detoxification mechanisms. A dietary strategy using simultaneous consumption of plants containing chemical chelators (e.g. tannin) and plants containing other toxins may provide a mechanism for reducing both pre- and post- absorption toxicity. As we can see in Figure 1 the second group of goats ate more biomass than the first group which indicates that *H. helix* positively influenced the consumption of shrubs like *A. unedo, Q. ilex and P. lentiscus* containing tannins. It seems obvious that combination of shrubs containing tannins with *H. helix* that contain saponins may enhance shrub intake, but likewise biochemical diversity has a positive influence on shrub intake.

Conclusion

Tannins and saponins are very common classes of phytochemicals (secondary compounds) in Mediterranean shrubby species. Goats offered a choice of shrubs each containing a different toxin

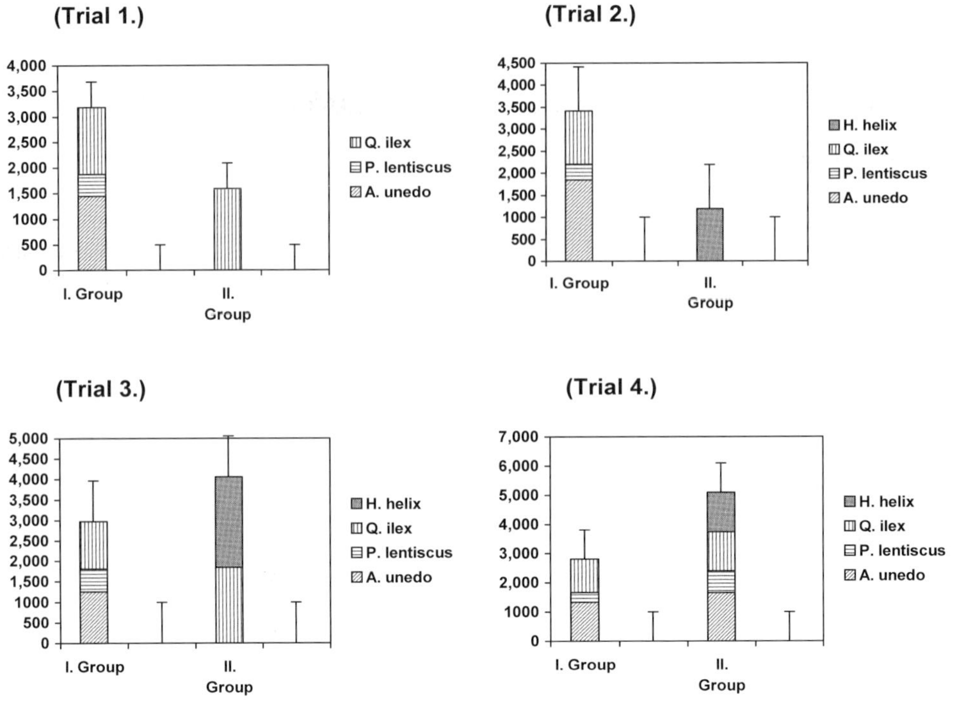

*Figure 1. Intake of goats (biomass (g)) fed different combinations of high-tannin (*Quercus ilex, Arbutus unedo, *and* Pistacia lentiscus*) and/or high-saponin* Hedera helix *shrubs; n=12 in each experiment.*

– tannins and saponins eat more forage than goats offered only one shrub. Goats offered rations of shrubs with tannins and saponins eat more biomass than sheep and goats offered shrubs with either tannins or saponins. Besides complementary tannin and saponin interactions, the variety of shrubs offered to goats is likely to be an important concern in effort to increase forage intake and animals performance on Mediterranean shrub lands.

References

Cheeke, P., 1998. Natural toxicants in feeds, forage and poisonous plants. Interstate Publ. Danville, IL.350 pp.

Freeland, W.J. and D.H. Janzen, 1974. Strategies in herbivory mammals: The role of plant secondary compounds. Am. Nat. 108: 269-288.

Freeland, W.J., P.H. Calcott and L.R. Andersen, 1985. Tannins and saponins: Interaction in herbivore diets. Biochem. Syste. Ecol. 13: 189-193.

Harshberger, J.W., 1920. Pastoral and Agricultural Botany. Blakiston's Son and Co, Philadelphia, 87 pp.

McArthur, C., A.E. Hagerman and C.T. Robbins, 1991. Physiological strategies of mammalian herbivores against plant defenses. pp. 103-114. in R.T. Palo and C.T. Robbins (eds.) Plant Defenses Against Mammalian Herbivory. CRC Press, Boca Raton, FL. 345 pp.

Provenza, F.D., J.J. Villalba, L.E. Dziba, S.B. Atwood and R.E. Banner, 2003. Linking herbivore experience, varied diets, and plant biochemical diversity. Small Rum. Res. 49:257-274.

Rogosic, J., 2000. Management of the Mediterranean Natural Resources. Skolska naklada Mostar. ISBN 9958-774-00-, COBISS/BiH – ID 8755974. 352 pp. (In Croatian).

Rogosic, J., J.A. Pfister and F.D. Provenza, 2003. Interaction of Tannins and Saponin in Herbivore Diet. VII International Rangelands congress: Rangelands in the New Millennium. Durban, South Africa 26 July - 1 Aug. pp. 104-105.

Round table

Traditions in Mediterranean livestock activities: obstacle or source of innovations?

A. Araba
Institut Agronomique et Vétérinaire Hassan II, Rabat, Maroc

This final round table has been organized to debate whether traditional knowledge in animal production is an obstacle or a source of innovation. I suggest you starting this discussion by a presentation of what 'traditional knowledge' means. Then, the several chairmen would mobilize the mains results of their sessions to feed the exchanges: Pr. Zervas is questioned about the relationship between tradition and innovation in terms of technical efficiency of production systems and activities, Dr. Brunori on the management of knowledge and innovation processing, then Dr. Faye on the relationships between social demand and animal production sectors. To finish, Dr. Boutonnet will give us his vision on how to structure between innovation and tradition and will conclude this round table.

Since the very beginning of this symposium, discussions have focused more on innovation, especially innovation in favor of traditional knowledge. Several issues, related to this topic, were discussed by various speakers and the audience. From all these, we could now consider the following question: Can traditional knowledge serve innovation? As a first step, I would just say a few words on what we could mean by traditional knowledge. It is a body of knowledge held by a community and built along the centuries for a better land use, a better management of natural resources and this in response to several nutritional, sanitary and other needs. This is as much related to technical farm management as social organizations, or sometimes crisis management (drought, epidemics, etc.). This knowledge has been built gradually over a long period of time, and could have changed also from moving circumstances. It is very often associated with a given human community. Each community may have their own 'traditional knowledge' which could not be found elsewhere. This knowledge has some common characteristics:

- It is not a static one, because the successive ecological circumstances shape this knowledge gradually.
- It is generally richer in pastoral systems than in intensive ones.
- It is owned with slight differences between women and men; some know-how being more owned by women than men.
- It is more owned by the elder than the younger.
- Although traditional knowledge is often rich, we should check its relevance and effectiveness before any attempt to include it in development projects. It is sometimes necessary to adjust or improve it.

It is important to pay attention to traditional knowledge in innovative projects, because it could lower the risks associated with dependence on external technologies, which are often very costly. Besides, by integrating it into innovative projects, it can involve local people and get confidence to make them project partners. This could give these projects more chance of success. In conclusion, I believe that in these conditions, traditional knowledge can serve innovation as much as innovation can serve tradition.

G. Zervas
Rector of the Agricultural University of Athens, Greece

My contribution aims to explore if the Traditional Livestock Farming Systems (TLFS) would be compatible with innovation. First of all, I want to underline that the TLFS have existed for many decades and it is well known that they are based on less favoured areas. Although they seem to be resistant to innovation, we observe that these TLFS have adopted a lot of incremental innovations.

I could quote several examples of the innovations more often related to planned conception, but not any related to 'innovating design' as such distinction has been provided in introduction of the symposium. Due to their resistance, the process of acceptance of innovation is slow, but farmers have accepted a lot of things, may be due to the necessity to survive:

- milking machine;
- milk tankers to reduce microbes and preserve milk much better;
- the artificial insemination has been largely adopted;
- new breeds and cross breeding have been introduced;
- synchronizing of Oestrus has been used for many years;
- artificial rearing now has been applied;
- identification systems from ear tags to electronic systems.

If we move to feeding, cereals, the only type of complementation some decades ago, are now replaced by balanced concentrates and minerals. We observe also a more rationalized utilization of natural resources. As an example, vaccinations, use of medicines (against parasites and mastitis) are now generalized.

All this happened for several reasons: legislation (European or national), demand of consumers asking for more safety and higher quality. The cheese making plants and ham processing have improved a lot due to these demands. Society demands also for organic products and a better organized production as mentioned during the symposium (SIEOL system of management for instance) are driving forces for the future of livestock activities.

As a conclusion, I underline that whatever their resistances, the TLFS have incorporated many innovations.

G. Brunori
Professor in Agricultural Economics and Agro-food economics, University of Pisa, Italy

When speaking about tradition and innovation, there is a problem because scientific knowledge has hardly given value to local knowledge, ignoring people who are in the field, and disregarding tradition. They have developed models totally detached from reality and from the perception of farmers. So this option has been an obstacle to the possibility of cross-fertilisation between local and scientific knowledge. From our perspective, we try to recover traditional knowledge because it is a source of diversity we need in order to make our system resilient.

I would like to mention biodiversity as an example, as biodiversity erosion is a product of modernisation of agriculture. But it has not been taken into account into scientific agriculture models because they were not efficient and realistic. For a long time, biodiversity has not been an issue. When we start to understand the role and the value of biodiversity, we recognize that people who are in the land have expertise to detect and to classify biodiversity. Furthermore, biodiversity conservation in most of the cases is based on social relations. It relies on local actors networking, for example for the exchange of seeds. Modernisation has reduced this capacity of improving biodiversity and now, we need to recover traditional knowledge.

Another very clear example is related to food. Most of the time, we have undergone a process of food standardisation. When we started to understand that quality is related to variety and diversity, the search began for food diversity, originating from traditional recipes. Such explorations became business models and now, you can find very little people that can deny the power of biodiversity based on marketing strategies and differentiation. Twenty years ago, it was impossible to speak about this, because a lot of people would say 'it's just a niche market, we can't feed the world with this'. Now it has been demonstrated that this is a viable business model especially for small farming.

So, I think that traditional knowledge remains a relevant source that should be taken into consideration within scientific knowledge. When scientific knowledge keeps local knowledge as an important part when developing models, researchers probably need to be more open about it. I think also that in the process of understanding what the role of tradition in innovation can be, it is important to be aware of the correct codification. In most cases, local knowledge is not codified. Thus, farmers have the know-how but they are often unable to convey their knowledge. So, we need experts capable of translating this knowledge into a language that can be circulated and shared. Otherwise the traditional knowledge is maintained as local skill and it dies with the owners. It is not kept from being transmitted because it is non-transmittable. Therefore, this is one important part of our job: try to codify traditional knowledge and conserve the value as much as possible. It is important to integrate this knowledge into new activities, new products and adapt it into different contexts.

B. Faye
Project leader in animal resources, CIRAD, France

I will probably repeat a little bit some ideas that have already been said. I would like to come back on several points.

First, a given innovation is only valid in the context in which it was established. This could look obvious: if it is valid here, it may not be valid elsewhere. In these differences, the weight of tradition is also important. But I do not use this term, 'the weight of tradition', in a pejorative way. I'll take a non-Mediterranean example.

Today we are observing a gradual disappearance of plough buffalo in Southeast Asia because a major technical innovation has been the mechanization of small farms with the introduction of mechanized farming. We realize a decrease in soil fertility and the slopes are becoming overgrown as consequences of these changes. Some people are now trying to restore the traditional role of buffalo in the agricultural systems of these areas. This example shows that although a technical innovation could improve productivity at short term, it can cause problems after a long period. And only tradition will solve this situation, an improved tradition with innovations to be brought not to mechanization but to buffalo breeding and management themselves. During the fourth session, we saw clearly that innovations are more and more related to societal pressure. But it was also underlined that this pressure must be managed by the producers themselves and that could be very difficult when they are facing crisis.

For example, sanitary crises are completely beyond the capacity of resilience of the systems. Within the production sectors or systems where irrationality is not absent, innovation must be based on some traditions to eliminate these irrationalities and find a balance.

Today we observe changes in nutrition habits, particularly those related to the growing urbanization of society. I think we are in a phase where we will switch to a majority of the world population becoming urban. We are witnesses of what I call the 'MacDonaldisation' of the young people. And at the same time, there is a growing societal demand for traditional products. As an example and argument of this point, in Central Asia after the fall of the Soviet Union, we saw the destruction of nearly all the traditional dairies existing in the country. But these dairies were using milk of unconventional species such as mare milk or camel milk. They were replaced by what they call over there the 'Western-style milk' from dairy cattle. This means they use now pasteurized milk or flavored European style yogurts. And the young people like this kind of products. But, we see also an increasing demand for products based on traditional camel and mare's milks. And we see also new product innovations to improve the attractiveness of these traditional yogurts regarding the now usual industrial ones (with fruits in it). So the consumers will not find mare or camel milk products, loose in the bazaars. They look for products with elegant or flavored packages, etc. Thus, innovation must also rely on traditional products that try to improve their image.

Finally, I want to conclude with a point related to my strong tropical experience: innovation is not a Western technical mirage. In the presentation of OECD data, we saw the dramatic increase in labor productivity. It will not be guaranteed to be the good solution everywhere in the world! To illustrate this Western technocratic mirage, we can quote the numerous failures in High Potential Holstein cow breeding projects in very warm conditions (until 50 degrees in the shade). These failures have also simultaneously contributed to genetic erosion, an important point often raised during the symposium. All this leads us to return to traditional species and animal genotypes that could be able to resist these extremely hard climates.

J.-P. Boutonnet
Scientist in INRA, SAD Department; UMR Innovation; SUPAGRO Montpellier; France

I also want to discuss the role of tradition in the emergence of innovations for livestock production in the Mediterranean. Traditions and innovations may appear contradictory. In fact, and we forget it often, they can be helped by one another: Innovation can support tradition, this has been said, but tradition too, can favor innovation. Firstly, innovation is seen as a process that aims to bring something new to improve while tradition is to do what we know how to do collectively. How can we abolish the gap between these two worlds?

First of all, what do we call tradition? Finally, is it not, what we do not understand? Traditions are not immutable and they result from the interaction between local actors, resources and learning processes. Tradition refers to heritage and can we live without heritage?

To answer these questions, I would say that the traditions should not be sacrosanct and we must not forget that some local resources will disappear when they are no longer needed. But tradition can also be a safety net for innovations and traditional organizations can be a support to implement innovations. On the other hand, there are many examples which show that innovation can be an important support to the development of traditional products (Internet marketing, new packaging, methods of food preservation).

In conclusion, I want to insist on several points. First of all, it is to enhance the synergy between tradition and innovation in the Mediterranean livestock sectors. I think it is necessary to review the practices of researchers who could better take into account local knowledge. On the other hand, traditions often lay down obligations of means (Geographic Origin, processes, methods of animal management) when innovation can focus both on means and results (fat or salt content, microbial flora).

Clear policy choices between best efforts obligations and obligatory results contracts or a combination of both, are needed. In the Mediterranean area, more perhaps than everywhere, traditional animal products are diverse and numerous and appreciated by consumers but their costs are not competitive compared to generic standardized products from Northern Europe. So and to finish, the implementation of these innovations by local stakeholders will only occur and create local development if public policies propose a clear regulatory framework to establish and enforce rules to coordinate (competition and/or cooperation) the local actors.

Conclusion

An overall view on recent innovations in Mediterranean Livestock Farming Systems

I. Casasús[1], D. Gabiña[2] and F. Casabianca[3]
[1]CITA-Aragón, Zaragoza, Spain
[2]IAMZ-CIHEAM, Zaragoza, Spain
[3]INRA, Corte, France

Livestock farming systems (LFS) in the Mediterranean basin are characterised by a high diversity in many different aspects, mainly associated to the particular resources and traditions specific of each area. A wide range of innovations have been conceived or introduced in these systems in the last decades; and while some of the operations have clearly increased their competitiveness, others have not been able to face new situations and have disappeared. The latter has often been the case when innovations led farms to an intensification level that could not be accomplished under their production environment at the time.

Some of these innovations have just been technical, with new incorporations into existing production units, or organizational, by the establishment of new relationships among different stakeholders. Some were related to implementing new procedures for obtaining the same products, while others have developed new products. Generally speaking, those of greater interest have dealt with production techniques, work organization, equipment, reproduction, feeding, health management, selection, product differentiation, marketing, etc. More recently, the increasing environmental and ethical concerns at all levels of society have motivated the appearance of innovations on these issues.

In the near future, the means and ways to innovate in European LFS, and very particularly in the Mediterranean, should be considered in the light of the latest Common Agricultural Policy reform, which will also apply in general terms for countries not included in the EU. These policies support LFS that can be defined as environmentally and territorially sustainable, and they highlight the role of agriculture and livestock farming in the preservation of natural resources but also, and very importantly, of the population in rural areas. The only means to maintain population is to ensure that farmers develop a profitable economic activity and reach their target quality of life, which can only be guaranteed by the development of systems that are technically sound, labour- and cost-efficient and that provide the market with products that meet consumer requirements on quality, ethical and environmental aspects, and at competitive prices. In the actual scenario of competition between food, feed and fuel, innovation may be the key to the survival of livestock farming that will have to adapt to an increasingly global situation.

Innovation in LFS can be implemented on some or many components of the production chain, with small or quite radical changes, but either way it is very important that innovation is given a multidisciplinary approach. This is because livestock is bred within a system, and innovation concerning any of its parts may have an effect on the rest. Sometimes even a 'butterfly effect', when a presumably small change finally generates big modifications in the rest of the system, and therefore it is fundamental that the chances, risks and effects of innovation are analysed from farm to fork.

In this context, the main objective of the X International Mediterranean Symposium 'New trends for innovation in the Mediterranean animal production', organized by the EAAP-FAO-CIHEAM Mediterranean Working group and hosted by INRA-Corte, was to analyse the innovations that can improve the efficiency of LFS, the adequacy of their products to consumer preferences, and the chances of farming to lead rural development by integrating economic, environmental and social sustainability throughout the process. In order to do so, this symposium was designed with contributions from experts at all the stages of the chain, experts from the whole Mediterranean basin who discussed and shared theoretical and practical approaches and experiences. Information was exchanged on the main innovations at different stages of the production chain, from animal genetics, nutrition, reproduction, health, etc. considered at the farm scale, to product differentiation and organization at different levels, taking into account ethical and environmental aspects.

How innovation should be conceived and developed

Innovation should arise from diagnosis of LFS in their production and market environment, analysing their strengths, weaknesses, opportunities and threats. The process should be collaborative, and involve the many different stakeholders who may have a direct or indirect role in the production chain. Experience and knowledge from different areas should be combined: farmers, scientists, engineers and, last but not least, society in a wide sense, because although consumers are at the end of the LFS chain, there are opinion groups that are not necessarily consumers of products provided by LFS but whose points of view need to be considered when designing production systems that meet societal demands. It is of major importance that stakeholders at all points are organized, so that their opinions and interests can appear in the agendas of current and future trends in LFS.

Public but also private investment in research and technology is crucial, as the effort on R+D+I provides an internal diagnosis of where in the operation innovations are needed to improve overall profitability. Integrative analyses were presented in the seminar by private companies and cooperatives involved in research, with clear objectives and definitions of their needs in order to improve production processes. For example, an innovative conception of the sheep production chain was presented, designed to be competitive both at the production and product commercialization levels. For this purpose, innovation coming from research was applied at all the stages of the chain: genetics, reproduction, management, feeding, commercialization, and development of new consumer-oriented products.

After innovations are conceived and technically tested, the success of their implementation depends on the appropriate transfer of this knowledge to the target users. Therefore, continuous updating of extension services and educational programmes for farmers and technicians involved in the agri-food industry is necessary to guarantee the objective is achieved.

Innovations to improve the efficiency of LFS in the Mediterranean area

In general terms, research in Mediterranean LFS should be focused on *improving the efficiency of use of the available resources in the area*, considering natural pastures and forage crops but also by-products that can be used for livestock feeding. Based on this criterion the performance of local breeds should be optimized, as well as technologies for producing and processing animal products, as well as their quality and safety. The wide diversity of breeds used and particularly adapted to these areas is doubtlessly a major reservoir of genetic diversity. However, in the current context of market globalization some of them may loose competitiveness in terms of quantitative output when compared with other improved breeds, and therefore may need specific management alternatives and market niches to guarantee their continuity. Modern reproductive technologies and genomic tools can significantly contribute to the *preservation of animal genetic resources*, particularly in the case of endangered breeds which are crucial for the survival of the production systems in which they are emplaced. Some of these may provide products of a high dietetic quality, giving them a chance to differentiate themselves from the general products in the markets, and even to be protected under official forms such as PGI or PDOs, as was presented in the Seminar.

Concerning *animal feeding*, particularly in arid areas, strategies are needed to face the wide range in forage availability and quality throughout the year. This can be achieved either by the improvement of potential productivity of pastures and the optimization of their use, or by using alternative by-products, obtained from agriculture (olive, fruit and orchard production) or from the biofuel industry.

As feeding is mostly based on grazed pastures, it seems indispensable to highlight the role of *pastoralism and traditional practices* in these systems, both of which play a major role in the sustainable development of rural areas. Although some innovations put into practice in areas of traditionally extensive production compromised the inherent multifunctionality of these systems, in the Seminar many examples were given on how new technologies could be easily implemented and supported the continuity of the farming activity. Some of them are widely used in our day-to-day life and are increasingly affordable, such as the combination of GPS-GPRS technologies, that permits real-time location of flocks and can significantly improve the shepherds' quality of life.

Some innovations based on the recovery of *traditional products or methods* may conceptually look like a step backwards, because sometimes traditions have been considered an obstacle to innovation. However, they may provide the way for technically- and economically-efficient diversification leading to products that meet consumers' demands concerning product quality and good practices during the production process (extensive farming, fattening animals on pastures, organic farming, etc.). There is increasing concern about animal welfare, and new EU-standardized tools were presented for the assessment of welfare quality throughout the chain, from farm to transport and slaughterhouses.

The development of *new sensing technologies* has provided the tools to collect and manage large data sets with information from animals and their products. Several sensors and applications were presented, used for identification (through radiofrequency, digital analysis of images from artificial vision, ultrasounds, thermal sensors), location, behaviour and performance records, and determination of product origin, composition or quality. Information is easily accessible in data bases, where the factors that can influence the system at different points can be identified almost immediately, as raw data are converted into practical indexes for management and decision-making. The latter is particularly important, because, doubtlessly, innovation can succeed or fail, and often the shadow of failure can prevent technically-sound innovations from being applied. Therefore, risk assessment is needed to detect how innovation can influence a particular LFS, which can be provided by modelization based on previous experience or experimental grounds, which was presented as a tool to support the decision-making process at several levels.

Conclusions

The future progress of Mediterranean LFS should be based upon revision of the main innovations that have significantly influenced these systems, analysing to what extent their adoption has improved techno-economic efficiency, has reinforced the socio-cultural basis of LFS or has reduced environmental impacts, determining the keys of success or failure.

Further innovation needs exploring and sharing, always using participatory procedures both in the diagnosis of needs and the analysis and extension of results, and therefore knowledge and experience from farmers, technicians, researchers, managers, social agents and policy makers should be considered. By this approach, a wide variety of solutions will be delivered for a large diversity of LFS, market niches and societal demands.

Keyword index

Author index